Systems & Control: Foundations & Applications

Alain Bensoussan
Giuseppe Da Prato
Michel C. Delfour
Sanjoy K. Mitter

Representation and Control of Infinite Dimensional Systems

Volume II

Birkhäuser
Boston · Basel · Berlin

Alain Bensoussan
Université Paris-Dauphine
UFR de Mathématiqes
Paris Cedex 16, France
and INRIA Domaine de Voluceau
B.P. 105
Rocquencourt,
78153 Le Chesnay, France

Michel C. Delfour
Université de Montréal
Département de Mathématiques
et de Statistique and
Centre de Recherches Mathématiques
Montréal, Québec, Canada H3C 3 J7

Giuseppe Da Prato
Scuola Normale Superiore
Piazza dei Cavalieri 7
56100 Pisa, Italy

Sanjoy K. Mitter
Dept. of Electrical Engineering
and Computer Science and Lab.
for Information & Decision Systems
MIT
Cambridge, MA 02139, USA
and Scuola Normale Superiore
Piazza dei Cavalieri 7
56100 Pisa, Italy

Library of Congress Cataloging-in-Publication Data

Representation and control of infinite dimensional systems.

(Systems & Control)
Includes bibliographical references and index.
1. Control theory. 2. Mathematical optimization. I. Bensoussan, Alain.
QA402.3.R47 1992 92-10323
003'.7--dc20 CIP

Printed on acid-free paper

ISBN-13: 978-1-4612-7651-7 e-ISBN-13: 978-1-4612-2750-2

DOI:10.1007/978-1-4612-2750-2

Camera-ready copy prepared by the Authors in LaTEX.

9 8 7 6 5 4 3 2 1

In memory of Andrée

Preface

Volume I of this two volume book has dealt with the question of modelling and representation of infinite-dimensional systems. Volume II is concerned with the optimal control of certain classes of infinite-dimensional systems with a quadratic cost criterion, both over a finite and an infinite-time horizon. The knowledge of Volume I is a prerequisite for reading Volume II.

Volume II consists of three parts. Chapter 1 of Part I is concerned with the theory of controllability and observability of finite-dimensional linear systems, and the role these concepts play in the study of the quadratic cost problem over an infinite-time horizon. This chapter is included so that the reader understands the background from which the infinite-dimensional theory developed and also obtains an appreciation of the substantial conceptual and technical difficulties which had to be overcome to give a satisfactory treatment of the subject in the infinite-dimensional context. Chapter 2 is devoted to the study of controllability for an infinite-dimensional abstract linear dynamical system which can be specialized to obtain results for controllability of parabolic and hyperbolic partial differential equations both when control is exercised in the interior of the domain and also when control is exercised through the boundary. The important problem of exact controllability of hyperbolic equations in appropriate spaces, which leads to stabilizability properties for these systems (and hence verification of the "finite cost" condition for infinite-time problems) is discussed in detail in this chapter. The systematic use of eigenvalues and eigenfunctions of appropriate differential operators to obtain the results is a somewhat novel aspect of this chapter.

Part II of the book is devoted to the Quadratic Cost Optimal Control Problem over a finite-time horizon. We develop the theory for an abstract dynamical model satisfying certain semi-group (group) assumptions and then we treat concrete situations by verifying these assumptions using differential equations (partial, functional-differential) methods. This chapter presents a reasonably complete treatment of the subject which includes both boundary control and boundary observation (not necessarily simultaneously) for parabolic and hyperbolic systems. Technically, these are the more difficult parts of the book since they involve unbounded control and observation operators. Many of the results presented here appear in book form for the first time.

The approach we adopt is Dynamic Programming which leads to a synthesis of the optimal control in feedback form via a study of an operator Riccati differential equation. The systematic use of Dynamic Programming gives a unified view of this topic.

Part III, the final part of the book is concerned with the Quadratic Cost Optimal Control Problem over a infinite-time horizon. Here the concepts of stabilizability and detectability play an essential role. The properties of stabilizability and detectability have to be verified, in some sense directly, for the parabolic case (which has finite-dimensional unstable part) and via the theory of exact controllability (for example) in the hyperbolic case. Thus, there is a close relation between Parts I and III of the book. The approach here is again via Dynamic Programming and a study of an algebraic Riccati equation. We also prove the stability of the closed-loop system. In many (but not all) situations the theory of the infinite-time quadratic cost problem in the infinite-dimensional case is as complete as for the finite-dimensional situation. Again, many results here appear for the first-time in book form.

Dynamic Programming is a deep conceptual idea. Its applications are manifold. For the philosophically-minded reader, we have provided a quotation from Soren Kierkegaard in the following page as an indication that philosophy might often anticipate developments in science.

Earlier versions of some chapters of this book have been used for graduate courses at the Scuola Normale Superiore and at the Université de Montréal. Other parts have been used for graduate seminars.

It is a pleasant task to thank the many institutions which have made this work possible—Université de Paris-Dauphine and INRIA, France; Scuola Normale Superiore, Pisa, Italy; Centre de recherches mathématiques and Département de mathématiques et de statistique, Université de Montréal, Canada; Department of Electrical Engineering and Computer Science and the Laboratory for Information and Decision Systems, Massachusetts Institute of Technology, Cambridge, Mass., U.S.A. The research of Giuseppe Da Prato has been partially supported by the Italian National Project MURST "Equazioni di Evoluzione e Applicazioni Fisico-Matematiche". The research of Michel Delfour has been supported by the Natural Sciences and Engineering Council operating grant OGP-8730 and infrastructure grant INF-7939, the "Ministère de l'Education du Québec" through a FCAR Team Grant, the France-Québec exchange program, and a Killam Fellowship from Canada Council. The research of Sanjoy Mitter has been supported by the U.S. National Science Foundation in the seventies, the U.S. Air Force Office of Scientific Research over a long period and the U.S. Army Research Office through the Center for Intelligent Control Systems during the last six years. This support is gratefully acknowledged.

It is a pleasure to thank Georg Schmidt for a critical reading of Part I, Chapter 2 of the book. His comments have led to many improvements in this chapter. We thank Edwin Beschler of Birkhäuser Boston Inc. for his cooperation, Ann Kostant also of Birkhäuser for expert help in word processing and Margaret Flaherty of the Laboratory for Information and Decision Systems (M.I.T.) for typing large parts of the manuscript with precision and care.

Alain Bensoussan *Paris, France*
Giuseppe Da Prato *Pisa, Italia*
Michel Delfour *Montréal, Canada*
Sanjoy Mitter *Cambridge, U.S.A.*

The Earth, March 26, 1993

It is perfectly true, as philosophers say, that life must be understood backwards. But they forget the other proposition, that it must be lived forwards ... And if one thinks over the proposition it becomes more and more evident that life can never really be understood in time simply because at no particular moment can I find the necessary resting-place from which to understand it – backwards.

From entry in Soren Kierkegaard's Journal for the Year 1843.
Quoted in Richard Wollheim, *Thread of Life*,
Harvard University Press, Cambridge, Mass. 1984.

Contents

Part I Qualitative Properties of Linear Dynamical Systems (continued)

Chapter 2
Controllability and Observability for a Class of Infinite Dimensional Systems 39

Part II Quadratic Optimal Control: Finite Time Horizon (continued)

Part III Quadratic Optimal Control: Infinite Time Horizon 257

Chapter 1
Systems with Bounded Control Operators: Control Inside the Domain 259

Chapter 2
Systems with Unbounded Control Operators: Parabolic Equations with Control on the Boundary 307

Part III Quadratic Optimal Control: Infinite Time Horizon (continued)

Part I

Qualitative Properties of Linear Dynamical Systems

Chapter 1

Control of Linear Finite Dimensional Differential Systems Revisited

1 Introduction

This part serves the purpose of an introduction to Volume II of Representation and Control of Infinite Dimensional Systems which is mainly concerned with the quadratic cost optimal control problem for distributed parameter systems and systems with time delay, both over a finite and and infinite time interval. For problems over a finite time interval, the main tool used is Dynamic Programming which leads to a Hamilton-Jacobi equation for the value function. For the class of control problems considered, the Hamilton–Jacobi equation can be explicitly solved via the study of an operator Riccati equation. The study of the operator Riccati equation when control is exercised through the boundary in the case of distributed parameter systems or when delays are present in the control in the case of systems with time delay poses additional technical difficulties. The results of Volume I are needed to overcome these difficulties. For problems over an infinite time interval the concepts of controllability and observability (and the weaker concepts of stabilizability and detectability) play an essential role in the development of the theory.

2 Controllability and observability of finite dimensional linear systems

In this section we present the theory of controllability and observability for finite dimensional linear systems in a manner which has implications in the study of controllability and observability for infinite dimensional systems. In particular, we show that if the system is controllable then the transfer from the zero-state to any other state can be carried out by means of minimum-energy controls. It is the same idea, which is later used in the theory of exact controllability of hyperbolic (second order) systems. The mathematical techniques needed to accomplish it however are far more difficult. Throughout this book $\mathbf{R}$ and $\mathbf{C}$ denote the respective fields of real and complex numbers.

Consider the linear finite-dimensional control system in the interval $[0,T]$:

$$\begin{cases} \dfrac{dx}{dt} = Ax(t) + Bu(t) \\ y(t) = Cx(t) \end{cases} \tag{2.1}$$

where the state $x(t) \in \mathbf{R}^n$, the control functions $(t \mapsto u(t)) \in L^2(0,T;\mathbf{R}^m)$, the output $y(t) \in \mathbf{R}^p$ and $A \in \mathcal{L}(\mathbf{R}^n;\mathbf{R}^n)$, $B \in \mathcal{L}(\mathbf{R}^m;\mathbf{R}^n)$ and $C \in \mathcal{L}(\mathbf{R}^n;\mathbf{R}^p)$. In this chapter we use $|\cdot|$ to denote the norm in various Hilbert spaces, $(\cdot,\cdot)$ to denote scalar product and * denotes adjoint of an operator (and also the dual space).

We assume that the initial state $x(0) = 0$.

The solution of the linear differential equation (2.1) can be written as

$$\begin{aligned} x(t;u) &= \int_0^t e^{A(t-s)} Bu(s)\,ds \\ &\triangleq (L_t u), \end{aligned} \tag{2.2}$$

where L_t is a linear bounded transformation from $L^2(0,t;\mathbf{R}^m)$ into $\mathbf{R}^n$.

2.1 Controllability

Definition 2.1 *We say that the system is* controllable *(from* $0 \in \mathbf{R}^n$*) in* $[0,T]$ *if given any* $\bar{x} \in \mathbf{R}^n$*, there exists a control function* $u(\cdot) \in L^2(0,T;\mathbf{R}^m)$ *such that*

$$x(T;u) = \bar{x}.$$

Controllability in $[0,T]$ is therefore equivalent to the surjectivity of the map $u \mapsto L_T u$. We now give necessary and sufficient conditions for controllability using elementary facts from the theory of linear transformations on Hilbert spaces.

Let H and K be Hilbert spaces and let $A \in \mathcal{L}(H;K)$. Let $A^* \in \mathcal{L}(K^*;H^*)$ be the adjoint operator. In the sequel we identify H and H^* and K and K^*. Let $\mathcal{R}(A)$ denote the range of the operator A and $\mathcal{N}(A)$ denote the null space of A. $\perp$ denotes the orthogonal complement of a closed subspace.

Proposition 2.1 *The following relations are true*

(i) $\mathcal{R}(A) \subseteq \mathcal{N}(A^*)^\perp$ *and* $\mathcal{R}(A) = \mathcal{N}(A^*)^\perp \Leftrightarrow \mathcal{R}(A)$ *is closed.*

(ii) $\mathcal{R}(A^*) \subseteq \mathcal{N}(A)^\perp$ *and* $\mathcal{R}(A^*) = \mathcal{N}(A)^\perp \Leftrightarrow \mathcal{R}(A)$ *is closed.*

($\mathcal{R}(A)$ *closed* $\Leftrightarrow \mathcal{R}(A^*)$ *is closed.*)

(iii) $\mathcal{N}(A) = \mathcal{R}(A^*)^\perp$

(iv) $\mathcal{N}(A^*) = \mathcal{R}(A)^\perp$

(v) Furthermore

$\mathcal{R}(A)$ *is closed* $\Leftrightarrow \exists c > 0$ *such that* $||h|| \le c||Ah||,\ \forall\, h \in H.$

Using the above, we can prove

Proposition 2.2 *The following are true:*

(i) $\mathcal{R}(A)$ *is dense in* $K \Leftrightarrow \mathcal{N}(A^*) = \{0\}$

(ii) $\mathcal{N}(A) = \{0\} \Leftrightarrow \mathcal{R}(A^*)$ *is dense in* H

(iii) rank $[B \vdots AB \vdots \cdots \vdots A^{n-1}B] = n$.

When the system is controllable we shall refer to (A, B) as a *controllable pair.*

It is of some interest to establish that if the system is controllable then we can transfer the state x_0 at time t_0 to the state x_1 at time t_1 by using a control u^* which has least energy

$$\int_{t_0}^{t_1} |u^*(s)|^2 \, ds$$

amongst all controls that transfer the phase (x_0, t_0) to the phase (x_1, t_1).

To establish this, let us use the notation

$$\Phi(t, s) = e^{A(t-s)}.$$

Introduce the adjoint system

$$\begin{cases} \dfrac{dp}{dt} = -A^* p(t), \quad t \in [t_0, t_1[\,. \\ p(t_1) = \eta, \end{cases} \tag{2.8}$$

where η will be specified later. The solution of (2.8) is given by

$$p(t) = \Phi^*(t_1, t)\eta. \tag{2.9}$$

Now choose the control

$$\begin{aligned} u^*(t) &= -B^* p(t) \\ &= -B^* \Phi^*(t_1, t)\eta. \end{aligned} \tag{2.10}$$

If the control is to transfer (x_0, t_0) to (x_1, t_1), we must have

$$x_1 = x(t_1) = \Phi(t_1, t_0)x_0 - \int_{t_0}^{t_1} \Phi(t_1, t)BB^*\Phi^*(t_1, t) \, dt \, \eta$$

and hence if the system is controllable and we choose

$$\eta = W^{-1}(t_0, t_1)[\Phi(t_1, t_0)x_0 - x_1]$$

2.1 Controllability

Definition 2.1 *We say that the system is* controllable *(from* $0 \in \mathbf{R}^n$*) in* $[0,T]$ *if given any* $\bar{x} \in \mathbf{R}^n$*, there exists a control function* $u(\cdot) \in L^2(0,T;\mathbf{R}^m)$ *such that*

$$x(T;u) = \bar{x}.$$

Controllability in $[0,T]$ is therefore equivalent to the surjectivity of the map $u \mapsto L_T u$. We now give necessary and sufficient conditions for controllability using elementary facts from the theory of linear transformations on Hilbert spaces.

Let H and K be Hilbert spaces and let $A \in \mathcal{L}(H;K)$. Let $A^* \in \mathcal{L}(K^*;H^*)$ be the adjoint operator. In the sequel we identify H and H^* and K and K^*. Let $\mathcal{R}(A)$ denote the range of the operator A and $\mathcal{N}(A)$ denote the null space of A. $\perp$ denotes the orthogonal complement of a closed subspace.

Proposition 2.1 *The following relations are true*

(i) $\mathcal{R}(A) \subseteq \mathcal{N}(A^*)^\perp$ *and* $\mathcal{R}(A) = \mathcal{N}(A^*)^\perp \Leftrightarrow \mathcal{R}(A)$ *is closed.*

(ii) $\mathcal{R}(A^*) \subseteq \mathcal{N}(A)^\perp$ *and* $\mathcal{R}(A^*) = \mathcal{N}(A)^\perp \Leftrightarrow \mathcal{R}(A)$ *is closed.*

($\mathcal{R}(A)$ *closed* $\Leftrightarrow \mathcal{R}(A^*)$ *is closed.)*

(iii) $\mathcal{N}(A) = \mathcal{R}(A^*)^\perp$

(iv) $\mathcal{N}(A^*) = \mathcal{R}(A)^\perp$

(v) Furthermore

$\mathcal{R}(A)$ *is closed* $\Leftrightarrow \exists c > 0$ *such that* $||h|| \le c||Ah||,\ \forall\, h \in H.$

Using the above, we can prove

Proposition 2.2 *The following are true:*

(i) $\mathcal{R}(A)$ *is dense in* $K \Leftrightarrow \mathcal{N}(A^*) = \{0\}$

(ii) $\mathcal{N}(A) = \{0\} \Leftrightarrow \mathcal{R}(A^*)$ *is dense in* H

(iii) $\mathcal{R}(A)$ *is dense in* $K \Leftrightarrow AA^* : K \to K$ *satisfies* $AA^* > 0$.

(iv) $\mathcal{N}(A) = \{0\} \Leftrightarrow A^*A : H \to H$ *satisfies* $A^*A > 0$

(v) $A \in \mathcal{L}(H; K)$ *is invertible* $\Leftrightarrow \mathcal{R}(A) = K, \mathcal{N}(A) = \{0\}$

$\Leftrightarrow \exists c > 0$, *such that* $||h|| \leq c||Ah||, \forall\, h \in H$

$\Leftrightarrow \exists c > 0$, *such that* $||k|| \leq c||AA^*k||, \forall\, k \in K$.

Remark 2.1 Much of the above extends to operators A which are densely defined and closed to spaces H and K, which are Banach spaces. For proofs of these facts, see, for example M. SCHECHTER [1]. For a use of these ideas in a systems context, see M.C. DELFOUR and S.K. MITTER [1]. ■

We can immediately use the above to get a necessary and sufficient condition for controllability. Using (2.2) and noting that

$$\begin{aligned} L_T^* \; &: \; \mathbf{R}^n \to L^2(0,T;\mathbf{R}^m) \\ &: \; y \mapsto B^* e^{(T-\cdot)} y \end{aligned}$$

we get

$$W(0,T) \triangleq L_T L_T^* = \int_0^T e^{A(T-s)} BB^* e^{A^*(T-s)}\, ds. \tag{2.3}$$

Hence we have proved.

Theorem 2.1 *The system* (2.1) *is controllable in* $[0,T]$ *if and only if* $W(0,T) > 0$ *(positive definite).*

We can transform this criteria of controllability to an algebraic criterion by noting that

$$\mathcal{N}(W(0,T)) = \mathcal{N}(\mathcal{C}) \tag{2.4}$$

where

$$\mathcal{C} = [B \vdots AB \vdots \cdots \vdots A^{n-1}B]\, [B \vdots AB \vdots \cdots \vdots A^{n-1}B]^* \tag{2.5}$$

This can be proved as follows.

Let $x \in \mathcal{N}(W(0,T))$. Then

$$0 = (x, W(0,T)x) = \int_0^T |B^* e^{A^*(T-t)} x|^2 \, dt,$$

and hence $B^* e^{A^*(T-t)} x = 0$, $\forall\, t \in [0,T]$.

Differentiating $(n-1)$-times with respect to t and setting $t = 0$, we get

$$B^* x = 0, B^* A^* x = 0, \ldots, B^* A^{*n-1} x = 0,$$

and hence

$$x \in \mathcal{N}(\mathcal{C}).$$

Conversely, suppose $x \in \mathcal{N}(\mathcal{C})$. By the Cayley-Hamilton theorem,

$$e^{A(T-t)} = \sum_{i=0}^{n-1} \alpha_i(T-t) A^i.$$

Hence

$$x^* W(0,T) = \int_0^T \left[\sum_{i=0}^{n-1} \alpha_i(T-t) x^* A^i B \right] B^* e^{A^*(T-t)} \, dt.$$

Now $x^* A^i B = 0$, $i = 0, \cdots, n-1$, since $x \in \mathcal{N}(\mathcal{C})$. Therefore $x^* W(0,T) = 0$ and since $W(0,T)$ is symmetric,

$$x \in \mathcal{N}(W(0,T)).$$

Hence using the fact that $W(0,T)$ is a symmetric matrix

$$\mathcal{R}(W(0,T)) = \mathcal{R}(\mathcal{C}). \tag{2.6}$$

But

$$\mathcal{R}(\mathcal{C}) = \mathcal{R}([B \vdots AB \vdots \cdots \vdots A^{n-1} B]), \tag{2.7}$$

and hence we have proved the following theorem.

Theorem 2.2 *The following conditions are equivalent:*

(i) System (2.1) *is controllable in* $[0,T]$

(ii) $W(0,T) > 0$ *(positive–definite)*

(iii) rank $[B \vdots AB \vdots \cdots \vdots A^{n-1}B] = n.$

When the system is controllable we shall refer to (A, B) as a *controllable pair.*

It is of some interest to establish that if the system is controllable then we can transfer the state x_0 at time t_0 to the state x_1 at time t_1 by using a control u^* which has least energy

$$\int_{t_0}^{t_1} |u^*(s)|^2 \, ds$$

amongst all controls that transfer the phase (x_0, t_0) to the phase (x_1, t_1).

To establish this, let us use the notation

$$\Phi(t, s) = e^{A(t-s)}.$$

Introduce the adjoint system

$$\begin{cases} \dfrac{dp}{dt} = -A^* p(t), \quad t \in [t_0, t_1[. \\ p(t_1) = \eta, \end{cases} \tag{2.8}$$

where η will be specified later. The solution of (2.8) is given by

$$p(t) = \Phi^*(t_1, t)\eta. \tag{2.9}$$

Now choose the control

$$\begin{aligned} u^*(t) &= -B^* p(t) \\ &= -B^* \Phi^*(t_1, t)\eta. \end{aligned} \tag{2.10}$$

If the control is to transfer (x_0, t_0) to (x_1, t_1), we must have

$$x_1 = x(t_1) = \Phi(t_1, t_0)x_0 - \int_{t_0}^{t_1} \Phi(t_1, t) B B^* \Phi^*(t_1, t) \, dt \, \eta$$

and hence if the system is controllable and we choose

$$\eta = W^{-1}(t_0, t_1)[\Phi(t_1, t_0)x_0 - x_1]$$

the desired transfer will be effected. Let us now prove that $u^*(t)$ is the minimum energy control.

Let $\bar{u}(t)$ be some other control which effects the desired transfer. Then

$$\begin{aligned} x_1 &= \Phi(t_1,t_0)x_0 + \int_{t_0}^{t_1} \Phi(t_1,t)B\bar{u}(t)\,dt \\ &= \Phi(t_1,t_0)x_0 + \int_{t_0}^{t_1} \Phi(t_1,t)Bu^*(t)\,dt. \end{aligned}$$

Therefore

$$\int_{t_0}^{t_1} \Phi(t_1,t)B[\bar{u}(t) - u^*(t)]\,dt = 0,$$

and hence

$$\int_{t_0}^{t_1} ([\bar{u}(t) - u^*(t)], -B^*\Phi^*(t_1,t)\eta)\,dt = 0$$

and hence

$$\int_{t_0}^{t_1} (\bar{u}(t) - u^*(t), u^*(t))\,dt = 0.$$

Therefore

$$\begin{aligned} \int_{t_0}^{t_1} |\bar{u}(t)|^2\,dt &= \int_{t_0}^{t_1} |\bar{u}(t) - u^*(t) + u^*(t)|^2\,dt \\ &= \int_{t_0}^{t_1} |u^*(t)|^2\,dt + \int_{t_0}^{t_1} |\bar{u}(t) - u^*(t)|^2\,dt \\ &> \int_{t_0}^{t_1} |u^*(t)|^2\,dt \quad \text{if } \bar{u} \neq u^*. \end{aligned}$$

Suppose now that U and X are infinite dimensional Hilbert spaces, $A: X \to X$ an unbounded linear operator with domain $D(A)$ and $B: U \to X$ a bounded linear operator. Then there is a difficulty with carrying through the arguments given earlier (for example, $\mathcal{R}(L_T)$ need not be a closed subspace of X). Nevertheless, $\mathcal{R}(L_T)$ can be given the structure of a Hilbert space with continuous injection in X (Proposition 2.1 in Chapter 2 of Part I). Furthermore, there is a need to make a distinction between approximate controllability ($\mathcal{R}(L_T)$ is dense in X) and exact controllability (whose definition is somewhat subtle, see Definition 2.2 in Chapter 2 of Part I); indeed $\mathcal{R}(L_T) = X$

will usually not hold). However, an abstract criterion for exact controllability is related to an estimate on $L_T L_T^*$ (see Proposition 2.2; this chapter). It is also of interest to note that the minimum energy control viewpoint of verifying exact controllability as outlined in this section extends to infinite dimensional situations, but of course at the cost of using much more elaborate technical machinery. Part I, Chapter 2 gives a treatment of exact controllability both for parabolic and hyperbolic partial differential equations when control is exercised in a distributed manner and also when control is exercised through the boundary.

2.2 Observability

We now discuss the dual concept of observability. For this purpose we may assume that the input is known and since we are dealing with linear time-invariant finite dimensional systems, we may take $u(t) \equiv 0$, $t \in [t_0, \infty[$. Consider therefore the system

$$\begin{cases} \dfrac{dx}{dt} = Ax(t) \\ y(t) = Cx(t), \end{cases} \tag{2.11}$$

and let the initial state $x(t_0) = x_0$, for some arbitrary x_0.

The solution of (2.11) can be written as

$$y(t) = Ce^{A(t-t_0)}x_0. \tag{2.12}$$

Definition 2.2 *The system* (2.1) *is said to be* observable *in* $[t_0, t_1]$ *if the map*

$$\mathcal{O} : \mathbf{R}^n \to L^2(t_0, t_1; \mathbf{R}^p) : x_0 \to y(\cdot) = Ce^{A(\cdot\, - t_0)}x_0$$

is injective.

The definition expresses the fact that we can recover uniquely the initial state from a knowledge of the output $y(\cdot)$ in the time interval $[t_0, t_1]$. The interval $[t_0, t_1]$ can in fact be taken to be arbitrarily small.

Now,

$$\begin{aligned}\mathcal{O} \text{ is injective} \quad &\Leftrightarrow \quad \mathcal{N}(\mathcal{O}) = \{0\}\\ &\Leftrightarrow \quad \mathcal{N}(\mathcal{O}^*\mathcal{O}) = \{0\}.\end{aligned}$$

An easy computation shows that

$$\Sigma(t_0, t_1) \triangleq \mathcal{O}^*\mathcal{O} = \int_{t_0}^{t_1} e^{(t_1-t)A^*} C^*Ce^{(t_1-t)A}\, dt,$$

and clearly

$$\mathcal{N}(\mathcal{O}^*\mathcal{O}) = \{0\} \Leftrightarrow \Sigma(t_0, t_1) > 0 \text{ (positive-definite)}.$$

It is also easy to show that,

$$\begin{aligned}\mathcal{N}(\mathcal{O}^*\mathcal{O}) &= \mathcal{N}([C^*\vdots A^*C^*\vdots \cdots \vdots A^{*n-1}C^*]^* [C^*\vdots A^*C^*\vdots \cdots \vdots A^{*n-1}C^*]),\\ &= \mathcal{N}([C^*\vdots A^*C^* \cdots A^{*n-1}C^*])\end{aligned}$$

and hence we have proved:

Theorem 2.3 *The system* (2.1) *is observable if and only if*

$$rank\ [C^*\vdots A^*C^*\vdots \cdots \vdots A^{*n-1}C^*] = n.$$

When the system is observable we shall refer to (A, C) as an observable pair.

2.3 Duality

There is a formal duality between the concepts of controllability and observability. For this purpose we introduce the dual system

$$\begin{cases} \dfrac{d\xi}{dt} = -A^*\xi(t) - C^*\gamma(t) \\ \eta(t) = -B^*\xi(t), \end{cases} \tag{2.1'}$$

which evolves backward in time.

Then mathematically,
The system (2.1) is observable $\Leftrightarrow$ The system (2.1') is controllable.

The system (2.1') is observable $\Leftrightarrow$ The system (2.1) is controllable.

In this book we do not study observability of partial differential equations but exploiting the above duality ideas much of the theory of controllability developed in Part I can be used to develop a theory of observability for partial differential equations (even with unbounded observation operators).

Conceptually, controllability of a system permits the choice of feedback controls resulting in certain desirable (e.g. stability) properties of closed-loop systems while observability of a system permits the design of state estimators with desirable properties. We shall discuss this in a later section.

2.4 Canonical structure for linear systems

In this section we wish to obtain a decomposition (structure) theorem for the system (2.1) which is not necessarily controllable and observable. Let us denote by

$$V := \mathbf{R}(\mathcal{C}) = \mathcal{R}([B \vdots AB \vdots \cdots \vdots A^{n-1}B])$$

V is the smallest A-invariant subspace containing the image of B.

Let us also denote by

$$W := \mathcal{N}(\mathcal{O}) = \mathcal{N}([C^* \vdots A^*C^* \cdots \vdots A^{*n-1}C^*])$$

and clearly W is the smallest A^*-invariant subspace containing the image of C^*.

Now we may write the state space $X := \mathbf{R}^n$ as

$$X = (V \cap W^\perp) \oplus (V^\perp \cap W^\perp) \oplus (V \cap W) \oplus (V^\perp \cap W).$$

Then there exists a basis in which we may write the system as

$$\begin{bmatrix} \frac{d\bar{x}_1}{dt} \\ \frac{d\bar{x}_2}{dt} \\ \frac{d\bar{x}_3}{dt} \\ \frac{d\bar{x}_4}{dt} \end{bmatrix} = \begin{bmatrix} A_{11} & A_{12} & A_{13} & 0 \\ 0 & A_{22} & A_{23} & A_{24} \\ 0 & 0 & A_{33} & 0 \\ 0 & 0 & A_{43} & A_{44} \end{bmatrix} \begin{bmatrix} \bar{x}_1(t) \\ \bar{x}_2(t) \\ \bar{x}_3(t) \\ \bar{x}_4(t) \end{bmatrix} + \begin{bmatrix} B_1 u_1(t) \\ 0 \\ B_3 u_3(t) \\ 0 \end{bmatrix} \tag{2.13}$$

$$y(t) = [C_1 \quad C_2 \quad 0 \quad 0] \begin{bmatrix} \bar{x}_1(t) \\ \bar{x}_2(t) \\ \bar{x}_3(t) \\ \bar{x}_4(t) \end{bmatrix} \tag{2.14}$$

In the above, (A_{11}, B_{11}), (A_{33}, B_3) are controllable pairs and (A_{11}, C_1), (A_{22}, C_2) are observable pairs.

2.5 The pole-assigment theorem

An important consequence of a linear system possessing the controllability property is that by means of linear state feedback control the poles of the corresponding closed-loop system can be placed in arbitrary locations.

By state feedback control we mean that the control law is of the form

$$u(t) = Kx(t) + v(t),$$

where $K : \mathbf{R}^n \to \mathbf{R}^m$ and $v(\cdot)$ is a reference input.

Theorem 2.4 *Let $\Sigma = \{\lambda_i \in \mathbf{C}, i = 1, 2, \ldots, n\}$ with the proviso that if λ_i is complex then its conjugate $\bar{\lambda}_i \in \Sigma$. Then the system is controllable iff there exists a feedback control $u(t) = Kx(t)$ with spectrum $(A + BK) = \Sigma$.*

Proof. We start by noting two facts:

(i) Controllability is invariant under non-singular linear transformations of state space. In case $m = 1$, that is we are dealing with the system:

$$\frac{dx}{dt} = Ax(t) + bu(t), \ u(t) \in \mathbf{R} \text{ and } b \in \mathbf{R}^n,$$

then assuming (A, b) is a controllable pair, there exists a non-singular linear transformation $T : \mathbf{R}^n \to \mathbf{R}^n$ such that the above system transforms to

$$\tilde{A} = \begin{bmatrix} 0 & 1 & 0 & \dots & 0 \\ & & \ddots & & \\ \vdots & & \ddots & \ddots & \\ & & & \ddots & 1 \\ \alpha_1 & \alpha_2 & \dots & & \alpha_n \end{bmatrix}, \quad \tilde{b} = \begin{bmatrix} 0 \\ \vdots \\ \vdots \\ \vdots \\ 1 \end{bmatrix}.$$

It is then easy to see that if the characteristic polynomial corresponding to the desired spectrum Σ is

$$\chi = s^n - \beta_n s^{n-1} \dots - \beta_2 s - \beta_1,$$

then the feedback law

$$\tilde{u}(t) = K\tilde{x}(t) + v(t), \quad K = \text{diag}\ (\beta_1 - \alpha_1, \dots, \beta_n - \alpha_n)$$

transforms

$$\frac{d\tilde{x}}{dt} = \tilde{A}\tilde{x}(t) + \tilde{b}\tilde{u}(t)$$

to

$$\frac{d\tilde{x}}{dt} = (\tilde{A} + \tilde{b}k)\tilde{x}(t) + \tilde{b}v(t)$$

such that the spectrum of $(\tilde{A} + \tilde{b}K)$ in Σ.

(ii) Controllability is not destroyed by state feedback. That is

$$\mathcal{R}[B \vdots (A + BK)B \vdots \cdots \vdots (A + BK)^{n-1}B] = \mathcal{R}[B \vdots AB \vdots \cdots \vdots A^{n-n}B]$$

Now the general case is reduced to the case $m = 1$ by virtue of the following lemma. ∎

Lemma 2.1 *Let* $0 \neq b \in \mathcal{R}(B)$. *If* (A, B) *is a controllable pair, there exists a* $K \in \mathcal{L}(\mathbf{R}^m; \mathbf{R}^n)$ *such that* $(A+BK, b)$ *is a controllable pair.*

Let $b_1 = b$ and let $n_1 =$ dimension of the cyclic subspace $\mathcal{X}_1$ generated by (A, b_1). Let $x_1 = b_1$ and define $x_j = Ax_{j-1} + b_1 (j = 2, \cdots, n_1)$. Then the x_j, $j = 1, \cdots, n_1$, forms a basis for the cyclic subspace generated by (A, b_1). If $n_1 < n$, choose $b_2 \in \mathcal{R}(B)$ such that $b_2 \notin \mathcal{X}_1$. Since (A, B) is controllable such a b_2 exists. Let n_2 be the largest integer such that

$$x_1, x_2, \cdots, x_{n_1}, b_2, Ab_2, \cdots, A^{n_2-1}b_2$$

are independent.

Define

$$x_{n_1+i} = Ax_{n_1+i-1} + b_2, \; i = 1, 2, \cdots, n_2.$$

Then $x_1, x_2, \cdots, x_{n_1+n_2}$ is a basis for the cyclic subspace generated by (A, b_1+b_2). Continuing in this way, we eventually get an independent set of vectors $x_1, \cdots, x_n$, and

$$x_{i+1} = Ax_i + \tilde{b}_i, \; i = 1, \cdots, n-1,$$

and $\tilde{b}_i \in \mathcal{R}(B)$.

Since $\tilde{b}_i = Bu_i$ for some $u_i \in \mathbf{R}^m$ (the control space) and the $x_1, \cdots, x_n$ are independent, there exists a K_1 such that

$$BK_1 x_i = \tilde{b}_i, \; i = 1, \cdots, n,$$

where $\tilde{b}_i \in \mathcal{R}(B)$ is arbitrary. Therefore

$$(A + BK_1)x_i = x_{i+1}, \; i = 1, 2, \cdots, n-1,$$

and hence

$$x_i = (A + BK_1)^{i-1} b, \; i = 1, \cdots, n.$$

Therefore $(A + BK_1, b)$ is a controllable pair.

Hence the problem has been reduced to the case of $n = 1$ and the control law $K = K_1 + bK^*$ achieves the desired property.

To prove the theorem in the other direction, let λ_i, $i = 1, \cdots, n$, be real and distinct with $\lambda_i \notin$ spectrum of A. Choose $K \in \mathcal{L}(\mathbf{R}^m; \mathbf{R}^n)$ such that spectrum $(A + BK) = \{\lambda_1, \cdots, \lambda_n\}$. Let x_i, $i = 1, \cdots, n$, be the corresponding eigenvectors. Hence

$$x_i = (\lambda_i I - A)^{-1} BKx_i, \quad i = 1, 2, \cdots, n.$$

Now

$$(\lambda I - A)^{-1} = \sum_{j=1}^{n} \rho_j(\lambda) A^{j-1},$$

where $\rho_j(\lambda)$ are rational functions defined on the complement $\mathbf{C} \backslash \sigma(A)$ of the spectrum $\sigma(A)$ of A. Hence

$$x_i = \sum_{j=1}^{n} \rho_j(\lambda_i) A^{j-1} BFx_i \in \mathcal{R}([B \vdots AB \vdots \cdots \vdots A^{n-1}B]).$$

Now the x_i's span $\mathbf{R}^n$ and hence $\mathcal{R}([B \vdots AB \vdots \cdots \vdots A^{n-1}B]) = n$.

No satisfactory result similar to the Pole-Assignment theorem is known in infinite dimensions. It would be interesting to prove a result of this kind for delay systems

$$\begin{cases} \dfrac{dx}{dt} = A_1 x(t) + A_2 x(t-h) + Bu(t) \\ x(0) = \phi_0 \in \mathbf{R}^n \\ x(0) = \phi_1(\theta) \text{ a.e. } \theta \in [-h, 0), \quad \phi_1 \in L^2(-h, 0; \mathbf{R}^n). \end{cases}$$

Corollary 2.1 *If the system is controllable then there exists a feedback control* $u(t) = Kx(t)$ *such that the closed-loop system*

$$\frac{dx}{dt} = (A + BK)x(t)$$

is asymptotically stable.

2.6 Stabilizability and detectability

The structure theorem of linear systems and the pole-assigment theorem motivates the introduction of the concepts of stabilizability and detectability. Consider the input-state part of the control system (2.1):

$$\frac{dx}{dt} = Ax(t) + Bu(t).$$

Then from the structure theorem, we know that there is a coordinate system in which the above system has a representation

$$\begin{bmatrix} \dfrac{d\bar{x}_1}{dt} \\ \\ \dfrac{d\bar{x}_2}{dt} \end{bmatrix} = \begin{bmatrix} \bar{A}_{11} & \bar{A}_{12} \\ \\ 0 & \bar{A}_{22} \end{bmatrix} \begin{bmatrix} \bar{x}_1(t) \\ \\ \bar{x}_2(t) \end{bmatrix} + \begin{bmatrix} \bar{B}_{11} \\ \\ 0 \end{bmatrix} u(t) \tag{2.15}$$

and $(\bar{A}_{11}, \bar{B}_{11})$ is a controllable pair. Hence from the corollary of the pole-assignment system we can stabilize the system by means of the control $u(t) = K\bar{x}_1(t)$, provided

$$\frac{d\bar{x}_2}{dt} = \bar{A}_{22}\bar{x}_2(t)$$

is asymptotically stable.

Definition 2.3 *(i) We say that the system* (2.1) *is* open-loop stabilizable *if*

$$\forall x_0 \in \mathbf{R}^n, \quad \exists u \in L^2(0, \infty; \mathbf{R}^m)$$

such that the solution $x(t; x_0, u)$ *of* (2.1) *satisfies*

$$\int_0^\infty |x(t; x_0, u)|^2 \, dt < \infty \tag{2.16}$$

(ii) We say that the system (2.1) *is* stabilizable by feedback *if there exists a feedback matrix* K *such that*

$$u(t) = Kx(t)$$

and

$$\frac{dx}{dt} = (A + BK)x(t)$$

is asymptotically stable.

Remark 2.2 For finite-dimensional linear systems if the system is open-loop stabilizable then it is stabilizable by feedback and conversely. Note that by the previous discussion, stabilizability is a weaker concept than controllability, since stabilizability only requires that the uncontrollable part of the system be asymptotically stable. ∎

The dual of the concept of stabilizability is that of detectability. This is best discussed by considering the dual system

$$\frac{d\xi}{dt} = -A^*\xi(t) - C^*y(t), \tag{2.17}$$

which involves only the state-output part of the system.

Definition 2.4 *The system* (2.1) *is said to be* detectable *if the dual system* (2.17) *is stabilizable by feedback.*

Remark 2.3 Detectability requires that the unobservable part of the system be asymptotically stable and hence is a weaker concept than observability. ∎

A matrix test for stabilizability and detectability can be given.

Theorem 2.5 *The following properties are equivalent:*
(i) (A, B) *is a controllable pair*

(ii) rank $[\lambda I - A \vdots B] = n, \quad \forall \lambda \in \mathbf{C}$

(iii) rank $[\lambda I - A \vdots B] = n$, *for each eigenvalue* λ *of* A.

Proof. Note that (ii) and (iii) are equivalent since the first $n \times n$ block of the $n \times (n+m)$ matrix $[\lambda I - A \vdots B]$ has full rank whenever λ is not an eigenvalue of A.

We now prove that (i) $\Rightarrow$ (ii). For this purpose, assume that rank $[\lambda I - A \vdots B] < n$ for some $\lambda \in \mathbf{C}$. Hence the space spanned by the rows of $[\lambda I - A \vdots B]$ has dimension less than n and therefore there exists some vector $z^* \in \mathbf{R}^n$ and some λ such that $z^*[\lambda I - A \vdots B] = 0$. Hence

$$z^*A = \lambda z^* \text{ and } z^*B = 0,$$

and therefore $z^* A^k B = \lambda^k z^* B = 0$, $\forall K$, and hence

$$z^*[B \vdots AB \vdots \cdots \vdots A^{n-1}B] = 0$$

which contradicts controllability.

To prove (i) $\Rightarrow$ (ii) assume (A, B) is not a controllable pair. Then from the Structure Theorem, there exists a non-singular linear transformation $T : \mathbf{R}^n \to \mathbf{R}^n$ such that

$$(A, B) \mapsto \left[\bar{A} = \begin{bmatrix} A_{11} & A_{12} \\ 0 & A_{22} \end{bmatrix}, \quad \bar{B} = \begin{bmatrix} B_1 \\ 0 \end{bmatrix}\right]$$

where A_{11}, B_1 is a controllable pair and A_{11} is $r \times r$ with $r < n$. Let λ be an eigenvalue of A_{22}^* with the corresponding eigenvector v so that

$v^*(\lambda I - A_{22}) = 0$. Then the n-vector $(\neq 0)$

$$w = \begin{bmatrix} 0 \\ v \end{bmatrix} \text{ is such that}$$

$w^* \bar{A} = \lambda w^*$ and $w^* \bar{B} = 0$. Hence $z = (T^*)^{-1} w \neq 0$ satisfies $z^*(\lambda I - A) T^{-1} \vdots B) = 0$ and therefore $z^*(\lambda I - A \vdots B) = 0$ contradicting (ii). ∎

Corollary 2.2 *(i)* (A, B) *is* stabilizable *if and only if*

$$\mathit{rank}\, [\lambda I - A \vdots B] = n, \ \forall\, \lambda \in \mathbf{C}, \ \mathit{Re}\, \lambda \geq 0.$$

(ii) Dually (A, C) *is detectable if and only if*

$$\mathit{rank} \begin{bmatrix} \lambda I - A^* \\ C^* \end{bmatrix} = n, \ \forall\, \lambda \in \mathbf{C} \text{ such that } \mathit{Re}\, \lambda \geq 0.$$

Corollary 2.2 is generalized to an infinite dimensional setting when A is the generator of a C_0-semigroup on a Hilbert space H with e^{tA} compact for $t > 0$ (and indeed for more general situations) and $B \in \mathcal{L}(U, H)$ and $C \in \mathcal{L}(H; Y)$ and U and Y are Hilbert spaces (Proposition 3.3, Part III, Chapter 1).

2.7 Applications of controllability and observability

The concepts of controllability, observability and the slightly weaker notions of stabilizability and detectability have important applications in diverse areas of systems theory and optimal control. In this section we present two such applications. The first application is to theory of stability and tests for stability via the Lyapunov equations. Our second application is to regulator theory where we show that if a system is controllable and observable then we can build a compensator to regulate the system in a satisfactory way.

2.7.1 Stability

Our first application is to the study of stability of linear differential equations via the Lyapunov method.

Consider the stability problem for the equation

$$\frac{dx}{dt} = Ax(t), \quad A : \mathbf{R}^n \to \mathbf{R}^n. \tag{2.18}$$

It is well-known (cf. Chapter 1 in Volume I) that the following are equivalent:

(i) (2.18) is asymptotically stable

(ii) (2.18) is exponentially stable

(iii) $\forall x_0 \in \mathbf{R}^n, \quad \int_0^\infty |e^{At}x_0|^2 \, dt < \infty.$

To test for stability of (2.18), we look for a Lyapunov function of the form

$$V(x) = (x, Px), \quad P^* = P > 0 \text{ positive definite}$$

Computing the derivative of V along trajectories of (2.18) we get

$$\frac{dV}{dt} = (x, A^*P + PAx)$$

and if it is to be negative we require that

$$A^*P + PA = -Q, \quad Q > 0 \text{ positive definite.} \tag{2.19}$$

It is well-know that if A is a stability matrix (all its eigenvalues have strictly negative real part), then the solution of (2.19) can be written as

$$P = \int_0^\infty e^{A^* t} Q e^{At}\, dt, \tag{2.20}$$

and conversely if (2.19) has a positive definite solution then A is a stability matrix. A generalization of these results in infinite dimensional spaces has been extensively discussed in Volume I, Chapter 1.

Now suppose that $Q \geq 0$ and factorize Q as $Q = C^*C$ $C \in \mathcal{L}(\mathbf{R}^p; \mathbf{R}^n)$. Introduce the state-output system:

$$\begin{cases} \dfrac{dx}{dt} = Ax(t) \\ y(t) = Cx(t). \end{cases} \tag{2.21}$$

Definition 2.5 *The state-output system is said to be* asymptotically output stable *if*

$$\forall x \in \mathbf{R}^n, \quad \int_0^\infty |Ce^{At}x|^2\, dt < \infty.$$

Now it is easy to see that if (A, C) is an observable pair and A is asymptotically stable then the equation

$$A^*P + PA = -C^*C$$

has a solution and

$$P = \int_0^\infty e^{A^* t} C^* C e^{At}\, dt > 0 \text{ (positive-definite)},$$

the positive-definiteness being a consequence of observability. Conversely if the state-output system is asymptotically output stable and if (A, C) is an observable pair then A is asymptotically stable. We can now weaken this further.

Theorem 2.6 *If the state-output system is asymptotically output stable (by some feedback matrix K) and if (A, C) is a detectable pair then the system*

$$\frac{dx}{dt} = (A + BK)\, x(t)$$

is asymptotically stable.

Proof. By hypothesis there exists a $K^* \in \mathcal{L}(\mathbf{R}^n; \mathbf{R}^p)$ such that $(A^* + C^*K^*)$ is a stability matrix and hence $(A + KC)$ is a stability matrix. Now write

$$\frac{dx}{dt} = Ax(t)$$

as

$$\frac{dx}{dt} = (A + KC)x(t) - KCx(t)$$

and we can write its solution as

$$x(t) = e^{(A+KC)t}x + \int_0^t e^{(A+KC)(t-s)}KCx(s)\,ds. \tag{2.22}$$

Hence

$$\begin{aligned} &[\int_0^\infty |x(t)|^2\,dt]^{\frac{1}{2}} \\ &\le [\int_0^\infty |e^{(A+KC)t}x|^2\,dt]^{\frac{1}{2}} \\ &\quad +[\int_0^\infty |\int_0^t e^{(A+KC)(t-s)}KCx(s)\,ds|^2\,dt]^{\frac{1}{2}}. \end{aligned} \tag{2.23}$$

Furthermore, since $(A+KC)$ is a stability matrix, there exists $\alpha > 0$ and $M \ge 1$ such that

$$\forall x \in \mathbf{R}^n, \quad |e^{(A+KC)t}x| \le M\,e^{-\alpha t}|x|,$$

and the second term of the right-hand side of inequality (2.23) can be majorized by

$$\int_0^\infty [\int_0^t M\,e^{-\alpha(t-s)}\,||K||\,|Cx(s)|\,ds]^2\,dt$$

Now introduce the function

$$f(s) = \begin{cases} M||K||e^{-\alpha s}, & s \ge 0 \\ 0, & \text{otherwise} \end{cases}$$

and

$$g(s) = \begin{cases} |Cx(s)|, & s \ge 0 \\ 0, & \text{otherwise.} \end{cases}$$

Now $f \in L^1(-\infty, \infty; \mathbf{R})$ and $g \in L^2(-\infty, \infty; \mathbf{R})$. Hence by Young's inequality

$$||f * g||_{L^2} \leq ||f||_{L^1}.||g||_{L^2},$$

where

$$(f * g)(t) = \int_{-\infty}^{\infty} f(t-s)g(s)\, ds = \int_0^t M\, ||K||\, e^{-\alpha(t-s)}\, |Hx(s)|\, ds.$$

This proves the theorem. ■

Remark 2.4 This theorem and its proof generalize to certain infinite dimensional Hilbert space situations and have implications in the study of the algebraic Riccati equation. See §3.4, Part III, Chapter 1. ■

2.7.2 Compensators for linear systems

Consider the linear control system

$$\begin{cases} \dfrac{dx}{dt} = Ax(t) + Bu(t) \\ y(t) = Cx(t). \end{cases} \tag{2.24}$$

We have seen previously that if (A, B) is a controllable pair then the spectrum of the closed-loop system can be made arbitrary by state-feedback. We now wish to discuss the situation where full-state feedback is not available and the control has to be of the form

$$u(t) = \int_0^t \Lambda(t-s)y(s)\, ds + v(t) \tag{2.25}$$

where the kernel Λ corresponds to some finite-dimensional system and $v(t)$ is a reference input. We would like to investigate the properties of the corresponding closed-loop system. We shall construct such a control by

(i) constructing a state estimator $\hat{x}(\cdot)$ from the observations $y(\cdot)$.

(ii) and by constructing an appropriate control as a linear function of the estimate $\hat{x}(t)$.

We first construct such an estimator in the form

$$\begin{cases} \dfrac{d\hat{x}}{dt} = A\hat{x}(t) + M\nu(t) + Bu(t), \\ \nu(t) = y(t) - C\hat{x}(t), \\ \text{and } M \text{ is to be chosen later.} \end{cases} \tag{2.26}$$

The vector $\nu(\cdot)$ has the interpretation as an "innovation" function, that is, $\nu(t)$ is the new information in the output $y(t)$ at time t not contained in $C\hat{x}(t)$. Let $e(t) = x(t) - \hat{x}(t)$ be the error between the state and its estimate. We see that

$$\frac{de}{dt} = (A - MC)e(t). \tag{2.27}$$

Now if we assume that (A, C) is an observable pair then the spectrum of $(A - MC)$ can be made arbitrary by suitable choice of M. In particular, by suitable choice of M, $(A - MC)$ can be made a stability matrix. Note that for the requirement of stability of the $(A - MC)$ it is enough to assume that (A, C) is a detectable pair.

Now consider a feedback control of the form

$$u(t) = -K\hat{x}(t), \tag{2.28}$$

giving us a closed-loop system,

$$\begin{aligned} \frac{dx}{dt} &= Ax(t) - BK\hat{x}(t) \\ &= (A - BK)x(t) + BKe(t). \end{aligned} \tag{2.29}$$

By the pole-assignment theorem, the spectrum of $(A - BK)$ can be made arbitrary by suitable choice of K, provided (A, B) is a controllable pair (stabilizability is enough).

Consider the pair of differential equations:

$$\begin{bmatrix} \dfrac{dx}{dt} \\ \\ \dfrac{de}{dt} \end{bmatrix} = \begin{bmatrix} A - BK & BK \\ \\ 0 & A - MC \end{bmatrix} \begin{bmatrix} x(t) \\ \\ e(t) \end{bmatrix}. \tag{2.30}$$

It is clear that the spectrum of the above block matrix is determined by that of $A - BK$ and $A - MC$ and hence can be made arbitrary by proper choice of K and M and in particular can be made a stability matrix. This is one of the fundamental results of linear feedback control.

We now turn to a discussion of quadratic-cost optimal control and show how this theory gives rise to optimal feedback controllers which render the closed-loop system asymptotically stable. It turns out that the theory of controllability and observability plays an important role in the study of optimal control over an infinite time horizon.

3 Optimal control

Consider the *linear control system*

$$\begin{cases} \dfrac{dx}{dt} = Ax(t) + Bu(t) \\ y(t) = Cx(t) \end{cases} \tag{3.1}$$

on the time interval $[s, T]$, where $0 < s < T$ with the initial condition $x(s) = x$; where x is arbitrary. The notation and terminology is the same as the ones defined in the section on controllability and observability.

We shall be primarily concerned with the following *optimal control problem*: to choose a control $\hat{u}(\cdot) \in L^2(s, T; \mathbf{R}^m)$ which minimizes the cost functional

$$J(u; s, x) = \frac{1}{2}\int_s^T [(u(t), Ru(t)) + (x(t), Qx(t)]\,dt + \frac{1}{2}(x(T), Sx(T)) \tag{3.2}$$

where $R = R^* > 0$, $Q = Q^* = C^*C \geq 0$ and $S = S^* \geq 0$. A control $\hat{u}(\cdot)$ minimizing $J(u; x)$ will be called an *optimal control.* We shall not be concerned with the question of existence and uniqueness of solutions but with the characterization of the optimal control $\hat{u}(\cdot)$ and the corresponding optimal trajectory $\hat{x}(\cdot)$.

Even though we are interested in solving the optimal control problem over a fixed interval $[0, T]$ and for the fixed initial condition x_0, it will turn out to be conceptually important to solve the problem for all initial points (s, x), $0 \leq s < T$.

We shall also be interested in the infinite time problem: find a control $\hat{u}(\cdot) \in L^2(s, \infty; \mathbf{R}^m)$ which minimizes:

$$J(u; s, x) = \frac{1}{2} \int_s^\infty [(u(t), Ru(t)) + (x(t), Qx(t))] \, dt \tag{3.3}$$

3.1 Optimal control in a finite time interval

We first discuss the finite-time situation. We are interested in obtaining the optimal control $\hat{u}(\cdot)$ in feedback form $\hat{u}(t) = K(t)x(t)$, where $K(t) \in \mathcal{L}(\mathbf{R}^n; \mathbf{R}^m)$. There are three basic approaches to the solution of this problem. The first is via the Calculus of Variations or equivalently the Maximum Principle and makes use of the adjoint equation. It can be shown that the optimal control is characterized by the solution of the following system:

$$\begin{cases} \dfrac{d\hat{x}}{dt} = A\hat{x}(t) + B\hat{u}(t) \\ \hat{x}(s) = x \end{cases} \tag{3.4}$$

$$\begin{cases} \dfrac{dp}{dt} = -A^*p(t) - Q\hat{x}(t) \\ p(T) = S\hat{x}(T) \end{cases} \tag{3.5}$$

$$R\hat{u}(t) + B^*p(t) = 0 \quad s \le t \le T. \tag{3.6}$$

We recognize this as a *two-point boundary value problem.* The system (3.4)–(3.6) can be decoupled as follows. It is here the idea of invariant embedding, namely, the idea of considering the optimal control problem for all initial (s, x) where $0 < s < T$ and $x \in \mathbf{R}^n$ arbitrary becomes important. One first shows that the system of equations

$$\begin{cases} \dfrac{d\xi}{dt} = A\xi(t) - BR^{-1}B^*\eta(t) \\ \xi(s) = x. \end{cases} \tag{3.7}$$

$$\begin{cases} \dfrac{d\eta}{dt} = -A^*\eta(t) - Q\xi(t) \\ \eta(T) = S\xi(T) \end{cases} \tag{3.8}$$

admits a unique solution. We can then prove that the mapping

$$x \mapsto (\xi(\cdot), \eta(\cdot)) : \mathbf{R}^n \to W(s, T; \mathbf{R}^n) \times W(s, T; \mathbf{R}^n) \tag{3.9}$$

where

$$W(s,T;\mathbf{R}^n) = \{z : z \in L^2(s,T;\mathbf{R}^n), \frac{dz}{dt} \in L^2(s,T;\mathbf{R}^n)\}$$

is an affine continuous map and finally that the mapping

$$x \mapsto \eta(s) : \mathbf{R}^n \to \mathbf{R}^n \text{ is an affine mapping.} \tag{3.10}$$

From this it follows that we can write

$$\eta(s) = P(s)x + r(s), \tag{3.11}$$

where

$$P(s) \in \mathcal{L}(\mathbf{R}^n;\mathbf{R}^n) \text{ and } r(s) \in \mathbf{R}^n. \tag{3.12}$$

Let $(x(\cdot), p(\cdot))$ be a solution of (3.4)–(3.6) in the interval $[0,T]$. Then

$$p(t) = P(t)x(t) + r(t), \quad \forall t \in [0,T],$$

and a calculation shows that $P(t)$ and $r(t)$ satisfy:

$$\begin{cases} \dfrac{dP}{dt} + A^*P(t) + P(t)A - P(t)BR^{-1}B^*P(t) + Q = 0 \\ P(T) = S \end{cases} \tag{3.13}$$

$$\begin{cases} \dfrac{dr}{dt} + A^*r(t) - P(t)BR^{-1}B^*r(t) = 0 \\ r(T) = 0. \end{cases} \tag{3.14}$$

Clearly $r(t)$ is identically zero and standard results of ordinary differential equations prove that (3.13) has a unique global solution.

Therefore we have obtained the optimal control in feedback from:

$$\hat{u}(t) = -R^{-1}B^*P(t)\hat{x}(t).$$

The same result can be obtained using the method of Dynamic Programming. Conceptually, the Calculus of Variations gives necessary conditions of optimality while the method of Dynamic Programming gives sufficient conditions of optimality. But for the strictly convex quadratic cost problem both methods give necessary and sufficient conditions of optimality.

Let
$$V(s,x) = \inf_u J(u;s,x).$$

Then an application of the Principle of Optimality leads to the following partial differential equation for V:

$$\begin{cases} \dfrac{\partial V}{\partial s}(s,x) + \min_u [\dfrac{1}{2}(u(s), Ru(s)) + \dfrac{1}{2}(x(s), Qx(s)) \\ \qquad\qquad + (\nabla_x V(x,s), Ax(s) + Bu(s))] = 0 \\ \\ V(T,x) = (Sx, x), \end{cases} \tag{3.15}$$

where $\nabla_x V(x,s)$ is the gradient of $V(x,s)$ with respect to the vector x.

Carrying out the minimization, we obtain

$$u(s) = -R^{-1}B^*\nabla_x V(x,s), \quad \forall\, s \in [0,T]. \tag{3.16}$$

Hence we obtain the so-called Bellman-Hamilton-Jacobi equation

$$\begin{aligned} &\frac{\partial V}{\partial s}(s,x) + \frac{1}{2}(\nabla_x V(x,s), BR^{-1}B^*\nabla_x V(x,s)) \\ &+\frac{1}{2}(x, Qx) + (A^*\nabla_x V(s,x), x) = 0 \end{aligned} \tag{3.17}$$

Now, equation (3.17) can be solved by looking for a solution of the form:
$$V(s,x) = \frac{1}{2}(x, P(s)x) + (x, r(s)). \tag{3.18}$$

We can check that $P(s)$ and $r(s)$ satisfies equations (3.13) and (3.14) previously obtained, and we recover the same results as obtained by Calculus of Variation arguments.

There is a third approach, in some sense related to the Dynamic Programming approach to the solution of quadratic cost optimal control problems. This is the method of completing the squares. The main idea here is to observe that if

$$\frac{dx}{dt} = Ax(t) + Bu(t),$$

and if $P(t) = P^*(t)$ be such that

$$\frac{dP}{dt}$$

exists in the interval $[s, T]$, then defining

$$z(t) = \begin{bmatrix} u(t) \\ x(t) \end{bmatrix} \text{ and } L(t) = \begin{bmatrix} 0 & R^{-1}B^*P(t) \\ P(t)B & \dfrac{dP}{dt} + A^*P(t) + P(t)A \end{bmatrix}$$

we get

$$\int_s^T (z(t), L(t)z(t))dt - (x(T), P(T)x(T)) + (x(s), P(s)x(s)) = 0 \tag{3.19}$$

Now let $P(t)$ be the global solution on the interval $[s, T]$ of the equation

$$\begin{cases} \dfrac{dP}{dt} + A^*P(t) + P(t)A - P(t)BR^{-1}B^*P(t) + Q = 0 \\ \\ P(T) = S. \end{cases} \tag{3.20}$$

Then adding the identity (3.20) to $J(u; s, x)$ and doing some algebraic manipulations, we obtain:

$$J(u; s, x) = \int_s^T |u(t) + R^{-1}B^*P(t)x(t)|^2 dt + (x(s), P(s)x(s)). \tag{3.21}$$

Hence the optimal control is given by

$$\hat{u}(t) = -R^{-1}B^*P(t)x(t), \tag{3.22}$$

and the optimal cost $V(s, x)$ by

$$V(s, x) = (x, P(s)x). \tag{3.23}$$

In this book, we use the latter two methods in infinite dimensional situations to obtain results on optimal control for the quadratic cost problem in a finite-time interval. Apparently, one obtains more general results using this methodology.

Part II, Chapter 1 to Chapter 3 develop the optimal control of- parabolic, hyperbolic and delay equations over a finite time interval. The observation operator is usually considered to be also bounded but the case of unbounded observation operators is also treated. Chapter 2 deals with parabolic equations with unbounded control operators while Chapter 3 deals with hyperbolic equations with unbounded control operators. The latter two chapters are the most technical and the full power of the theory developed in Volume I needs to be used.

3.2 Optimal control over an infinite time interval

The study of the quadratic cost problem over a finite and an infinite time interval is essentially a study of the Riccati differential equation over the finite time interval $[s,T]$

$$\frac{dP}{dt} + A^*P(t) + P(t)A - P(t)BR^{-1}B^*P(t) + C^*C = 0 \tag{3.24}$$

subject to the terminal condition

$$P(T) = P_T \geq 0, \tag{3.25}$$

its asymptotic behaviour and the corresponding matrix quadratic equation

$$A^*P + PA - PBR^{-1}B^*P + C^*C = 0. \tag{3.26}$$

The following theorem is proved (in certain infinite dimensional situations) in Part II, Chapter 1, §2 and Part III, Chapter 1, §2 to 4.

Theorem 3.1 *There exists a matrix $P(t)$ satisfying the following:*
(i) $P(\cdot)$ is defined and belongs to $C^1(s,T;\mathcal{L}(\mathbf{R}^n))$ and satisfies (3.24) *and* (3.25).
(ii) $P(t) \geq 0, s \leq t \leq T$ and is the unique solution of (3.24)–(3.25).
(iii) Let $\tilde{K}(t)$ be a continuous function on $[s,T]$ and let $\tilde{P}(t)$ be the solution of the linear differential equation

$$\begin{aligned} &\frac{d\tilde{P}}{dt} + (A - B\tilde{K}(t))^*\tilde{P}(t) + \tilde{P}(t)(A - B\tilde{K}(t)) \\ &+ C^*C + \tilde{K}(t)^*R\tilde{K}(t) = 0 \end{aligned} \tag{3.27}$$

$$\tilde{P}(T) = P_T \geq 0. \tag{3.28}$$

If $P(t)$ is a solution (3.24) and (3.25), then

$$P(t) \leq \tilde{P}(t), \; s \leq t \leq T.$$

(iv) Consider the equation (3.24) with $s \to -\infty$ and $T = 0$. If (A, B) is stabilizable, then $P(t)$ is bounded on $]-\infty, 0]$. If (A, C) is detectable then

$$P_\infty = \lim_{t \to -\infty} P(t)$$

exists and is positive semi-definite.

In that case P_∞ is the unique positive semi-definite solution of (3.26) and

$$A - BR^{-1}B^* P_\infty$$

is a stability matrix.

The optimal control over the infinite time interval is then given by

$$\hat{u}(t) = -R^{-1}B^* P_\infty \hat{x}(t). \tag{3.29}$$

The study of these control problems when control is exercised through the boundary is most complicated for hyperbolic equations and is taken up in Chapter 3 of Part III. The study of exact controllability comes into the picture here to ensure that the space of admissible controls is nonempty.

4 A glimpse into H^∞-theory: state feedback case

4.1 Introduction

In many control problems the quadratic criterion is not the most appropriate. We consider a *disturbance attenuation problem* when the full state vector can be measured and introduce the so-called $\mathcal{H}^\infty$-optimal control problem. There is now a vast literature on this topic. For two textbook presentations the reader is referred to T. Basar and P. Bernhard [1] and B.A. Francis [1].

Consider the finite-dimensional linear time-invariant system

$$\begin{aligned} \dot{x}(t) &= Ax(t) + Lw(t) + Bu(t) \\ z(t) &= Cx(t) + Du(t), \end{aligned} \tag{4.1}$$

where u is interpreted as the controls and z as controlled outputs which are to be made small in the presence of exogenous disturbances w. The matrices and vectors have appropriate dimensions.

Let $S(A, B)$ denote the set of all constant gain stabilizing state feedback matrices. That is

$$S(A, B) = \{K : \mathrm{Re}\,[\lambda_i(A + BK)] < 0\}.$$

Then for $u \in S(A, B)$, the closed loop dynamics take the form

$$\dot{x}(t) = (A + BK)x(t) + Lw(t) \tag{4.2}$$
$$z(t) = (C + DK)x(t). \tag{4.3}$$

or in transfer function form

$$z(s) = T_K(s)w(s)$$

where

$$T_K(s) = (C + DK)(sI - (A + BK))^{-1}L.$$

The problem of $\mathcal{H}^\infty$-optimal state feedback is now stated as

$$\inf_{K \in S(A,B)} ||T_K||_{\mathcal{H}^\infty},$$

where

$$||T_K||_{\mathcal{H}_\infty} = \sup_\omega \sigma_{\max}||T(j\omega)||$$

and $\sigma_{\max}$ denotes the maximum singular value of the matrix $T(j\omega)$.

This $\mathcal{H}^\infty$-optimal state feedback problem represents a special case of the more general $\mathcal{H}^\infty$- optimal output feedback disturbance rejection problem.

4.2 Main results

The following assumptions are made on (4.1)
(A1) The pair (A, B) is stabilizable,
(A2) The pair (A, C) is observable, and
(A3) $D^*[C\ D] = [0\ I]$.

(A1) is necessary for the stabilization of (4.1). Assumption (A2) is a technical assumption which guarantees the invertibility of certain algebraic Riccati equation solutions. The orthogonality assumption (A3) is analogous to no cross weighting between the state and control in the standard LQ problem. In case (A3) is violated, the change of control variables from u to v given by

$$u(t) = (D^*D)^{-\frac{1}{2}}v(t) - (D^*D)^{-1}Cx(t)$$

yields a new system which satisfies (A3). Furthermore, it can be shown that a controlled output of the form

$$z(t) = Cx(t) + Du(t) + Eu(t)$$

may be transformed to the form of (4.1). Thus, (A3) is made with minimal loss of generality.

The main result is now stated.

Theorem 4.1 *Consider the linear system* (4.1) *under assumptions* (*A*1) *to* (*A*3). *Under these conditions, there exists a* $K \in S(A, B)$ *such that* $||T_K||_{\mathcal{H}^\infty} < \gamma$ *if and only if there exists an* $X = X^* > 0$ *which satisfies*

$$XA + A^*X + C^*C + X(\frac{1}{\gamma^2}LL^* - BB^*)X = 0 \tag{4.4}$$

with $A + (\frac{1}{\gamma^2}LL^* - BB^*)X$ *stable (i.e. all eigenvalues in the open left half complex plane).*

The remainder of this section is devoted to the proof of Theorem 4.1 for $\gamma = 1$. By linearity, this simplification is made without loss of generality.

First, some preliminary results from linear quadratic optimization theory are stated for the linear system

$$\begin{aligned} \dot{x}(t) &= Ax(t) + Bu(t), \quad x(0) = x_0 \\ y(t) &= Cx(t) \end{aligned}$$

Lemma 4.1 *Let (A, B) be stabilizable and (A, C) detectable (resp. observable). Then there exists a unique $P = P^* \geq 0$ (resp. > 0) which satisfies*

$$PA + A^*P + C^*C - PBB^*P = 0$$

with $A - BB^*P$ stable. Furthermore, the state feedback control law $u(t) = -B^*Px(t)$ is a minimizing solution for the problem

$$\inf_u \frac{1}{2} \int_0^\infty (|y(t)|^2 + |u(t)|^2)\, dt.$$

This is Theorem 3.1 restated.

Lemma 4.2 *Let $G(s) = C(sI - A)^{-1}B$ with A stable. Then $||G||_\infty < 1$ if and only if there exists an $X = X^* \geq 0$ which satisfies*

$$XA + A^*X + C^*C + XBB^*X = 0$$

with $A + BB^*X$ stable. Furthermore, the state feedback $u(t) = BB^*Xx(t)$ solves

$$\sup_u \frac{1}{2} \int_0^\infty (|y(t)|^2 - |u(t)|^2)\, dt.$$

For a proof of the above lemma see J.C. WILLEMS [1].

The proof of Theorem 4.1 is now presented. To prove the "if" direction, let $X = X^* > 0$ satisfies (4.4) with $A + (LL^* - BB^*)X$ stable. A manipulation of (4.4) yields

$$\begin{aligned} &X(A - BB^*X) + (A - BB^*X)^*X \\ &+(C - DB^*X)^*(C - DB^*X) + XLL^*X = 0. \end{aligned} \tag{4.5}$$

Now assumptions (A2) to (A3) imply that the pair $(A - BB^*X, C - DB^*X)$ is observable. This observability, (4.5), and $X > 0$ together imply that the matrix $A - BB^*X$ is stable via standard Lyapunov stability theory. Thus, $K = -B^*X \in S(A, B)$. Finally, using (4.5) and the stability of $A + (LL^* - BB^*)X$ in Lemma 4.2 implies that $||T_K||_{\mathcal{H}^\infty} < 1$, which is the desired result.

To prove the "only if" direction requires the following preliminary result:

Lemma 4.3 *There exists an $X = X^* > 0$ which satisfies*

$$XA + A^*X + C^*C + X(LL^* - BB^*)X = 0 \tag{4.6}$$

with $A+(LL^*-BB^*)X$ stable if and only if there exists a $P = P^* > 0$ which satisfies

$$AP + PA^* + PC^*CP + LL^* - BB^* = 0 \tag{4.7}$$

with $-A^* - C^*CP$ stable.

Proof. Let $P = X^{-1}$. This establishes an equivalence of the existence of positive definite solutions to either (4.6) or (4.7). To show equivalence of the stability conditions, a manipulation of either (4.6) or (4.7) yields the similarity condition

$$A + (LL^* - BB^*)X = P(-A^* - C^*CP)P^{-1}.$$

■

Thus to prove the "only if" portion of Theorem 4.1, it suffices to prove the equivalent condition associated with (4.7).

Towards this end, let $K \in S(A, B)$ be such that $||T_K||_{\mathcal{H}^\infty} < 1$. Then from Lemma 4.2, there exists an $\tilde{X} = \tilde{X}^* \geq 0$ which satisfies

$$\tilde{X}(A + BK) + (A + BK)^*\tilde{X} + (C + DK)^*(C + DK) + \tilde{X}LL^*\tilde{X} = 0 \tag{4.8}$$

with $A + BK + LL^*\tilde{X}$ stable. A straightforward manipulation of (4.8) yields

$$\begin{aligned} &\tilde{X}A + A^*\tilde{X} + C^*C + \tilde{X}(LL^* - BB^*)\tilde{X} \\ &= -(K + B^*\tilde{X})^*(K + B^*\tilde{X}) \end{aligned} \tag{4.9}$$

Let $\tilde{P} = \tilde{X}^{-1}$. Note that assumptions (A2)–(A3) imply observability of the pair $(A + BK, C + DK)$ which in turn implies $\tilde{X} > 0$ thus allowing the inversion of $\tilde{X}$. In terms of $\tilde{P}$, (4.9) now takes the form

$$A\tilde{P} + \tilde{P}A^* + \tilde{P}C^*C\tilde{P} + LL^* - BB^* = -\tilde{V}^*\tilde{V} \tag{4.10}$$

where

$$\tilde{V} = K\tilde{P} + B^*.$$

Now (4.10) is not quite in the desired form of (4.7). Thus, as in J.C. WILLEMS [1], let $\Delta P = P - \tilde{P}$. Then subtracting (4.10) from (4.7) yields (after some straightforward manipulations)

$$(-\tilde{A})^*\Delta P + \Delta P(-\tilde{A}) + \tilde{V}^*\tilde{V} - \Delta P C^* C \Delta P = 0 \tag{4.11}$$

where $\tilde{A} = A^* + C^*C\tilde{P}$. Clearly, if one can find a symmetric positive semidefinite solution, ΔP, to (4.11), then $P = \tilde{P} + \Delta P$ satisfies the desired (4.7). However, (4.7) is of the form of a standard LQ Riccati equation as in Lemma 4.1. The requisite stabilizability and detectability in Lemma 4.1 is now shown as follows:

(1) The matrix pair $(-\tilde{A}, C^*)$ is controllable. This follows directly from assumption (A2) and that $-\tilde{A} = -A^* - C^*C\tilde{P}$.

(2) The matrix pair $(-\tilde{A}, \tilde{V})$ is detectable. To see this, a straightforward manipulation of (4.10) yields the similarity condition

$$-\tilde{A} - K^*\tilde{V} = \tilde{P}(A + BK + LL^*\tilde{X})\tilde{P}^{-1}.$$

However, recall that $\tilde{X}$ is such that $A + BK + LL^*\tilde{X}$ is stable, thus proving the desired detectability.

These observations along with Lemma 4.1 guarantees the existence and uniqueness of a positive definite ΔP which satisfies (4.11). Thus, $P = \tilde{P} + \Delta P > 0$ and satisfies (4.7).

It then remains to be shown that $-A^* - C^*CP$ is stable. However, from Lemma 4.1, one has that $-\tilde{A} - C^*C\Delta P$ is stable. But

$$-\tilde{A} - C^*C\Delta P = -A^* - C^*CP$$

which, via Lemma 4.3, completes the proof.

It is worthwhile remarking that the present proof exploits the fact that the condition $||T_K|| \leq 1$ is a characterization of dissipativity in terms of scattering variables. It is well known that the dissipative property in turn can be characterized in terms of a quadratic Lyapunov function obtained from an indefinite quadratic cost problem (cf. Lemma 4.2). The exposition here essentially exploits these ideas to obtain the desired result.

It should be remarked that the state space theory of $\mathcal{H}^\infty$ should only be considered as a computational device and not as a substitute for the operator theory solution of $\mathcal{H}^\infty$-problems. The important questions of approximation, and robustness should be treated from the input-output viewpoint and it is unclear whether it is meaningful in the state-space description. Finally, the characterization of optimality in the state-space framework is apparently an unsolved problem.

5 Final remarks

If one were to write a Volume III it should be concerned with a generalization of the following finite dimensional problem to infinite dimensions:

$$\inf_u \int_s^T w(x(t), u(t))\, dt \tag{5.1}$$

where $w(x,u) = \frac{1}{2}[(x,Qx) + (u,Ru) + 2(u,\mathbf{C}x)]$ and $Q = Q^*$ but without any definiteness condition, $R = R^* \geq 0$, subject to the dynamical constraint

$$\begin{cases} \dfrac{dx}{dt} = Ax(t) + Bu(t) \\ x(s) = x \end{cases} \tag{5.2}$$

and the related infinite time problem

$$\inf_u \int_s^\infty w(x(t), u(t))\, dt \tag{5.3}$$

with the dynamical constraint (5.2) and final value conditions $x(\infty) = 0$ or $x(\infty) = $ free.

The paper of J.C. WILLEMS [1] is the subject of this problem in finite dimensions. A variety of problems such as linear quadratic differential games, the KALMAN–YACUBOVICH–POPOV lemma of stability theory and H^∞-theory (in state space form) is related to this topic. Undoubtedly a large part of Willem's results can be generalized to infinite dimensions using the ideas presented in this work.

Notes

The results of this chapter are for the most part classical. The concepts of controllability and observability were introduced by R.E. KALMAN [1], [2] where the criteria of controllability and observability and the structure theorem were developed. In this connection see also the early paper of E. GILBERT[1]. The minimum energy control for transfer from the origin to a final state was first discussed in R.E. KALMAN, Y.C. HO and K.S. NARENDRA [1]. For a text-book study of these questions as well as a treatment of stability theory, see R.W. BROCKETT [1]. The concepts of stabilizability and detectability were introduced by Wonham in his study of the Algebraic Riccati equation (cf. (W.M. WONHAM [1])). The criteria for stabilizability and detectabiity are due to M.L.J. HAUTUS [1]. The pole-assignment theorem in the real case was proved by W.M. WONHAM [2] and independently by J.D. SIMON and S.K. MITTER [1]. The proof presented here follows W.M. WONHAM [3]. Theorem 2.6 is due to J. ZABCZYK [1]. The idea of a state estimator was first introduced by D.G. LUENBERGER [1]. For the treatment of compensators for linear systems as estimator-controller and the use of the pole-assigment theorem in this context, see J.D. SIMON and S.K. MITTER [1]. For a discussion of the quadratic cost optimal control problem see the book of R.W. BROCKETT [1] where the completing the square argument is developed. The idea of invariant embedding used in §3.1 is due to R.E. BELLMAN [1]. Its use in decoupling the Hamiltonian equations in the form described here is due to J.-L. LIONS [4]. It was formally used in the same way by S.K. MITTER [1]. Theorem 3.1 is due to W.M. WONHAM [2] (under the slightly stronger hypothesis of observability). The results of §4 follows an unpublished manuscript of S.K. MITTER and J.S. SHAMMA [1] but are essentially an application of the results given in the paper of J.C. WILLEMS [1]. There is now a vast literature on the so-called H^∞-approach to the control of linear systems. For two textbook presentations see T. BASAR and P. BERNHARD [1] and B.A. FRANCIS [1].

Chapter 2

Controllability and Observability for a Class of Infinite Dimensional Systems

1 Introduction

In §2.1 and §2.2 of Chapter 1 we have discussed criteria for controllability and observability for finite dimensional systems and have also shown that when the system is controllable we can transfer the state $z_0 \in H$ at time t_0 to the state $z_1 \in H$ at time t_1 using minimum energy controls. These results were obtained by considering the controllability operator

$$\begin{aligned} L_T &: L^2(0,T;U) \to H \\ &: u \longmapsto \int_0^T e^{(T-s)A} Bu(s)ds, \end{aligned}$$

and its adjoint

$$\begin{aligned} L_T^* &: H \to L^2(0,T;U) \\ &: y \longmapsto B^* e^{(T-\cdot)A^*} y, \end{aligned}$$

and studying the relation between the ranges and null spaces of these two operators and by showing that controllability is equivalent to invertibility of $L_T L_T^*$. As we have remarked (see Remark 2.1), Chapter 1) in some sense the same ideas can be used to obtain characterizations of controllability when the spaces U and X are infinite-dimensional Hilbert spaces, but at the expense of using much elaborate technical machinery. In this chapter we discuss questions of controllability for parabolic and second-order hyperbolic equations, plate equation and Maxwell's equation. We first deal with controllability of the abstract linear dynamical system

$$\begin{cases} z'(t) = Az(t) + Bu(t) \\ z(0) = 0, \end{cases} \tag{1.1}$$

evolving in a Hilbert space H and where A is the infinitesimal generator of a strongly continuous semi-group e^{tA} on H and $B \in \mathcal{L}(U, H)$, where U is a Hilbert space, the control space. In an infinite-dimensional setting there are at least two concepts of controllability: approximate controllability and exact controllability. These concepts are introduced in Definitions 2.1 and 2.2 and criteria of approximate controllability and exact controllability for the above abstract linear dynamical system are presented in Proposition 3.1 and §3.2 (see criterion (3.4)). Approximate controllability and null controllability for parabolic equations (distributed, boundary and pointwise control), are studied in §5 by converting it to the abstract model (1.1) in appropriate spaces. The exact controllability problem for hyperbolic equations (Neumann and Dirichlet boundary control), the plate equation and Maxwell's equation is studied by first studying exact controllability of the model (1.1) by specializing A to the case of skew-symmetric operators (§6 and §7) and then converting the various partial differential equations to this abstract model in appropriate spaces where the assumptions of the abstract results are verified. The method of proof for the abstract model makes essential use of the eigenvalue-eigenfunction structure of an appropriate partial differential operator related to the generator A and hence assumptions need to be made on the control operator B in relation to these eigenfunctions (assumptions which clearly have to be verified in concrete cases), so that lower estimates on boundary operators can be obtained.

For partial differential equations, the study of controllability has some intrinsic interest. Indeed, it appears that the most natural approach to the exact controllability problem for hyperbolic equations is obtained by using the theory of pseudo-differential operators and micro-local analysis as developed by Melrose and Sjöstrand (see C. BARDOS, G. LEBEAU and J. RAUCH [1]). For our purposes the theory of controllability (in particular exact controllability in suitably defined spaces) provides a criterion for stabilizability, a requirement for infinite-time quadratic control problems to be well-posed (see Part III of this volume). The interplay between the choice of state and control spaces in the formulation of infinite-time quadratic control problems, regularity properties for solutions of partial differential equations and verifying exact controllability properties in these spaces (leading to stabilizability and uniform stabilizability) is subtle and has only been worked out in the eighties (see, in particular the recent work of I. LASIECKA and R. TRIGGIANI [4], J.-L. LIONS [2] and Part III, Chapter 2 and 3, of this volume). A general methodological approach to these issues is via the Hilbert Uniqueness Method (HUM) of J.-L. Lions.

The scheme of development adopted in this chapter, although not necessarily leading to the sharpest results, has considerable appeal (especially from an engineering viewpoint) and provides a bridge between the finite-dimensional and infinite-dimensional theory. It uses the controllability operator and minimum energy transfer between states in an essential way. It is therefore of some interest in placing it both in historical context and also in the context of most recent developments in the subject. For this purpose we cite three sources on which the remainder of this section is based (see, I. LASIECKA and R. TRIGGIANI [4], J.-L. LIONS [2] and D.L. RUSSELL [1]). For finite-dimensional systems, the concept of controllability first arose as a technical condition related to normality and abnormality of Calculus of Variations problems and in obtaining stuctural results on the reachable set (see J.P. LASALLE [1]). In the same way, the concept of approximate controllability and its characterization first arose in the work of YU.V. EGOROV [1] again as a technical requirement in the study of time-optimal control problems for parabolic systems. An important early contribution to controllability is H.O. FATTORINI [1]. There is a close relation between approximate controllability and observability (see Definition 3.1 and Proposition 3.1 of this chapter). In concrete situations involving parabolic partial differential equations, approximate controllability is verified via observability using uniqueness and unique continuation theorems for the adjoint system (for an early use of this idea see J.-L. LIONS [4] who used a uniqueness theorem of S. MIZOHATA [1] to achieve the desired objective). Our results in §5, especially Theorem 5.2 and Lemma 5.1 implicitly use uniqueness and unique continuation ideas. For other work related to §5, see H.O. FATTORINI [1], [2], [3], R.C. MCCAMY, V.J. MIZEL and T.I. SEIDMAN [1], E.J.P.G. SCHMIDT and N. WECK [1] and Part III, Chapter 2, §3.3 of this volume.

A deeper issue in the context of parabolic systems is the concept of null controllability (exact controllability to the origin), that is, the ability of transferring an arbitrary state $z_0 \in D(A)$ to the zero state in time $T > 0$ using admissible controls. This question and its observability counterpart have been investigated by several authors (see, for example, H.O. FATTORINI and D.L. RUSSELL [1], V.J. MIZEL and T.I. SEIDMAN [1], [2], T.I. SEIDMAN [1], [2] and W. LITTMAN [1]). Most of the results known here are for parabolic equations in one-dimension (the work of Seidman being an exception) and are based on deep results in Harmonic Analysis (excepting the work of Littman which uses different ideas) related to the independence of exponentials $e^{-\lambda_k t}$, where λ_k are eigenvalues of appropriate partial differential operators entering the dynamical system. The work in the early seventies started in one-dimension but both T.I. Seidman and H.O. FATTORINI and D.L. RUSSELL [2] proved the result for balls, and it was well known that this gave the result for general domains by extension to a ball containing the domain and then restriction to the domain and its boundary. Null controllability is not discussed in this chapter but we conjecture that the methods that we adopt would lead to results for parabolic equations.

The focus of this chapter is on exact controllability (which is often a mean of verifying stabilizability) for second order hyperbolic equations, plate equations and Maxwell's equations. The reason for this is that there is an essential difference between the problem of stabilizability (uniform stabizability) for parabolic systems and for hyperbolic systems. In the parabolic case, we are generally dealing with a semigroup which is analytic with a compact resolvent and hence A (the generator) has only finitely many unstable eigenvalues with finite multiplicity. In this case we have to check whether the projection of A on the unstable subspace is controllable or not and this can be done via the Pole-Assignment Theorem (see Theorem 2.4, Chapter 1, Part I). These issues are dealt with in §3.3, Chapter 1, Part III and §3.3, Part III, Chapter 2. Concrete situations, such as the heat equation with control exercised through Dirichlet or Neumann boundary conditions can be handled using these ideas (see Remark 3.1, Chapter 2, Part III for bibliographical references).

For second order hyperbolic equations (and for plate and Maxwell's equation) the stabilizability question is far more difficult since these systems have an infinite-dimensional unstable (marginally stable) part. It is here that the study of exact controllability is most important and this has been carried out in J.-L. LIONS [2], [3] and I. LASIECKA and R. TRIGGIANI [1], [2], [3] and others (see the bibliographical references cited in the above- mentioned works). We remark here that the key to these results are estimates on boundary operators.

Let us examine this in the context of the wave equation with Dirichlet control (see Part III, Chapter 3, Example 2.1).

In this case we denote by Λ the operator

$$\Lambda h = -\Delta h, \quad \mathcal{D}(\Lambda) = H^2(\Omega) \cap H_0^1(\Omega).$$

and by D the Dirichlet map

$$Dv = y \Leftrightarrow \{\Delta y = 0 \text{ in } \Omega \text{ and } y|_{\partial\Omega} = v\}.$$

and the state space $H = L^2(\Omega) \times H^{-1}(\Omega)$ and the control space $U = L^2(\partial\Omega)$. Then the abstract model is

$$A = \begin{bmatrix} 0 & I \\ -\Lambda & 0 \end{bmatrix}, \quad Bu = \begin{bmatrix} 0 \\ \Lambda D u \end{bmatrix}.$$

We may then compute

$$B^* \begin{bmatrix} y_1 \\ y_2 \end{bmatrix} = D^* y_2 = -\frac{\partial}{\partial \nu} \Lambda^{-1} y_2 \text{ since } D^* \Lambda = -\frac{\partial}{\partial \nu}.$$

Hence

$$B^* e^{A^* t} \begin{bmatrix} y_1 \\ y_2 \end{bmatrix} = \frac{\partial \varphi}{\partial \nu}(t), \quad (y_1, y_2) \in L^2(\Omega) \times H^{-1}(\Omega).$$

and φ is the solution of the homogeneous equation

$$\begin{cases} \varphi_{tt} = \Delta\varphi \quad \text{in } \Omega \times]0,T[\\ \varphi = 0 \quad \text{in } \partial\Omega \times]0,T[\\ \left.\begin{array}{l} \varphi(\cdot,0) = \varphi_0 \\ \varphi_t(\cdot,0) = \varphi_1 \end{array}\right\} \text{in } \Omega \end{cases}$$

Now

$$\varphi_0 = \Lambda^{-1} y_2 \in \mathcal{D}(\Lambda^{1/2}) = H_0^1(\Omega)$$
$$\varphi_1 = y_1 \in L^2(\Omega).$$

The two relevant estimates are (see Part II, Chapter 3, Example 4.3 and Part III, Chapter 3, Example 2.1):

$$\text{(i)} \quad \int_\Sigma \left| \frac{\partial \varphi}{\partial \nu} \right|^2 d\Sigma \le c(T) \| \{\varphi_0, \varphi_1\} \|^2_{H_0^1(\Omega) \times L^2(\Omega)}$$
$$\text{(ii)} \quad \int_\Sigma \left| \frac{\partial \varphi}{\partial \nu} \right|^2 d\Sigma \ge c(T) \| \{\varphi_0, \varphi_1\} \|^2_{H_0^1(\Omega) \times L^2(\Omega)}$$

at least for large T, where $\Sigma = \partial\Omega \times]0,T[$. By virtue of the fact that

$$B^* e^{A^* t} \begin{bmatrix} y_1 \\ y_2 \end{bmatrix} = \frac{\partial \varphi}{\partial \nu}(t),$$

the inequality (ii) above is equivalent to the invertibility of the controllability operator

$$\int_0^T e^{tA} B B^* e^{tA^*} \, dt$$

and this is the criterion for exact controllability.

In the case of control through Neumann Boundary conditions it is difficult to identify the space on which the above estimates on boundary operators hold.

For a discussion of the above examples in the abstract setting developed in this chapter, see Theorem 6.1 and §8.1 and §8.2.

2 Main definitions

2.1 Notation

Let H be a complex Hilbert space, identified with its dual. More specifically H is the complexified version of a real Hilbert space, also denoted by H to save notation. This means that an element of H complexified is of the form

$$h = h_1 + ih_2, \text{ with } h_1, h_2 \in H.$$

The norm in H real is denoted by $|\cdot|$ and the scalar product by $(\cdot,\cdot)$. In H (complexified) the scalar product of h and h' is denoted by $(h, \overline{h'})$ where $\overline{h'}$ is the conjugate of h. The norm of h is $|h| = (h, \overline{h})^{\frac{1}{2}}$.

We consider a linear operator $A : D(A) \to H$ and $B \in \mathcal{L}(U; H)$, where U is also identified with its dual. We call H the state space and U the control space. The norm and scalar product in U is denoted in the same way as for H, and possibly with a subscript U when there is a risk of confusion. A and B are defined when H and U use scalars which are real valued, and are real valued. They are extended to the complexified versions of H and U by setting

$$Ah = Ah_1 + iAh_2$$

and similar formula for B. In the sequel, the complexified versions are needed because of eigenvalues (in fact only starting with §6). Note that the operator

$$A^*h = A^*h_1 + iA^*h_2$$

satisfies the condition

$$\forall\, h, h', \quad (A^*h, \overline{h'}) = (h, \overline{Ah'})$$

and is the adjoint of A considered as a linear operator in H complexified.

We consider the linear dynamical system

$$\begin{aligned} &z'(t) = Az(t) + Bv(t), \quad t \in (0,T) \\ &z(0) = 0. \end{aligned} \tag{2.1}$$

We assume that A is the infinitesimal generator of a strongly continuous semigroup e^{At} in H. The control function $v(\cdot)$ belongs to $L^2(0,T;U)$ (U is the ordinary real version). The solution of (2.1) is denoted by $z(t;v)$ and naturally

$$z(t;v) = \int_0^t e^{(t-s)A} Bv(s)\, ds. \tag{2.2}$$

2.2 Definitions

We set

$$F_T = \{z(T;v) \,:\, \forall\, v \in L^2(0,T;U)\}$$

we can provide F_T with the structure of a Hilbert space thanks to the following.

Proposition 2.1 *The space F_T can be structured as a Hilbert space, with continuous injection in H.*

Proof. Consider the closed subspace of $L^2(0,T;U)$ made of $v(\cdot)$ such that $z(T;v) = 0$. Let N_T be this subspace, and $L^2(0,T;U)/N_T$ be the Hilbert quotient space of $L^2(0,T;U)$ by N_T. If $u(\cdot)^\circ$ is an element of $L^2(0,T;U)/N_T$, and $u(\cdot)$ is a representative of $u(\cdot)^\circ$, then the quotient norm is given by

$$|u^\circ|^2 = \min\{J(v) \,:\, v \in L^2(0,T;U),\ v - u \in N_T\} \tag{2.3}$$

where we have set

$$J(v) = \int_0^T |v(t)|^2\, dt.$$

If $\hat{u}$ realizes the minimum in (2.3), then the scalar product in

$$L^2(0,T;U)/N_T$$

is defined by

$$(u^\circ, v^\circ) = \int_0^T (\hat{u}(t), \hat{v}(t))\, dt$$

We define the map Ψ_T from $L^2(0,T;U)/N_T$ to F_T by setting

$$\Psi_T(u^\circ) = z(T;u)$$

and Ψ_T is clearly a bijection. We define on F_T a structure of Hilbert space by setting

$$((\xi, \eta)) = (\Psi_T^{-1}\xi, \Psi_T^{-1}\eta). \tag{2.4}$$

By construction Ψ_T is an isometry. It remains to show that the injection of F_T into H is continuous. We show that

$$\frac{||\xi||}{|\xi|} \geq c, \quad \forall \xi \in F_T. \tag{2.5}$$

Indeed if (2.5) does not hold, then there exists a sequence ξ_n in F_T, such that $|\xi_n| = 1$, and $\xi_n \to 0$ in F_T. Let $u_n^\circ = \Psi_T^{-1}\xi_n$, and $\hat{u}_n$ the representative of u_n° of minimum norm, then $\hat{u}_n \to 0$ in $L^2(0,T;U)$ and $\xi_n = z(T;\hat{u}_n)$, therefore $\xi_n \to 0$ in H, which contradicts the fact that $|\xi_n| = 1$. The desired result has been proved. ∎

Let Ψ_T^* be the adjoint of Ψ_T. It is an isometry from F_T' to $L^2(0,T;U)/N_T$. Therefore the map $\Psi_T\Psi_T^*$ is an isometry from F_T' to F_T. If $\xi \in F_T$, defining $\zeta_* \in F_T'$ by the equation

$$\Psi_T\Psi_T^*\zeta_* = \xi \tag{2.6}$$

and considering the class $\Psi_T^*\zeta_* \in L^2(0,T;U)/N_T$, then any representative of this class will realize ξ. The canonical isomorphism from F_T into F_T' is $(\Psi_T\Psi_T^*)^{-1} = (\Psi_T^*)^{-1}(\Psi_T)^{-1}$. The norm product in F_T' is $||\zeta_*|| = |\Psi_T^*\zeta_*|$.

Consider now the operator

$$\Gamma_T = \int_0^T e^{tA}BB^*e^{tA^*}\,dt$$

which belongs to $\mathcal{L}(H;H)$. We call Γ_T the *controllability operator.* We can decompose Γ_T as a product of an operator with its transpose. For that, introduce π the injection of F_T in H and π^* its transpose. Note that π^*H is dense in F_T'. We can then give an interpretation of $\psi_T^*\pi^*$. We have

$$\psi_T^*\pi^*h(t) = \text{equivalence class of } B^*e^{(T-t)A^*}h. \tag{2.7}$$

Let $u^0 \in L^2(0,T;U)/N_T$. We have

$$(\psi_T^*\pi^*h, u^0) = (h, z(T;\hat{u}))$$

where $\hat{u}$ is the representative of minimum norm of u^0. Hence,

$$(\psi_T^* \pi^* h, u^0) = \int_0^T (B^* e^{(T-t)A^*} h, \hat{u}(t))\, dt.$$

Therefore what remains to do is to show that $B^* e^{(T-t)A^*} h$ minimizes the norm of elements which belong to its equivalence class. According to (2.3) we look for the element $\hat{v}$ which minimizes

$$J(v) = \int_0^T |v(t)|^2\, dt$$

among all v such that

$$\int_0^T e^{(T-t)A} B(v(t) - B^* e^{(T-t)A^*} h)\, dt = 0. \tag{2.8}$$

The optimal $\hat{v}$ satisfies (2.8) and

$$\int_0^T (\hat{v}(t), v(t))\, dt = 0$$

for all v such that

$$\int_0^T e^{(T-t)A} Bv(t)\, dt = 0.$$

It is clear that $\hat{v}(t) = B^* e^{(T-t)A^*} h$ satisfies this necessary condition and thus the property (2.7) has been demonstrated.

From (2.7) we deduce at once the decomposition

$$\Gamma_T = \pi \psi_T \psi_T^* \pi^*. \tag{2.9}$$

We now introduce the following.

Definition 2.1 *The pair* (A, B) *is* approximately controllable at time T *whenever* F_T *is* densely *embedded in* H.

If the pair (A, B) is approximately controllable, then the map π^* is injective. In that case we can identify an element h of H, with its image $\pi^* h$, and $\pi^* H$ with H. In that case we have

$$F_T \subset H \subset F_T' \tag{2.10}$$

with continuous and dense embedding of each space in the following one.

Moreover from (2.9) we have

$$\Gamma_T = \psi_T \psi_T^* \tag{2.11}$$

and Γ_T extends as an isometry from F_T' to F_T. If $h \in H$, then

$$||h||_{F_T'} = (\Gamma_T h, h)^{\frac{1}{2}} \tag{2.12}$$

and F_T' appears as the completion of H with respect to the norm (2.12).

Suppose that the pair (A, B) is approximately controllable, then $D(A)$ and F_T are two dense subspaces of H. An important question is the comparison between these two spaces. This justifies the

Definition 2.2 *The pair (A, B) is* exactly controllable at time T *whenever*

$$D(A) \subset F_T$$

with continuous and dense injection.

One may wonder why we do not use as definition of exact controllability at time T, the property $F_T = H$. The reason is that as soon as H is infinite dimensional, this property will very rarely occur. To partially support this assertion, we can state the following

Proposition 2.2 (R. Triggiani [1]) *If H is infinite dimensional and B is compact, the property $F_T = H$ cannot hold for finite time T.*

Proof. The operator B is compact, there exists a non unique sequence of orthonormal vectors w_k of U and positive numbers α_k, which tends monotonically to 0 as k tends to infinity, such that

$$w_k' = \frac{1}{\alpha_k} B w_k \text{ is an orthonormal sequence of } H \tag{2.13}$$

$$Bv = \sum_{k=0}^{\infty} \alpha_k (v, w_k) w_k' \tag{2.14}$$

Let $\{v_m\}$ be a sequence in $L^2(0,T;U)$ which tends to 0 weakly and $z_m(T) = z(T;v_m)$. We shall show that

$$z_m(T) \to 0 \text{ strongly in } H. \tag{2.15}$$

If (2.15) is proved, then defining F_T^n = subset of F_T obtained with v restricted to the ball of center 0 and radius n, it is clear that F_T^n has a compact closure in H. Since H is infinite dimensional $\overline{F_T^n}$ has an empty interior. It is nowhere dense.

Now

$$F_T \subset \bigcup_{n=1}^{\infty} \overline{F_T^n}.$$

Therefore, according to Baire Category theorem, F_T cannot coincide with H. Similarly

$$\overline{\bigcup_{T>0} F_T} = \overline{\bigcup_{m>0} F_m} \subset \bigcup_{n,m=1}^{\infty} \overline{F_m^n}$$

cannot coincide either with H.

It remains to prove (2.15). Using (2.13) one has

$$z_m(T) = \sum_{k=0}^{\infty} \alpha_k \int_0^T e^{(T-t)A} w_k'(v_m(t), w_k)\, dt. \tag{2.16}$$

Let

$$z_{m,N}(T) = \sum_{k=0}^{N} \alpha_k \int_0^T e^{(T-t)A} w_k'(v_m(t), w_k)\, dt.$$

Since v_m tends to 0 weakly in $L^2(0,T;U)$, its norm remains bounded. Then it is easy to check that

$$|z_m(T) - z_{m,N}(T)| \leq C\alpha_N$$

independently of m. Since α_N tends to 0 as N tends to infinity, to prove (2.15) it is sufficient to prove that $z_{m,N}(T)$ tends to 0 as m tends to infinity, for fixed N. For this it is enough to prove that

$$\int_0^T e^{(T-t)A} h f_m(t)\, dt \to 0 \text{ in } H$$

whenever f_m tends to 0 weakly in $L^2(0,T)$ and h is a fixed element of H. This is an easy consequence of the fact that $t \mapsto e^{(T-t)A}h$ is a continuous function on $[0,T]$ with values in H. The proof has been completed. ∎

In the sequel, we shall be interested in obtaining criteria for approximate and exact controllability.

3 Criteria for approximate and exact controllability

3.1 Criterion for approximate controllability

We introduce the following definition.

Definition 3.1 *Given $T > 0$, we say that the pair (A^*, B^*) is observable on $(0,T)$ when*

$$B^* e^{tA^*} h = 0 \text{ on } (0,T) \quad \Rightarrow \quad h = 0. \tag{3.1}$$

We can state the following result.

Proposition 3.1 *The following statements are equivalent:*

a) The pair A^*, B^* is observable on $(0,T)$,

b) $(\Gamma_T h, h)^{\frac{1}{2}}$ is a norm on H,

c) the pair (A, B) is approximately controllable at Γ.

Proof. If the observability property holds, then $(\Gamma_T h, h) \geq 0$ and $(\Gamma_T h, h) = 0$ implies $B^* e^{tA^*} h = 0$ on $(0,T)$, therefore $h = 0$.

If $(\Gamma_T h, h)^{\frac{1}{2}}$ is a norm on H, then if h satisfies $B^* e^{tA^*} h = 0$ on $(0,T)$, $(\Gamma_T h, h) = 0$ hence $h = 0$. Hence a) and b) are equivalent.

Now approximate controllability at T is equivalent to

$$(z(T;v), h) = 0, \ \forall v \quad \Rightarrow \quad h = 0. \tag{3.2}$$

But

$$\begin{aligned} (z(T;v), h) &= \left(\int_0^T e^{(T-t)A} Bv(t)\, dt, h\right) \\ &= \int_0^T (v(t), B^* e^{(T-t)A^*} h)\, dt. \end{aligned}$$

Therefore

$$\forall v, \quad (z(T;v),h)=0 \quad \Leftrightarrow \quad B^* e^{(T-t)A^*} h = 0$$

and (3.2) is thus the same thing as observability. Hence a) and c). ∎

3.2 Criteria for exact controllability and continuous observability

From our definition of exact controllability we deduce the sequence

$$D(A) \subset F_T \subset H \subset F_T' \subset (D(A))' \tag{3.3}$$

each space being densely and continuously embedded in the following one.

From (2.12) we deduce

$$(\Gamma_T h, h) \geq c_T ||h||^2_{D(A))'} \tag{3.4}$$

for any $h \in H$.

Conversely, if (3.4) holds then (3.3) holds. Indeed necessarily

$$F_T' \subset (D(A))' \tag{3.5}$$

To prove (3.5) consider the vector space $F_T' \cap (D(A))'$ provided with the norm

$$||\xi||_{F_T' \cap (D(A))'} = \max \left\{ ||\xi||_{F_T'}, ||\xi||_{(D(A))'} \right\}.$$

Let $\xi \in F_T'$ and consider a sequence $\xi_n \in H$, $\xi_n \to \xi$ in F_T'. Note that ξ_n is a Cauchy sequence in F_T' and from (3.4) a Cauchy sequence in $(D(A))'$. Therefore ξ_n is a Cauchy sequence in $F_T' \cap (D(A))'$. It follows that $\xi \in F_T' \cap (D(A))'$, in particular to $(D(A))'$. Hence (3.5). From (3.4) the injection is clearly continuous. Since H is dense in $(D(A))'$, F_T' is also densely embedded in $(D(A))'$. By duality we deduce (3.3). Therefore, the estimate (3.4) is a *criterion of exact controllability*.

Remark 3.1 There is an analogue of Proposition 3.1 for exact controllability. For this purpose, define the dual system as follows: first introduce the observation equation for the system (2.1)

$$y(t) = Cz(t),$$

where $C \in \mathcal{L}(H; Y)$, and the observation space Y is a Hilbert space identified with its dual. Then introduce the dual system (backward equation)

$$\begin{cases} p'(t) = -A^*p(t) + C^*y(t) \\ p(T) = 0. \end{cases}$$

and the dual observation equation

$$u(t) = B^*p(t).$$

We then say that the pair (A, C) is continuously observable if and only if the pair (A^*, C^*) is exactly controllable at time 0 (starting at time T). Therefore we obtain a criterion analogous to (3.4) for continuous observability. ■

3.3 Approximation

Let h be an element of H. There exists a sequence $\{\xi_n\}$ of elements of F_T which converges to h in H. Solving the equation

$$\Psi_T \Psi_T^* \zeta_n^* = \xi_n$$

defines a unique ζ_n^* in F_T'. Any control of the equivalence class $\Psi_T^* \zeta_n^*$ realizes ξ_n. In this way we can construct a sequence of controls whose corresponding state at time T is as close as wishable to h. But this procedure is not very constructive. A constructive approach is obtained in the following way: define

$$J_\varepsilon(v) = J(v) + \frac{1}{\varepsilon}|z(T; v) - h|^2 \tag{3.6}$$

with

$$z(T; v) = \int_0^T e^{(T-t)A} Bv(t)\, dt \tag{3.7}$$

and minimize $J_\varepsilon(v)$ over all $v \in L^2(0,T;U)$. This problem has a unique solution u_ε. We have

$$|z(T;u_\varepsilon)-h|^2 \le |z(T;v)-h|^2 + \varepsilon J(v), \quad \forall v,$$

and thus

$$\limsup_{\epsilon\to 0} |z(T;u_\varepsilon)-h|^2 \le \inf |z(T;v)-h|^2 = 0.$$

Therefore the sequence $z(T;u_\varepsilon)$ belongs to F_T and converges to h in H as ε tends to 0. Now we may write the necessary conditions of optimality for u_ε. It is easy to check that if p_ε is the adjoint state defined by

$$-p'_\varepsilon = A^* p_\varepsilon, \quad p_\varepsilon(T) = \frac{1}{\varepsilon}(x_\varepsilon(T)-h) \tag{3.8}$$

then

$$\begin{aligned} u_\varepsilon(t) &= -B^* p_\varepsilon(t) \\ &= -B^* e^{(T-t)A^*} p_\varepsilon(T) \\ &= -\Psi_T^* p_\varepsilon(T). \end{aligned} \tag{3.9}$$

Therefore

$$z(T;u_\varepsilon) = -\Gamma_T\, p_\varepsilon(T). \tag{3.10}$$

Since $p_\varepsilon(T) \in H$ and not just to F'_T, we have an even better result than that explained at the beginning of the paragraph.

If h belongs to F_T, then we can say something more. Indeed let $\hat{u}$ be such that

$$\Psi_T \hat{u} = h$$

and $\hat{u}$ has minimum L^2 norm. Since

$$J(u_\varepsilon) \le J(\hat{u}) \tag{3.11}$$

it follows that u_ε remains bounded in L^2. From the formula (3.9), it follows that $p_\varepsilon(T)$ remains in a bounded subset of F'_T. We pick a subsequence, still denoted by $p_\varepsilon(T)$ which converges weakly to ζ_* in F'_T. Going to the limit in (3.10) yields

$$h = -\Gamma_T \zeta_* \tag{3.12}$$

From (3.11) and the minimality of $\hat{u}$, it is easy to check that u_ε tends to $\hat{u}$ in $L^2(0,T;U)$. From (3.9) we deduce also

$$\hat{u} = -\Psi_T^* \zeta_* \tag{3.13}$$

and thus we have an approximation procedure for the element ζ_*.

We refer to A. BEL FEKIH [1] for related topics.

4 Finite dimensional control space

We shall consider in this section the case when U is finite dimensional, $U = \mathbf{R}^m$, and make precise the conditions of controllability (clearly approximate and exact controllability coincide in that case).

4.1 Finite dimensional case

Assume here that $H = \mathbf{R}^n$ and $U = \mathbf{R}^m$, $A \in \mathcal{L}(\mathbf{R}^n;\mathbf{R}^n)$, $B \in \mathcal{L}(\mathbf{R}^m;\mathbf{R}^n)$. We have the following classical result due to R.E. KALMAN [1], for which see Theorems 2.1 and 2.2, Part I, Chapter 1.

Proposition 4.1 *The pair (A,B) is controllable at any time $T > 0$ if and only if the matrix $[B\ AB \ldots A^{n-1}B] \in \mathcal{L}(\mathbf{R}^{nm};\mathbf{R}^n)$ has rank n.*

Consider now the case of diagonalizable operators

$$A = \sum_{j=1}^{J} \lambda_j P_j \tag{4.1}$$

where λ_j is complex valued, P_j is a projector, satisfying

$$P_j = P_j^*, \quad P_j P_k = \delta_{jk}, \quad \sum_{j=1}^{J} P_j = I.$$

Note that we have complexified the set up, so P_j^* means

$$(P_j^* h, k) = (h, \overline{P_j k}) = (P_j h, k)$$

hence

$$A^* = \sum_{j=1}^{J} \overline{\lambda_j} P_j.$$

We then state the following proposition.

Proposition 4.2 *Assume* (4.1), *then the pair* (A, B) *is controllable at any time* $T > 0$, *if and only if for any* j, *the operator* $P_j B B^* P_j$ *has full rank, in the subspace* $P_j H$.

Proof. The condition is necessary.
Suppose otherwise that there exists some h, with $P_j h \neq 0$, and $B^* P_j h = 0$, then

$$\begin{bmatrix} B^* \\ B^* A^* \\ \cdots \\ \cdots \\ B^* A^{*n-1} \end{bmatrix} P_j h = \begin{bmatrix} B^* P_j h \\ \overline{\lambda_j} B^* P_j h \\ \cdots \\ \cdots \\ \overline{\lambda_j}^{n-1} B^* P_j h \end{bmatrix} = 0$$

which contradicts the fact that the pair (A, B) is controllable.

The condition is sufficient.
Consider an element h such that $(\Gamma_T h, h) = 0$, hence $B^* e^{A^* t} h = 0$, i.e.,

$$\sum_{j=1}^{J} \exp(\overline{\lambda_j} t) B^* P_j H = 0, \quad t \in (0, T). \tag{4.2}$$

From the analycity of the function to the left of (4.2), it follows that (4.2) holds for any $t \in (-\infty, +\infty)$. It first follows that $B^* P_j h = 0$, for any j such that $\operatorname{Re} \lambda_j \neq 0$. Indeed, suppose λ_1 is such that $|\operatorname{Re} \lambda_1| > |\operatorname{Re} \lambda_j|$, $\forall\, j \neq 1$. We can write

$$B^* P_1 h + \sum_{j \neq 1} \exp((\overline{\lambda_j} - \overline{\lambda_1}) t) B^* P_j h = 0.$$

Taking the limit as t tends to $+$ or $-$ infinity, according to the fact that $\operatorname{Re} \lambda_1$ is positive or negative, we deduce $B^* P_1 h = 0$. More generally, there may be a set J_0 of indices j such that $\operatorname{Re} \lambda_j = \operatorname{Re} \lambda_1$, for

$j \in J_0$, and $|\mathrm{Re}\,\lambda_1| > |\mathrm{Re}\,\lambda_j|$, $\forall j \notin J_0$. Dividing (4.2) by $\exp(\mathrm{Re}\,\lambda_1 t)$ we deduce

$$\sum_{j\in J_0} e^{-i\gamma_j t} B^* P_j h + \sum_{j\notin J_0} \exp(\overline{\lambda_j} - \mathrm{Re}\,\lambda_1)t B^* P_j h = 0, \tag{4.3}$$

where $\gamma_j = \mathrm{Im}\,\lambda_j$. Suppose to fix the ideas that $\mathrm{Re}\,\lambda_1 > 0$. We derive from (4.3)

$$\begin{aligned} 0 &= \int_0^T |\sum_{j\in J_0} e^{-i\gamma_j t} B^* P_j h|^2\, dt \\ &+ \int_0^T (\sum_{j\in J_0} e^{+i\gamma_j t} B^* \overline{P_j} h, \sum_{k\notin J_0} \exp(\overline{\lambda_k} - \mathrm{Re}\,\lambda_1)t B^* P_k h)\, dt, \end{aligned}$$

hence

$$\begin{aligned} 0 &= T \sum_{j\in J_0} (P_j B B^* P_j h, h) \\ &+ \sum_{j\in J_0} \sum_{\substack{k\,\in\, J_0 \\ k\,\neq\, j}} \frac{\exp(i(\gamma_k - \gamma_j)T) - 1}{i(\gamma_k - \gamma_j)} (P_k B B^* P_j h, h) \\ &+ \sum_{j\in J_0} \sum_{k\neq J_0} \frac{\exp(\overline{\lambda}_k - \mathrm{Re}\,\lambda_1 + i\gamma_j)T - 1}{\overline{\lambda}_k - \mathrm{Re}\,\lambda_1 + i\gamma_j} (P_j B B^* P_k h, h). \end{aligned}$$

Dividing by T, and letting T tend to ∞, we deduce

$$\sum_{j\in J_0} (P_j B B^* P_j h, h) = 0,$$

or $B^* P_j h = 0$, $\forall j \in J_0$. If $\mathrm{Re}\,\lambda_1 < 0$, one should consider a similar integral between $-T$ and 0, to deduce a similar result.

Successively, we can treat in the same way all j such that $\mathrm{Re}\,\lambda_j \neq 0$. So we have

$$B^* P_j h = 0, \quad \forall j \text{ with } \mathrm{Re}\,\lambda_j \neq 0.$$

There remains the case when (4.2) reduces to

$$\sum_{j\in J_0} e^{-i\gamma_j t} B^* P_j h = 0, \quad \forall t \in (-\infty, +\infty)$$

where $\lambda_j = i\gamma_j, \gamma_j$ real, and J_0 is a subset of $\{1, \dots, J\}$. We make a calculation similar to that done after (4.3) to prove that

$$\sum_{j \in J_0} (P_j BB^* P_j h, h) = 0,$$

or $B^* P_j h = 0, \ \forall j \in J_0$.

Hence, we have proved that $B^* P_j h = 0, \ \forall j$, and this implies $h = 0$, from the assumption. ∎

4.2 General state space

We now turn to the case of a general state space, with finite dimensional control space $U = R^m$. We have

$$Bu = \sum_{i=1}^{m} b_i u_i, \text{ with } b_1, b_2, \dots, b_m \in H, \ u \in \mathbf{R}^m. \tag{4.4}$$

We state (see L. Markus [1]) the

Theorem 4.1 *Assume that*

$$b_1, b_2, \dots, b_m \in D_\infty(A) = \bigcap_{k>0} D(A^k).$$

If the linear set generated by the vectors

$$A^k b_j, \ j = 1, \dots, m, \ k = 0, 1, \dots$$

is dense in H, then the pair (A, B) is approximately controllable at any time $T > 0$. Conversely, assume that $b_j = e^{t_j A}\overline{b_j}$, with $\overline{b_j} \in D_\infty(A), t_j > 0$, and the trajectories $(e^{tA}\overline{b_j}, z)$ for any $z \in H$ are real analytic, then if the pair (A, B) is approximately controllable at some $T > 0$ necessarily the linear set generated by the vectors $\{A^k b_j, \ j = 1 \dots m, \ k = 0, 1, \dots\}$ is dense in H.

Proof. If A, B is not approximately controllable at $T > 0$, then A^*, B^* is not observable on $(0, T)$ (see Proposition 3.1). Hence

$$\exists h \neq 0, \text{ such that } \quad \forall t \in (0, T), \quad B^* e^{tA^*} h = 0,$$

hence also

$$\forall\, u \in \mathbf{R}^m, \quad (e^{tA}Bu, h) = 0, \quad \text{in } (0,T).$$

This implies

$$\forall\, j = 1, \ldots, m, \quad (e^{tA}b_j, h) = 0 \text{ in } (0,T). \tag{4.5}$$

From the assumption on the b_j's, we can differentiate in t as many times as we wish and set $t = 0$. We deduce

$$\forall\, k > 0,\, j = 1, \ldots, m, \quad (A^k b_j, h) = 0. \tag{4.6}$$

Since $h \neq 0$, this contradicts the fact that the set generated by the vectors $\{A^k b_j,\, j = 1, \ldots, m,\, k = 0, 1, \ldots\}$. is dense in H.

Let us now prove the second part of the statement. Assume that the set generated by the vectors $\{A^k b_j,\, j = 1, \ldots m,\, k = 0, 1, \ldots\}$ is not dense in H. There exists $h \neq 0$ such that (4.6) holds. This means also

$$(A^k e^{t_j A}\overline{b_j}, h) = 0. \tag{4.7}$$

Set

$$f_j(t) = (e^{tA}\overline{b_j}, h)$$

which is real analytic. Then (4.7) means that $f_j(t_j) = 0$, $f_j^{(k)}(t_j) = 0$, $\forall\, k > 0$. Therefore, $f_j(t) = 0$, $\forall\, t > 0$. Writing $f_j(t + t_j) = 0$, we have also (4.5). Therefore the pair (A^*, B^*) is not observable on $(0,T)$, which implies that the pair cannot be approximately controllable at T. ∎

Let us give a variant of Theorem 4.1 in the case when $-A$ has a discrete spectrum $\lambda_1 < \lambda_2 < \ldots \lambda_k \ldots$ with $\lambda_k \uparrow \infty$ and λ_k has multiplicity r_k. It bears similarity with Proposition 4.2. Using the projector P_k on the finite dimensional eigenspace corresponding to λ_k, the semigroup e^{tA} is then represented by the expansion

$$e^{tA}h = \sum_k e^{-\lambda_k t} P_k h \tag{4.8}$$

We consider P_k as an element of $\mathcal{L}(H; H)$. It is self-adjoint and $P_k^2 = P_k$. We have

$$z(T; u) = \sum_k \int_0^T e^{-(T-t)\lambda_k} P_k B u(t)\, dt \tag{4.9}$$

$$(\Gamma_T h, h) = \int_0^T |\sum_k e^{-\lambda_k t} B^* P_k h|^2 \, dt. \tag{4.10}$$

We shall prove the following result.

Theorem 4.2 *Assume that (4.8) is verified. Then the pair (A, B), is approximately controllable at T if and only if $P_k BB^* P_k$ is full rank, for any k.*

Proof. From Proposition 3.1 we know that the pair (A, B), is approximately controllable at T if and only if $(\Gamma_T h, h)$ is a norm on H. From the expression (4.10), we deduce that $(\Gamma_T h, h)$ is a norm on H if and only if

$$\sum_k e^{-\lambda_k t} B^* P_k h = 0 \text{ on } (0, T) \quad \Rightarrow \quad h = 0. \tag{4.11}$$

The function

$$t \mapsto \sum_k e^{-\lambda_k t} B^* P_k h$$

being analytic, the left part of the statement (4.11) implies also

$$\sum_k e^{-\lambda_k t} B^* P_k h = 0, \quad \forall t \geq 0.$$

Multiplying by $e^{\lambda_1 t}$ and letting t tend to infinity, we deduce $B^* P_1 h = 0$, and successively $B^* P_k h = 0$. Therefore (4.11) means also

$$B^* P_k h = 0, \quad \forall k \geq 0 \quad \Rightarrow \quad h = 0. \tag{4.12}$$

It is easy to check that (4.12) is equivalent to $P_k BB^* P_k$ has full rank for any k. ∎

Remark 4.1 The restriction of the system to E_k (eigenspace corresponding to λ_k) is given by the dynamics

$$\begin{aligned} z_k' &= -\lambda_k z_k + B_k u, \\ z_k(0) &= 0 \end{aligned} \tag{4.13}$$

where $B_k = P_k B$. The state z_k lies in the finite dimensional space E_k. The pair $(-\lambda_k I, B_k)$ is controllable. ∎

Remark 4.2 Note that the condition (4.12) holds also when U is not finite dimensional. Note also that from the formula (4.10) it follows that there exists a positive constant c such that

$$(\Gamma_T h, h) \leq c \sum_k \frac{(1 - e^{-2\lambda_k T})|P_k h|^2}{2\lambda_k}. \tag{4.14}$$

■

5 Controllability for the heat equation

5.1 Distributed control

We consider a bounded domain of $\mathbf{R}^n$, denoted Ω, assumed to be smooth to simplify a little. Let Γ be the boundary of Ω, and $\Sigma = \Gamma \times (0, T)$. Let O be a subdomain of Ω. We denote by χ_O the characteristic function of the subdomain O. We consider the dynamic system

$$\begin{aligned} &z' - \Delta z = v(x,t)\chi_O \\ &z|_\Sigma = 0 \\ &z(x,0) = 0 \end{aligned} \tag{5.1}$$

with v the control belongs to $L^2(O \times (0,T))$. In fact, (5.1) can be written under the general framework (2.1). Indeed, let $H = L^2(\Omega)$ and $D(A) = H^2(\Omega) \cap H_0^1(\Omega)$. We take $A = \Delta$, and $D(A)$ is equipped with the norm $|Az|, z \in D(A)$. Note that A is an isometry from $D(A)$ to H. Let next $U = L^2(O)$ and if u belongs to U,

$$Bu \equiv \begin{cases} u(x), & \text{if } x \in O \\ 0, & \text{if } x \notin O. \end{cases}$$

It is well known that there exists a sequence $0 < \lambda_1 < \lambda_2 < \ldots \lambda_k \uparrow \infty$ of eigenvalues and w_{kj}, $j = 1, \ldots, r_k$, corresponding eigenvectors where r_k is the multiplicity of λ_k, with

$$\begin{aligned} &-\Delta w_{kj} = \lambda_k w_{kj} \\ &w_{kj} \in H_0^1(\Omega), \quad |w_{kj}|_{L^2} = 1. \end{aligned} \tag{5.2}$$

Moreover (4.5) hold true. We can then state the

Theorem 5.1 *The pair (A, B) defined above is approximately controllable at any time $T > 0$*

Proof. From Remark 4.2 it is sufficient to check that the condition (4.12) holds. Note that $B^*h = h_{|O}$ and thus since $B_k^* = B^*P_k$, condition (4.12) means

$$\sum_{j=1}^{r_k} (w_{kj}, h) w_{kj|O} = 0 \quad \forall\, k \quad \Rightarrow \quad h = 0. \tag{5.3}$$

The function

$$\sum_{j=1}^{r_k} (w_{kj}, h) w_{kj}$$

is analytic in Ω. From (5.3) it follows that it is identically 0. Hence the coefficients $(w_{kj}, h) = 0$. Since this holds for any k and $j = 1 \dots r_k$, necessarily $h = 0$. The desired result is thus proved. ∎

Remark 5.1 From (4.14) we have

$$\begin{aligned}(\Gamma_T h, h) &\leq C \sum_k \frac{1 - e^{-2\lambda_k T}}{2\lambda_k} \sum_{j=1}^{r_k} (w_{kj}, h)^2 \\ &\leq C \sum_k 1/\lambda_k \sum_{j=1}^{r_k} (w_{kj}, h)^2 \\ &= ||h||^2_{H^{-1}}.\end{aligned}$$

This means that

$$H^{-1}(\Omega) \subset F_T'$$

with dense and continuous injection. By duality, it follows that

$$F_T \subset H_0^1(\Omega)$$

with dense and continuous injection. Clearly F_T increases with the Lebesgue measure of O, and $F_T = H_0^1(\Omega)$, when $O = \Omega$. In that case, exact controllability holds true. ∎

5.2 Boundary control

Consider the following situation, using a formal write up

$$\begin{aligned} &\zeta' - \Delta\zeta = 0 \\ &\zeta|_\Gamma = v \\ &\zeta(z;0) = 0 \end{aligned} \tag{5.4}$$

where $v(t) \in U = L^2(\Gamma)$. We need to clarify in what space $\zeta(T)$ lies and to what extent we can reduce (5.4) to the general framework (2.1).

In fact, we shall show that $\zeta(T;v)$ lies in $(H^2 \cap H_0^1)'$, and $\zeta \in C(0;T;(H^2 \cap H_0^1)')$. One defines $\zeta(T)$ by the transposition method of J.-L. Lions and E. Magenes [1]. Let h be an element of $L^2(\Omega) = H$, then $(-\Delta)^{-1}h \in H^2 \cap H_0^1$. Consider the solution ψ of

$$\begin{aligned} &-\psi' - \Delta\psi = 0 \\ &\psi|_\Gamma = 0 \\ &\psi(x;T) = -(-\Delta)^{-1}h = A^{-1}h \\ &\text{(recalling the notation } A = \Delta) \end{aligned} \tag{5.5}$$

and we note that $\psi = -(-\Delta)^{-1}\phi$, where

$$\begin{aligned} &-\phi' - \Delta\phi = 0 \\ &\phi|_\Gamma = 0 \\ &\phi(x;T) = h \end{aligned} \tag{5.6}$$

Clearly $\phi \in C(0;T;H)$ and $\psi \in C(0;T;H^2 \cap H_0^1)$. If we perform a formal integration by parts between (5.4) and (5.5) we obtain

$$< \zeta(T), \psi(T) >= -\int_0^T \int_\Gamma v \frac{\partial\psi}{\partial\nu}\, d\Gamma\, dt \tag{5.7}$$

and this constitutes the definition of $\zeta(T)$ as an element of $(H^2 \cap H_0^1)' = (D(A))'$.

Consider now the operator $n \in \mathcal{L}(H^2; L^2(\Gamma))$ defined by

$$n\phi = \frac{\partial\phi}{\partial\nu} \tag{5.8}$$

We set

$$B = (-\Delta)^{-1}n^* = -A^{-1}n^* \tag{5.9}$$

which belongs to $\mathcal{L}(L^2(\Gamma); H)$. We consider

$$\begin{aligned} z' &= Az + Bv \\ z(0) &= 0 \end{aligned} \tag{5.10}$$

Noting that (5.6) is equivalent to

$$\begin{aligned} -\phi' &= A\phi \\ \phi(T) &= h \end{aligned} \tag{5.11}$$

then from (5.10) and (5.11) we deduce

$$\begin{aligned} (z(T), h) &= \int_0^T (\phi, Bv)\, dt = \int_0^T (B^*\phi, v)\, dt \\ &= -\int_0^T (nA^{-1}\phi, v)\, dt = -\int_0^T (\frac{\partial \psi}{\partial \nu}, v)\, dt \end{aligned} \tag{5.12}$$

which is exactly the right-hand side of (5.7). Hence

$$(z(T), h) = (\zeta(T), -(-\Delta)^{-1}h). \tag{5.13}$$

Therefore

$$z(T) = -(-\Delta)^{-1}\zeta(T). \tag{5.14}$$

Similarly for any t, $z(t) = -(-\Delta)^{-1}\zeta(t)$, and since $z \in C(0, T; H)$ we deduce that $\zeta \in C(0; T; (H^2 \cap H_0^1)')$. Now (5.10) is exactly equivalent to (2.1).

We shall prove the following theorem.

Theorem 5.2 *The pair (A, B) is approximately controllable at any time $T > 0$. The range of $\zeta(T; v)$ is dense in $(H^2 \cap H_0^1)'$.*

Proof. It is clear that it is sufficient to prove that the pair (A, B) is approximately controllable at any time $T > 0$. For that we must check the condition (4.12), i.e.

$$\sum_{j=1}^{r_k} (h, w_{kj}) B^* w_{kj} = 0, \quad \forall\, k > 0 \quad \Rightarrow \quad h = 0. \tag{5.15}$$

This means also

$$\sum_{j=1}^{r_k}(h, w_{kj})\frac{\partial w_{kj}}{\partial \nu} = 0, \quad \Rightarrow \quad (h, w_{kj}) = 0, \quad \forall j = 1, \ldots, r_k. \quad (5.16)$$

But from the following lemma we have

$$\sum_{j=1}^{r_k}(h, w_{kj})w_{kj} = 0.$$

Hence the desired result. ∎

Lemma 5.1 *Let w be an eigenfunction of the Laplace operator corresponding to the eigenvalue λ, i.e.*

$$\begin{aligned} -\Delta w &= \lambda w, \\ w|_\Gamma &= 0 \end{aligned} \quad (5.17)$$

then one has

$$\int_\Gamma (\frac{\partial w}{\partial \nu})^2 m \cdot \nu \, d\Gamma = 2\lambda \int_\Omega w^2 \, dx \quad (5.18)$$

where $m(x) = x - x_0$.

Proof. We multiply the equation (5.17) by

$$m_\alpha \frac{\partial w}{\partial x_\alpha}$$

(using summation convention) and perform integration by parts. The desired result follows. ∎

Remark 5.2 Let Γ_0 be a part of Γ. The control is exerted on Γ_0 only, which means that it is 0 on $\Gamma_1 = \Gamma - \Gamma_0$. To treat this case, let us define

$$n_0\phi = \left.\frac{\partial \phi}{\partial \nu}\right|_{\Gamma_0}.$$

Take $B = -A^{-1}n_0^*$. The result of Theorem 5.2 extends to the pair (A, B) provided that for an eigenfunction w as in (5.17)

$$\left.\frac{\partial w}{\partial \nu}\right|_{\Gamma_0} = 0$$

implies $w = 0$. From the formula (5.18) it is sufficient that $m \cdot \nu \geq 0$ on Γ_0. ∎

Remark 5.3 From (4.14) again we deduce

$$\begin{aligned}(\Gamma_T h, h) &\leq C\sum_k \frac{1-e^{-2\lambda_k T}}{2\lambda_k}\sum_{j=1}^{r_k}(w_{kj}, h)^2 \\ &\leq C(||h||_{H^{-1}})^2.\end{aligned} \tag{5.19}$$

Therefore $H^{-1} \subset F'_T$, with continuous and dense injection. Hence also $F_T \subset H_0^1(\Omega)$, with continuous and dense injection. It follows that the range of $\zeta(T; v)$, when v lies in $L^2(0,T; L^2(\Gamma))$ is dense in H^{-1}. ∎

5.3 Neumann boundary control

We introduce some notation. We take

$$H = L^2(\Omega), \quad V = H^1(\Omega), \quad A = \Delta - I,$$

associated with the Neumann boundary condition. Hence

$$D(A) = \{z \in H^2(\Omega) \,:\, \frac{\partial z}{\partial \nu} = 0\}.$$

Again $D(A)$ is equipped with the norm $|Az|$, $z \in D(A)$. The eigenvectors are defined by

$$\begin{aligned}-\Delta w + w &= \lambda w \\ \frac{\partial w}{\partial \nu} &= 0.\end{aligned} \tag{5.20}$$

The first eigenvalue is 1. It is isolated hence $1 < \lambda_2 \leq \lambda_3 \ldots$ and $w_1 = 1/\sqrt{|\Omega|}$. We set $m(x) = x - x_0$ and define

$$\Gamma_0 = \{x \in \Gamma \,:\, m \cdot v > 0\}, \quad \Gamma_1 = \Gamma - \Gamma_0. \tag{5.21}$$

Let γ be the trace operator on Γ. Note the $\gamma \in \mathcal{L}(V; L^2(\Gamma)) \cap \mathcal{L}(H^2(\Omega);\, H^1(\Gamma))$. Let also σ_j be the tangential operators on $\Gamma \in \mathcal{L}(H^1(\Gamma); L^2(\Gamma))$, such that

$$\gamma\frac{\partial \phi}{\partial x_j}|_\Gamma = \nu_j\frac{\partial \phi}{\partial \nu} + \sigma_j\gamma\phi = \nu_j n\phi + \sigma_j\gamma\phi, \quad \forall\, \phi \in H^2(\Omega). \tag{5.22}$$

Note that $n\phi = \nu \cdot \gamma D\phi$, and $\nu \cdot \sigma = 0$.

Similarly, we define by $\gamma_0, \sigma_{j,0}$ $(\gamma_1, \sigma_{j,1})$ the respective restrictions of γ, σ_j to Γ_0 and Γ_1. We also use the notation

$$D_\sigma = (\sigma_1, \dots, \sigma_n) \in \mathcal{L}(H^1(\Gamma); (L^2(\Gamma))^n).$$

The operators $\gamma, \sigma_1\gamma, \sigma_2\gamma, \dots \sigma_n\gamma \in \mathcal{L}(D(A); L^2(\Gamma))$ and not identically 0, whereas $n = 0$ on $D(A)$. This should be compared to the situation in the Dirichlet case, where $\gamma, \sigma_1\gamma, \sigma_2\gamma, \dots \sigma_n\gamma$ are 0 on $D(A)$ and n is not 0. This explains why the analogy with the Dirichlet case implies the control to be an element of $U = (L^2(\Gamma))^{n+1}$, instead of just $L^2(\Gamma)$. The operator B is defined by

$$\begin{aligned} B &= A^{-1}(\gamma^* \ \gamma^*\sigma_1^* \ \gamma^*\sigma_2^* \dots \gamma^*\sigma_n^*) \\ &= A^{-1}(\gamma^* \ \gamma^* D_\sigma^*) \in \mathcal{L}(U; H). \end{aligned} \tag{5.23}$$

We can also view U as $(L^2(\Gamma_0))^{n+1} \times (L^2(\Gamma_1))^{n+1}$. Some of the components might be taken equal to 0, in which case they are omitted and we use γ_0^* or γ_1^* instead of γ^* (or $\sigma_{1,0}^*, \sigma_{1,1}^* \dots ; D_{\sigma,0}^*, D_{\sigma,1}^*$). We are now interested in the pair A, B. Let us first interpret the dynamic system (5.10), corresponding to this pair,

$$\begin{aligned} z' &= Az + Bv \\ z(0) &= 0. \end{aligned} \tag{5.24}$$

Consider also (5.11), i.e.

$$\begin{aligned} -\phi' &= A\phi \\ \phi(T) &= h \end{aligned} \tag{5.25}$$

then we have

$$\begin{aligned} (z(T), h) &= \int_0^T (B^*\phi, v)\, dt \\ &= \int_0^T [(\gamma\psi, v_0) + \sum_{i=1}^n (\sigma_i\gamma\psi, v_i)]\, dt \\ &= \int_0^T [(\gamma\psi, v_0) + \sum_{i=1}^n (\sigma_i\gamma\psi, v_i)]\, dt, \end{aligned} \tag{5.26}$$

where $\psi = A^{-1}\phi$, and $v = (v_0, v_1, \ldots, v_n)$. Since $n\psi = 0$, we can also write the above relation as

$$(z(T), h) = \int_0^T \int_\Gamma (\psi v_0 + \sum_{i=1}^n \frac{\partial \psi}{\partial x_i} v_i) \, d\Gamma \tag{5.27}$$

where we omit to indicate explicitly the operator γ.

Note that $z(\cdot) \in C(0, T; H)$. Let $\zeta(t) = Az(t) \in C(0, T; (D(A))')$. Then from (5.27) we deduce

$$< \zeta(T), \psi(T) >= \int_0^T \int_\Gamma (\psi v_0 + \sum_{i=1}^n \frac{\partial \psi}{\partial x_i} v_i) \, d\Gamma. \tag{5.28}$$

Note that if $v \in (H^1(\Gamma))'$ and $\psi \in D(A)$ (hence $\gamma\psi \in H^1(\Gamma)$), we can write

$$< v, \gamma\psi >= \int_\Gamma (\psi v_0 + \sum_{i=1}^n \frac{\partial \psi}{\partial x_i} v_i) \, d\Gamma$$

where $(v_0, v_1, \ldots, v_n)$ is a representation of v.

Therefore (5.28) reads also

$$< \zeta(T), \psi(T) >= \int_0^T < v(t), \gamma\psi(t) > \, dt. \tag{5.29}$$

This is exactly the formula which follows from the definition of the solution of

$$\begin{aligned} &\zeta' - \Delta\zeta = 0 \\ &\frac{\partial \zeta}{\partial \nu} = v \\ &\zeta(x, 0) = 0 \end{aligned} \tag{5.30}$$

obtained by the transposition method of J.-L. LIONS and E. MAGENES [1] (see also Chapter 2 in Volume I), using the dual equation

$$\begin{aligned} &-\psi' - \Delta\psi = 0 \\ &\frac{\partial \psi}{\partial \nu} = 0 \\ &\psi(T) \in D(A). \end{aligned} \tag{5.31}$$

We are interested in the density of the range of $\zeta(T;v)$, solution of (5.30) at some time T, in the space $(D(A))'$. This is equivalent to the density of $z(T;v)$ in H, where z is the solution of (5.24). We are in the framework of the model (2.1), with a semigroup satisfying the condition (4.5), and thus the problem is that of the approximate controllability of the pair (A,B) at some time T. The general theory implies that the pair (A,B) is approximately controllable at any positive time T, if and only if the condition (4.12) holds.

We write the condition (4.12)

$$\sum_{j=1}^{r_k}(h,w_{kj})B^*w_{kj}=0, \quad \forall\, k>0 \quad \Rightarrow \quad h=0. \tag{5.32}$$

This implies that

$$w=\sum_{j=1}^{r_k}(h,w_{kj})w_{kj}$$

satisfies $\gamma w=0$, $D_\sigma w=0$. Clearly $w=0$, hence the coefficients $(h,w_{kj})=0$, for $j=1,\dots,r_k$. Hence $h=0$.

It is clear from the above that if we control the whole boundary, we can take $v(t)$ in $L^2(\Gamma)$ instead of $(H^1(\Gamma))'$, i.e. $v(t)=v_0(t)$ and $v_1(t)=\dots=v_n(t)=0$.

On the other hand, we need $(H^1(\Gamma))'$ if we control only a part of the boundary. Let us make this precise. We first state a lemma.

Lemma 5.2 *The eigenvector w solution of* (5.20) *satisfies the relation*

$$\begin{aligned}&\int_\Gamma m\cdot\nu\,|D_\sigma w|^2\,d\Gamma+2(\lambda-1)\int_\Omega w^2\,dx\\&=(\lambda-1)\int_\Gamma m\cdot\nu\,w^2\,d\Gamma\end{aligned} \tag{5.33}$$

Proof. As usual, we multiply (5.20) by

$$m_\alpha\frac{\partial w}{\partial x_\alpha}$$

and perform integration by parts. Details are left to the reader. ∎

We then state the

Theorem 5.3 *Let $A = \Delta - I$ associated with the Neumann boundary condition, and the two following cases for U, B. Either $U = L^2(\Gamma)$ and $B = A^{-1}\gamma^*$ or $U = L^2(\Gamma_0)\times(L^2(\Gamma_1))^n$ and $B = A^{-1}(\gamma_0^*\ \gamma_1^* D_{\sigma,1}^*)$. Then the pair (A, B) is approximately controllable at any $T > 0$. The range of $\zeta(T; v)$ solution of (5.40) is dense in $(D(A))'$.*

Proof. The first case has already been discussed. Consider the 2nd case. We have to prove (5.32). Note first that for $k = 1, r_k = 1$, and w_1 is a constant. We necessarily have $(h, w_1) = 0$. Starting with $k = 2$, we have $\lambda_k > 1$. From the hypothesis of the statement (5.32) we have, writing

$$\begin{aligned} w &= \sum_{j=1}^{r_k} (h, w_{kj})\, w_{kj}, \\ \gamma_0\, w &= 0, \quad D_{\sigma,1} w = 0. \end{aligned} \tag{5.34}$$

Splitting the surface integrals in (5.33) in two parts Γ_0 and Γ_1 and using the sign properties of $m.\nu$ on Γ_0, Γ_1 and the fact that $\lambda > 1$ we deduce that

$$\int_\Omega w^2\, dx = 0.$$

Therefore the coefficients $(h, w_{kj}) = 0$, for $j = 1, \ldots, r_k$. The desired result has been proved. ∎

Remark 5.4 Apparently, it is not sufficient to take $U = L^2(\Gamma_0)$, unlike the Dirichlet case, see Remark 5.2. ∎

Remark 5.5 Consider again $(\Gamma_T h, h)$. As usual from (4.14), we can prove that $(\Gamma_T h, h) \le C(||h||_{V'})^2$, recalling that $V = H^1$. Hence $V' \subset F_T'$ and $F_T \subset V$, with continuous and dense injection. In the case $U = L^2(\Gamma)$ and $B = A^{-1}\gamma^*$, one can obtain an additional result, namely $(D(A))' \subset F_T'$, therefore $F_T \subset D(A)$. It follows that the range of $\zeta(T; v)$, solution of (5.30) as $v(\cdot)$ varies in $L^2(0, T; L^2(\Gamma))$ is dense in H. ∎

5.4 Pointwise control

We consider the system

$$\begin{aligned} &\zeta' - \Delta\zeta = \sum_{i=1}^{m} v_i(t)\delta(x - \alpha_i) \\ &\zeta|_\Gamma = 0 \\ &\zeta(x;0) = 0 \end{aligned} \tag{5.35}$$

where the points α_i belong to Ω and represent actuators. We assume that the dimension of the space is $n \leq 3$. We shall reduce the formal model (5.35) to our general set up.

We take $A = \Delta$ with Dirichlet conditions. We define $b_i \in H$ by the equation

$$(b_i, h) = (A^{-1}h)(\alpha_i). \tag{5.36}$$

This makes sense since for $h \in H = L^2, A^{-1}h$ belongs to $H^2 \subset C(\overline{\Omega})$ with continuous injection. Hence the right-hand side of (5.36) is defined and represents a linear continuous functional on H. Consider the dynamic system

$$\begin{aligned} &z' = Az + \sum_{i=1}^{m} v_i(t)b_i \\ &z(0) = 0. \end{aligned} \tag{5.37}$$

Let us check that

$$\zeta(T) = Az(T) \tag{5.38}$$

coincides with the definition of (5.35) through the transposition method. Indeed consider

$$\begin{aligned} -\phi' &= A\phi \\ \phi(T) &= h \end{aligned} \tag{5.39}$$

Then from (5.37) and (5.39) it follows taking account the definition of b_i

$$(z(T), h) = \sum_{i=1}^{m} \int_0^T v_i(t)\psi(\alpha_i, t)\, dt \tag{5.40}$$

where $\psi = A^{-1}\phi$. But from the transposition method we define $\zeta(T)$ by

$$(\zeta(T), \psi(T)) = \sum_{i=1}^{m} \int_0^T v_i(t)\psi(\alpha_i, t)\, dt. \tag{5.41}$$

From (5.40) and (5.41), using the fact that $\psi(T) = A^{-1}h$, we deduce that (5.38) holds. Now the system (5.37) has already been studied in §4.2. Consider the matrices

$$F_k, \text{ with } F_{k,ji} = w_{kj}(\alpha_i), \quad j = 1, \ldots, r_k,\ i = 1, \ldots, m, \tag{5.42}$$

then we can state the

Theorem 5.4 *The pair (A, B), $B = (b_1, b_2, \ldots, b_m)$ is approximately controllable at any $T > 0$, if and only if rank $F_k = r_k$ for any k.*

Remark 5.6 We refer to A. El Jai and T. Pritchard [1], [2] for general results of controllability in the case of pointwise control. ∎

Remark 5.7 We assume that the assumption of Theorem 5.4 is satisfied. Using condition (4.14) we obtain again that $F_T \subset H_0^1$ with dense and continuous injection. Hence the range of $\zeta(T; v)$ is dense in H^{-1}. ∎

6 Controllability for skew-symmetric operators

6.1 Notation and general comments

We shall need to work here with complex spectrum. We assume that A has a purely imaginary spectrum of the form

$$\{i\sqrt{\lambda_j}, -i\sqrt{\lambda_j} : j = 1, \ldots\},$$

where

$$0 < \lambda_1 \leq \lambda_2 \leq \ldots \leq \lambda_j \leq \ldots, \qquad \lambda_j \uparrow \infty,$$

with only a finite number of successive eigenvalues possibly equal. If ϕ_j is the eigenvector corresponding to $i\sqrt{\lambda_j}$, then $\overline{\phi_j}$ is the eigenvector corresponding to $-i\sqrt{\lambda_j}$. We assume that

$$\phi_j, \overline{\phi_j} \tag{6.1}$$

form an orthonormal basis of H (complexified).

This implies with the notation of the scalar product in H complexified

$$\begin{aligned}(\phi_j, \overline{\phi_k}) &= \delta_{jk} \\ (\phi_j, \phi_k) &= 0.\end{aligned} \tag{6.2}$$

By definition

$$\begin{aligned}A\phi_j &= i\sqrt{\lambda_j}\,\phi_j \\ A\overline{\phi_j} &= -i\sqrt{\lambda_j}\,\overline{\phi_j}.\end{aligned} \tag{6.3}$$

If ϕ is an element of H (real valued), then it can be represented with the expansion

$$\phi = \sum_j (c_j\overline{\phi_j} + \overline{c_j}\phi_j) \tag{6.4}$$

with

$$c_j = (\phi, \phi_j), \quad \overline{c_j} = (\phi, \overline{\phi_j}), \tag{6.5}$$

$$|\phi|^2 = 2\sum_j |c_j|^2. \tag{6.6}$$

It is easy to check that the operator A is skew adjoint,

$$A^* = -A. \tag{6.7}$$

In fact we can even characterize the type of operators we are considering, thanks to the following proposition.

Proposition 6.1 *Assume that A is a linear operator from a dense subspace $D(A)$ of H into H, satisfies (6.7) and*

$$\textit{Null space of } A = \{0\} \tag{6.8}$$

$$\begin{aligned}&\textit{the mapping } A + I \textit{ from } D(A) \textit{ to } H \textit{ is onto} \\ &\textit{(and thus invertible since already one to one)}\end{aligned} \tag{6.9}$$

$$\textit{the map } T = (A+I)^{-1} \textit{ is compact.} \tag{6.10}$$

Then the spectrum of A is of the form $\pm i\sqrt{\lambda_j}$, with

$$0 < \lambda_1 \leq \lambda_2 \leq \ldots \leq \lambda_j \leq \ldots, \qquad \lambda_j \uparrow \infty,$$

and the corresponding eigenvectors ϕ_j and $\overline{\phi_j}$ form an orthonormal basis for H. Conversely if A has these spectral properties, then it satisfies (6.7) to (6.10).

Proof. Let us assume (6.7), then if $z \in D(A)$, we can write

$$(Az, \overline{z}) = -(A^* z, \overline{z}) = -(z, \overline{Az}) = -\overline{(Az, \overline{z})}$$

and thus $(Az, \overline{z})$ is purely imaginary. Hence the eigenvalues are necessarily purely imaginary. Since -1 cannot be an eigenvalue, the map $A + I$ is necessarily one to one. Moreover if λ is an eigenvalue corresponding to the eigenvector ϕ, then $\overline{\phi}$ is an eigenvector corresponding to the eigenvalue $\overline{\lambda}$.

Assume now (6.8), (6.9), (6.10). Let us check the T is a normal operator on H, i.e.

$$T^* T = T T^* \tag{6.11}$$

Note that $T^* = (A^* + I)^{-1}$. Let $f \in H$ and define

$$\phi = \tfrac{1}{2}\{(A + I)^{-1} f + (-A + I)^{-1} f\}$$

$$\psi = \tfrac{1}{2}\{-(A + I)^{-1} f + (-A + I)^{-1} f\}$$

then clearly $\psi = A\phi$, thus

$$\begin{aligned} A\phi + \phi &= (A^* + I)^{-1} f \\ A^*\phi + \phi &= (A + I)^{-1} f \end{aligned}$$

which proves (6.11).

From the spectral theory of normal compact operators, see T. Kato [1, p. 260], we can state that T has a discrete spectrum $\{\mu_1, \mu_2, \ldots\}$ with $|\mu_1| \geq |\mu_2| \geq \ldots \geq |\mu_n| \geq \ldots$, $|\mu_n|$ tends to 0 as n tends to infinity. Each of these eigenvalues has a finite multiplicity. Since moreover the null space of T is 0, the eigenvectors form an orthonormal basis of H. But eigenvectors of T are eigenvectors of A, with eigenvalues obtained by the transformation

$$\lambda = -1 + 1/\mu.$$

The spectral properties of A are easily deduced.

Let us check the reverse. The set $D(A)$ is made of elements

$$\phi = \sum_j (c_j \overline{\phi_j} + \overline{c_j} \phi_j)$$

such that

$$\sum_j |c_j|^2 \lambda_j < \infty.$$

It is clear that $D(A)$ is dense in H. If $\phi \in D(A)$ then

$$A\phi = \sum_j (-ic_j\sqrt{\lambda_j}\,\overline{\phi_j} + i\,\overline{c_j}\sqrt{\lambda_j}\,\phi_j).$$

Therefore we can solve the equation

$$A\phi + \phi = f \tag{6.12}$$

as follows

$$c_j = \frac{f_j}{(1 - i\sqrt{\lambda_j}\,)}$$

where f_j denotes the components of f. Therefore the equation (6.12) has one and only one solution. To prove (6.10) consider a sequence f^n which tends to 0 in H weakly, and let ϕ^n the corresponding solution of (6.12), we must prove that ϕ^n tends to 0 in H strongly. Let f_j^n be the components of f^n and

$$c_j^n = \frac{f_j^n}{(1 - i\sqrt{\lambda_j}\,)}$$

be the corresponding components of ϕ^n.

We have

$$\begin{aligned} \sum_j |c_j^n|^2 &= \sum_j \frac{|f_j^n|^2}{1+\lambda_j} \\ &\leq \sum_{j=1\ldots N} |f_j^n|^2 + \frac{|f^n|^2}{\lambda_N}. \end{aligned}$$

Since f^n tends to 0 weakly, $|f^n|^2$ remains bounded and f_j^n tends to 0 for any fixed j. Moreover λ_n tend to 0 as N tends to infinity. It follows from the above inequality that

$$\sum_j |c_j^n|^2$$

tends to 0 as n tends to infinity. This means that ϕ^n tends to 0 in H. ■

The solution of

$$\begin{aligned} z' &= Az \\ z(0) &= \phi \end{aligned} \tag{6.13}$$

is given by

$$z(t) = e^{tA}\phi = \sum_j (c_j e^{-i\sqrt{\lambda_j}\,t}\,\overline{\phi_j} + \overline{c_j} e^{+i\sqrt{\lambda_j}\,t}\phi_j) \tag{6.14}$$

where c_j are the components of ϕ.

In the sequel we shall use the notation

$$\phi = \operatorname{Re}\phi + i\operatorname{Im}\phi$$

to express the decomposition of an element of H complexified.

Note also that if $\phi \in D(A)$ then

$$||\phi||^2_{D(A)} = |A\phi|^2 = 2\sum_j \lambda_j |c_j|^2$$

and

$$\frac{\phi_j}{\sqrt{\lambda_j}}, \frac{\overline{\phi_j}}{\sqrt{\lambda_j}}$$

form an orthonormal basis of $D(A)$. Similarly $\phi_j\sqrt{\lambda_j}$, $\overline{\phi_j}\sqrt{\lambda_j}$ form an orthonormal basis of $(D(A))'$.

6.2 Dynamical system

Consider $B \in \mathcal{L}(U; H)$. We shall make several assumptions on B in relation with the eigenvectors ϕ_j.

There exist N integer ≥ 0 and operators $M \in \mathcal{L}(H; H)$, $\Lambda \in \mathcal{L}(H; D(A)')$ such that

$$|B^* z|^2 \geq (Mz, z), \quad \forall\, z \in H, \tag{6.15}$$

for all $j, k \geq N + 1$,

$$(M \operatorname{Re}\phi_j, \operatorname{Im}\phi_k) = 0, \tag{6.16}$$

for all $j, k \geq N+1$, $j \neq k$

$$(M \operatorname{Re} \phi_j, \operatorname{Re} \phi_k) = \frac{(\Lambda \operatorname{Re} \phi_j, \operatorname{Re} \phi_k)}{\lambda_k} + \frac{(\Lambda \operatorname{Re} \phi_k, \operatorname{Re} \phi_j)}{\lambda_j} \tag{6.17}$$

for all $j, k \geq N+1$, $j \neq k$

$$(\Lambda \operatorname{Re} \phi_j, \operatorname{Re} \phi_k) + (\Lambda \operatorname{Re} \phi_k, \operatorname{Re} \phi_j) + \sqrt{\lambda_j \lambda_k}\,(M \operatorname{Im} \phi_j, \operatorname{Im} \phi_k) = 0. \tag{6.18}$$

There exists $c_0 > 0$ such that for all $j \geq N+1$,

$$(M \operatorname{Re} \phi_j, \operatorname{Re} \phi_j) + (M \operatorname{Im} \phi_j, \operatorname{Im} \phi_j) \geq \frac{c_0}{\lambda_j}, \tag{6.19}$$

for all $j \geq N+1$

$$\left|(M \operatorname{Re} \phi_j, \operatorname{Re} \phi_j) - (M \operatorname{Im} \phi_j, \operatorname{Im} \phi_j) - 4\frac{(\Lambda \operatorname{Re} \phi_j, \operatorname{Re} \phi_j)}{\lambda_j}\right| \leq \frac{K}{\sqrt{\lambda_j}}, \tag{6.20}$$

and

$$\lambda_{N+1} > \lambda_N.$$

For $j \leq N$, we consider only the λ_j having different values and call $P_j(\overline{P_j})$ the projector on the finite dimensional eigensubspace of H corresponding to the conjugate pair of eigenvalues $\{i\sqrt{\lambda_j}, -i\sqrt{\lambda_j}\}$, then

$$|B^* P_j z|^2 \geq c_1 \frac{|P_j z|^2}{\lambda_j}, \quad \forall\, z \in H,\ j \leq N. \tag{6.21}$$

With our convention we have

$$0 < \lambda_1 < \lambda_2 < \ldots < \lambda_N < \lambda_{N+1} \leq \lambda_{N+2} \leq \ldots.$$

An element ϕ of H is represented by the expansion

$$\phi = \sum_{j=1}^{N} (P_j \phi + \overline{P_j} \phi) + \sum_{j \geq N+1} (c_j \overline{\phi_j} + \overline{c_j} \phi_j). \tag{6.22}$$

Naturally if $N = 0$, the condition (6.21) is void.

We now consider the dynamic system corresponding to the pair (A, B)

$$\begin{aligned} z' &= Az + Bv \\ z(0) &= 0. \end{aligned} \tag{6.23}$$

Our main result is the following

Theorem 6.1 *We assume that A satisfies (6.7) to (6.10) and B satisfies (6.15) to (6.21), then the pair (A, B) is exactly controllable at T, for T sufficiently large.*

Proof. We shall consider the controllability operator

$$\Gamma_T = \int_0^T e^{tA} BB^* e^{tA^*} \, dt,$$

and prove that

$$\forall\, h \in H, \quad (\Gamma_T h, h) \geq c_T ||h||^2_{(D(A))'} \tag{6.24}$$

for T sufficiently large.

According to §3.2, this implies the desired result. Consider

$$h = \sum_{j=1}^{N} (P_j h + \overline{P_j} h) + \sum_{j \geq N+1} (c_j \overline{\phi_j} + \overline{c_j} \phi_j)$$

with $c_j = (h, \phi_j)$, then

$$\begin{aligned} (\Gamma_T h, h) &= \int_0^T |B^* e^{tA^*} h|^2 \, dt \\ &= 4 \int_0^T |B^* (\sum_{j=1}^{N} \mathrm{Re}\,(e^{ti\sqrt{\lambda_j}} P_j h) \\ &\qquad + \sum_{j \geq N+1} \mathrm{Re}\,(e^{-ti\sqrt{\lambda_j}} \overline{c_j} \phi_j))|^2 \, dt. \\ &= X_1 + X_2 + X_3, \end{aligned} \tag{6.25}$$

where

$$X_1 = 4\int_0^T |B^*[\sum_{j=1}^{N} \mathrm{Re}\,(e^{-ti\sqrt{\lambda_j}}P_j h)]\,|^2\,dt,$$

$$X_2 = 8\int_0^T (B^*[\sum_{j=1}^{N} \mathrm{Re}\,(e^{-ti\sqrt{\lambda_j}}P_j H)], \\ B^*[\sum_{j\geq N+1} \mathrm{Re}\,(e^{-ti\sqrt{\lambda_j}}\overline{c_j}\phi_j)])\,dt,$$

$$X_3 = 4\int_0^T |B^*[\sum_{j\geq N+1} \mathrm{Re}\,(e^{-ti\sqrt{\lambda_j}}\overline{c_j}\phi_j)]\,|^2\,dt.$$

By an easy computation

$$X_1 = 2T\sum_{j=1}^{N} |B^*P_j h|^2 + Y_1$$

with

$$\begin{aligned} Y_1 &= 2\,\mathrm{Re}\sum_{j,k=1}^{N} (B^*P_j h, B^*P_k h)\frac{1-e^{-iT(\sqrt{\lambda_j}+\sqrt{\lambda_k})}}{i(\sqrt{\lambda_j}+\sqrt{\lambda_k})} \\ &\quad +2\,\mathrm{Re}\sum_{j=1}^{N}\sum_{\substack{k=1\\k\neq j}}^{N} (B^*P_j h, B^*\overline{P_k}h)\frac{1-e^{-iT(\sqrt{\lambda_j}-\sqrt{\lambda_k})}}{i(\sqrt{\lambda_j}-\sqrt{\lambda_k})} \end{aligned}$$

$$\begin{aligned} X_2 &= 2\,\mathrm{Re}\sum_{j=1}^{N}\sum_{k\geq N+1} (B^*P_j h, B^*\phi_k\overline{c}_k)\frac{1-e^{-iT(\sqrt{\lambda_j}+\sqrt{\lambda_k})}}{i(\sqrt{\lambda_j}+\sqrt{\lambda_k})} \\ &\quad +2\,\mathrm{Re}\sum_{j=1}^{N}\sum_{k\geq N+1} (B^*P_j h, B^*\overline{\phi_k}c_k)\frac{1-e^{-iT(\sqrt{\lambda_j}-\sqrt{\lambda_k})}}{i(\sqrt{\lambda_j}-\sqrt{\lambda_k})}. \end{aligned}$$

Now according to (6.15)

$$X_3 \geq 4\int_0^T (M(\sum_{j\geq N+1} \mathrm{Re}\,(e^{-ti\sqrt{\lambda_j}}\overline{c_j\phi_j})), \sum_{j\geq N+1} \mathrm{Re}\,(e^{-ti\sqrt{\lambda_j}}\overline{c_j}\phi_j))\,dt.$$

At this stage it is convenient to introduce the following notation which will help reduce the length of the following equations and estimates:

$$\begin{aligned} m_{ij} &= (M \operatorname{Im} \phi_i, \operatorname{Im} \phi_j) \\ l_{ij} &= (\Lambda \operatorname{Re} \phi_i, \operatorname{Re} \phi_j). \end{aligned}$$

From (6.16).

$$\begin{aligned} &X_3 \geq \\ &4 \sum_{j,k \geq N+1} \int_0^T (M \operatorname{Re} \phi_j, \operatorname{Re} \phi_k) \operatorname{Re}(e^{-ti\sqrt{\lambda_j}} \overline{c_j}) \operatorname{Re}(e^{-ti\sqrt{\lambda_k}} \overline{c_k})\, dt \\ &+4 \sum_{j,k \geq N+1} \int_0^T m_{jk} (e^{-ti\sqrt{\lambda_j}} \overline{c_j}) \operatorname{Im}(e^{-ti\sqrt{\lambda_k}} \overline{c_k})\, dt \end{aligned}$$

and from (6.17)

$$X_3 \geq Y_3 + Z_3,$$

where

$$\begin{aligned} Y_3 \;=\; & 4 \sum_{j \geq N+1} \int_0^T (M \operatorname{Re} \phi_j, \operatorname{Re} \phi_j) |\operatorname{Re}(e^{-ti\sqrt{\lambda_j}} \overline{c_j})\,|^2\, dt \\ & +4 \sum_{j \geq N+1} \int_0^T (M \operatorname{Im} \phi_j, \operatorname{Im} \phi_j) \,|\operatorname{Im}(e^{-ti\sqrt{\lambda_j}} \overline{c_j})|^2\, dt \end{aligned}$$

$$\begin{aligned} Z_3 = 4 & \sum_{\substack{k \geq N+1 \\ j \neq k}} \int_0^T \left[\frac{l_{jk}}{\lambda_k} + \frac{l_{kj}}{\lambda_j} \right] \operatorname{Re}(e^{-ti\sqrt{\lambda_j}} \overline{c_j}) \operatorname{Re}(e^{-ti\sqrt{\lambda_k}} \overline{c_k})\, dt \\ +\, 4 & \sum_{\substack{k \geq N+1 \\ j \neq k}} \int_0^T m_{jk} \operatorname{Im}(e^{-ti\sqrt{\lambda_j}} \overline{c_j}) \operatorname{Im}(e^{-ti\sqrt{\lambda_k}} \overline{c_k})\, dt. \end{aligned}$$

Set

$$a_j = \frac{c_j}{\sqrt{\lambda_j}}.$$

Then we can write

$$Z_3/4 = \sum_{\substack{k \geq N+1 \\ j \neq k}} \int_0^T \left[\frac{l_{jk}\sqrt{\lambda_j}}{\sqrt{\lambda_k}} + \frac{l_{kj}\sqrt{\lambda_k}}{\sqrt{\lambda_j}} \right] \mathrm{Re}\,(e^{-ti\sqrt{\lambda_j}}\,\overline{a_j})\,\mathrm{Re}\,(e^{-ti\sqrt{\lambda_k}}\,\overline{a_k})\,dt$$

$$+ \sum_{\substack{k \geq N+1 \\ j \neq k}} \int_0^T \sqrt{\lambda_j\lambda_k}\, m_{jk}\,\mathrm{Im}\,(e^{-ti\sqrt{\lambda_j}}\,\overline{a_j})\,\mathrm{Im}\,(e^{-ti\sqrt{\lambda_k}}\,\overline{a_k})\,dt.$$

Then

$$Z_3 = -4 \sum_{\substack{k \geq N+1 \\ j \neq k}} \int_0^T \frac{l_{jk}}{\sqrt{\lambda_k}} \frac{d}{dt} \mathrm{Im}\,(e^{-ti\sqrt{\lambda_j}}\overline{a_j})\,\mathrm{Re}\,(e^{-ti\sqrt{\lambda_k}}\overline{a_k})\,dt$$

$$-4 \sum_{\substack{k \geq N+1 \\ j \neq k}} \int_0^T \frac{l_{kj}}{\sqrt{\lambda_j}} \mathrm{Re}\,(e^{-ti\sqrt{\lambda_j}}\overline{a_j}) \frac{d}{dt}\,\mathrm{Im}\,(e^{-ti\sqrt{\lambda_k}}\overline{a_k})\,dt$$

$$+4 \sum_{\substack{k \geq N+1 \\ j \neq k}} \int_0^T \sqrt{\lambda_j\lambda_k}\, m_{jk}\,\mathrm{Im}\,(e^{-ti\sqrt{\lambda_j}}\overline{a_j})\,\mathrm{Im}\,(e^{-ti\sqrt{\lambda_k}}\overline{a_k})\,dt.$$

Integrating by parts and using (6.18) obtains

$$Z_3 = -4 \sum_{\substack{k \geq N+1 \\ j \neq k}} \frac{(\Lambda\,\mathrm{Re}\,\phi_j, \mathrm{Re}\,\phi_k)}{\sqrt{\lambda_k}} (\,\mathrm{Im}\,(e^{iT\sqrt{\lambda_j}}\overline{a_j}) \cdot \mathrm{Re}\,(e^{iT\sqrt{\lambda_k}}\overline{a_k}) - \mathrm{Im}\,(\overline{a_j})\,\mathrm{Re}\,(\overline{a_k}))$$

$$-\,4 \sum_{\substack{k \geq N+1 \\ j \neq k}} \frac{(\Lambda\,\mathrm{Re}\,\phi_k, \mathrm{Re}\,\phi_j)}{\sqrt{\lambda_j}} (\mathrm{Im}\,(e^{iT\sqrt{\lambda_k}}\overline{a_k}) \cdot \mathrm{Re}\,(e^{iT\sqrt{\lambda_j}}\overline{a_j}) - \mathrm{Im}\,(\overline{a_k})\,\mathrm{Re}\,(\overline{a_j})).$$

Hence also

$$Z_3 =$$

$$- 8(\Lambda(\sum_{j\geq N+1} \operatorname{Re}\phi_j \operatorname{Im}(e^{-iT\sqrt{\lambda_j}}\overline{a_j})), \sum_{k\geq N+1} \operatorname{Re}\phi_k \frac{\operatorname{Re}(e^{-iT\sqrt{\lambda_k}}\overline{a_k})}{\sqrt{\lambda_k}})$$

$$+ 8(\Lambda(\sum_{j\geq N+1} \operatorname{Re}\phi_j \operatorname{Im}(\overline{a_j})), \sum_{k\geq N+1} \operatorname{Re}\phi_k \frac{\operatorname{Re}(\overline{a_k})}{\sqrt{\lambda_k}})$$

$$+ 8 \sum_{j\geq N+1} (\Lambda \operatorname{Re}\phi_j, \operatorname{Re}\phi_j) \operatorname{Im}(e^{-iT\sqrt{\lambda_j}}\overline{a_j}) \frac{\operatorname{Re}(e^{-iT\sqrt{\lambda_j}}\overline{a_j})}{\sqrt{\lambda_j}}$$

$$- 8 \sum_{j\geq N+1} (\Lambda \operatorname{Re}\phi_j, \operatorname{Re}\phi_j) \operatorname{Im}(\overline{a_j}) \frac{\operatorname{Re}(\overline{a_j})}{\sqrt{\lambda_j}}.$$

Next as easily seen

$$Y_3 = 2T \sum_{j\geq N+1} |c_j|^2((M \operatorname{Re}\phi_j, \operatorname{Re}\phi_j) + (M \operatorname{Im}\phi_j, \operatorname{Im}\phi_j)) + Y_3'$$

with

$$Y_3' = \sum_{j\geq N+1} \{(M \operatorname{Re}\phi_j, \operatorname{Re}\phi_j) - m_{jj}\} \frac{\operatorname{Im}(c_j^2(e^{2iT\sqrt{\lambda_j}} - 1))}{\sqrt{\lambda_j}}.$$

Collecting results we deduce that

$$\begin{aligned}(\Gamma_T h, h) \geq\ & 2T \sum_{j=1}^{N} |B^* P_j h|^2 \\ & +2T \sum_{j\geq N+1} |c_j|^2[(M \operatorname{Re}\phi_j, \operatorname{Re}\phi_j) + m_{jj}] \\ & +Y_1 + X_2 + Z_3 + Y_3'.\end{aligned} \tag{6.26}$$

Using the assumptions (6.19), (6.20) we get, setting $c = \min\{c_0, c_1\}$:

$$\begin{aligned}(\Gamma_T h, h) \quad &\geq \quad cT\left\{\sum_{j=1}^{N}\frac{|P_j h|^2}{\lambda_j} + \sum_{j\geq N+1}|a_j|^2\right\} \\ &\qquad + Y_1 + X_2 + Z_3 + Y_3' \\ &\geq \quad cT||h||^2_{(D(A))'} + Y_1 + X_2 + Z_3 + Y_3'.\end{aligned} \tag{6.27}$$

We now estimate the remainder $+Y_1 + X_2 + Z_3 + Y_3'$.

We first check

$$\begin{aligned} &Y_1 + X_2 \\ &= \; 2\,\mathrm{Re}\,(\sum_{j=1}^{N} BB^* P_j h, \; \sum_{\substack{k=1 \\ k\neq j}}^{N} [P_k h \frac{1 - e^{iT(\sqrt{\lambda_j}+\sqrt{\lambda_k})}}{i(\sqrt{\lambda_j}+\sqrt{\lambda_k})} \\ &\qquad + \overline{P_k} h \frac{1 - e^{-iT(\sqrt{\lambda_j}-\sqrt{\lambda_k})}}{i(\sqrt{\lambda_j}-\sqrt{\lambda_k})}] \\ &\qquad + \sum_{k\geq N+1} [\phi_k \overline{c_k} \frac{1 - e^{-iT(\sqrt{\lambda_j}-\sqrt{\lambda_k})}}{i(\sqrt{\lambda_j}+\sqrt{\lambda_k})} \\ &\qquad + \overline{\phi_k} c_k \frac{1 - e^{-iT(\sqrt{\lambda_j}-\sqrt{\lambda_k})}}{i(\sqrt{\lambda_j}-\sqrt{\lambda_k})}]).\end{aligned} \tag{6.28}$$

Hence

$$|Y_1 + X_2| \leq \beta_N ||h||^2_{(D(A))'} \tag{6.29}$$

where

$$\beta_N = \sqrt{2}\sup_h \sum_{j=1}^{N} \sqrt{\lambda_j}\,\Lambda_j \frac{|BB^* P_j h|}{|P_j h|}$$

and

$$\Lambda_j = \max\left\{\frac{\lambda_{j+1}^{\frac{1}{2}}(\lambda_j + \lambda_{j+1})^{\frac{1}{2}}}{\lambda_{j+1} - \lambda_j}, \frac{\lambda_{j-1}^{\frac{1}{2}}(\lambda_j + \lambda_{j-1})^{\frac{1}{2}}}{\lambda_j - \lambda_{j-1}})\right\}.$$

We interpret $\lambda_{j-1} = 0$, if $j = 1$. Now

$$Z_3 + Y_3' = T_3 + T_3'$$

where

$$T_3 =$$

$$-8(\Lambda(\sum_{j \geq N+1} \mathrm{Re}\,\phi_j \,\mathrm{Im}\,(e^{-iT\sqrt{\lambda_j}}\overline{a_j})), \sum_{k \geq N+1} \mathrm{Re}\,\phi_k \frac{\mathrm{Re}\,(e^{-iT\sqrt{\lambda_k}}\overline{a_k})}{\sqrt{\lambda_k}})$$

$$+ 8(\Lambda(\sum_{j \geq N+1} \mathrm{Re}\,\phi_j \,\mathrm{Im}\,(\overline{a_j})), \sum_{k \geq N+1} \mathrm{Re}\,\phi_k \frac{\mathrm{Re}\,(\overline{a_k})}{\sqrt{\lambda_k}})$$

and

$$T_3' = \sum_{j \geq N+1} \{(M\mathrm{Re}\,\phi_j, \mathrm{Re}\,\phi_j) - (M\,\mathrm{Im}\,\phi_j, \mathrm{Im}\,\phi_j)$$

$$- 4\frac{(\Lambda\,\mathrm{Re}\,\phi_j\,\mathrm{Re}\,\phi_j)}{\lambda_j}\} \frac{\mathrm{Im}\,(c_j^2(e^{2iT\sqrt{\lambda_j}} - 1))}{\sqrt{\lambda_j}}.$$

Since $\Lambda \in \mathcal{L}(H; (D(A))')$ and using the fact that

$$\frac{\phi_j}{\sqrt{\lambda_j}}, \frac{\overline{\phi_j}}{\sqrt{\lambda_j}}$$

are orthonormal in $D(A)$ we can write the estimate

$$|T_3| \leq 4||\Lambda||\;||h||^2_{(D(A))'} \tag{6.30}$$

Similarly thanks to (6.20)

$$|T_3'| \leq K||h||^2_{(D(A))'} \tag{6.31}$$

From (6.27), (6.29), (6.30), (6.32) we deduce the desired result. ∎

Remark 6.1 We have proved that

$$(\Gamma_T h, h) \geq c(T - T_0)||h||^2_{(D(A))'}$$

with

$$T_0 = \frac{\beta_N + B + 4||\Lambda||}{c}$$

where $c = \min(c_0, c_1)$ ($= c_0$ if $N = 0$). Therefore there is exact controllability if $T > T_0$. Moreover this estimate implies the sequence

$$D(A) \subset F_T \subset H \subset F'_T \subset (D(A))' \tag{6.32}$$

with continuous and dense injection, as seen in §3.2. ∎

Remark 6.2 In finite dimension, Theorem 6.1 reduces to Proposition 4.2. ∎

There is some flexibility in the type of assumption we can make to achieve exact controllability. We give next a variant of Theorem 6.1.

Theorem 6.2 *We make the assumptions* (6.7) *to* (6.10), (6.15), (6.16), (6.21),

$$BB^* \in \mathcal{L}((D(A))'; D(A)) \tag{6.33}$$

and $\forall\, j, k \geq N+1,\ j \neq k$

$$\begin{aligned} &\lambda_j \lambda_k (M \operatorname{Re}\phi_j, \operatorname{Re}\phi_k) \\ = \ &(\Lambda \operatorname{Re}\phi_j, \operatorname{Re}\phi_k) + (\Lambda \operatorname{Re}\phi_k, \operatorname{Re}\phi_j). \end{aligned} \tag{6.34}$$

There exists $Q \in \mathcal{L}(H; H) \geq 0$, *self adjoint such that* $\forall\, j, k \geq N+1$, $j \neq k$

$$\begin{aligned} &\frac{(\Lambda \operatorname{Re}\phi_j, \operatorname{Re}\phi_k)}{\lambda_k} + \frac{(\Lambda \operatorname{Re}\phi_k, \operatorname{Re}\phi_j)}{\lambda_j} \\ &+ \sqrt{\lambda_j \lambda_k}\, (M \operatorname{Im}\phi_j, \operatorname{Im}\phi_k) \\ &= (Q \operatorname{Re}\phi_j, \operatorname{Re}\phi_k). \end{aligned} \tag{6.35}$$

There exists $c_0 >$ *such that* $\forall\, j \geq N+1$

$$\begin{aligned} &(M \operatorname{Re}\phi_j, \operatorname{Re}\phi_j) + (M \operatorname{Im}\phi_j, \operatorname{Im}\phi_j) \\ &- \frac{(Q \operatorname{Re}\phi_j, \operatorname{Re}\phi_j)}{\lambda_j} \geq \frac{c_0}{\lambda_j^2}, \end{aligned} \tag{6.36}$$

and $\forall\, j,k \geq N+1$,

$$\begin{aligned}
&|(M\operatorname{Re}\phi_j,\operatorname{Re}\phi_j)-(M\operatorname{Im}\phi_j,\operatorname{Im}\phi_j)\\
&+\frac{(Q\operatorname{Re}\phi_j,\operatorname{Re}\phi_j)}{\lambda_j}-4\frac{\Lambda\operatorname{Re}\phi_j,\operatorname{Re}\phi_j)}{\lambda_j^2}| \qquad (6.37)\\
&\leq \frac{K}{\lambda_j^{\frac{3}{2}}}.
\end{aligned}$$

Then the pair (A,B) *is exactly controllable for* T *sufficiently large, and* $D(A^2)\subset F_T$, *with dense and continuous injection.*

Proof. One proves that

$$(\Gamma_T h,h)\geq c_T||h||^2_{(D(A^2))'},\quad \forall\, h\in H. \tag{6.38}$$

Of course this estimate is not as good as (6.24) but it is sufficient to establish that $(\Gamma_T h,h)$ is a norm on H and the rest of the statement. We again write

$$(\Gamma_T h,h)=X_1+X_2+X_3$$

where

$$X_1=4\int_0^T|B^*(\sum_{j=1}^N \operatorname{Re}(e^{-ti\sqrt{\lambda_j}}P_jh))|^2\,dt$$

$$X_2=$$

$$8\int_0^T(B^*(\sum_{j=1}^N\operatorname{Re}(e^{-ti\sqrt{\lambda_j}}P_jh)),B^*(\sum_{j\geq N+1}\operatorname{Re}(e^{-ti\sqrt{\lambda_j}}\overline{c_j}\phi_j)))\,dt$$

$$X_3=4\int_O^T|B^*(\sum_{j\geq N+1}\operatorname{Re}(e^{-ti\sqrt{\lambda_j}}\overline{c_j}\phi_j))|^2\,dt.$$

As in Theorem 6.1

$$X_1=2T\sum_{j=1}^N|B^*P_jh|^2+Y_1.$$

We can combine

$$X_2 + Y_1 = \text{ right-hand side of (6.28).} \tag{6.39}$$

We can state the estimate

$$|X_2 + Y_1| \leq \beta'_N ||h||^2_{(D(A^2))'} \tag{6.40}$$

where

$$\beta'_N = \sqrt{2} \sup_h \sum_{j=1}^{N} \sqrt{\lambda_j}\, \Lambda_j \frac{||BB^* P_j h||_{D(A)}}{||P_j h||_{(D(A))'}} \tag{6.41}$$

and

$$\Lambda_j = \max \left\{ \frac{\lambda_{j+1}^{\frac{1}{2}}(\lambda_j + \lambda_{j+1})^{\frac{1}{2}}}{\lambda_{j+1} - \lambda_j}, \frac{\lambda_{j-1}^{\frac{1}{2}}(\lambda_j + \lambda_{j-1})^{\frac{1}{2}}}{\lambda_j - \lambda_{j-1}} \right\}.$$

Moreover again as in the proof of Theorem 6.1.

$$\begin{aligned} X_3 \ \geq \ & 4 \sum_{j,k \geq N+1} \int_0^T (M \operatorname{Re} \phi_j, \operatorname{Re} \phi_k) \operatorname{Re} (e^{-ti\sqrt{\lambda_j}} \overline{c_j}) \\ & \qquad \cdot \operatorname{Re} (e^{-ti\sqrt{\lambda_k}} \overline{c_k})\, dt \\ & + 4 \sum_{j,k \geq N=1} \int_0^T (M \operatorname{Im} \phi_j, \operatorname{Im} \phi_k) \operatorname{Im} (e^{-ti\sqrt{\lambda_j}} \overline{c_j}) \\ & \qquad \cdot \operatorname{Im} (e^{-ti\sqrt{\lambda_k}} \overline{c_k})\, dt. \end{aligned}$$

Let us write $b_j = \frac{c_j}{\lambda_j}$, then from (6.34)

$$X_3 \geq Y_3 + Z_3$$

where

$$\begin{aligned} Y_3 = \ & 4 \sum_{j \geq N+1} \int_0^T \lambda_j^2 (M \operatorname{Re} \phi_j, \operatorname{Re} \phi_j) |\operatorname{Re} (e^{-ti\sqrt{\lambda_j}} \overline{b_j})|^2\, dt \\ & + 4 \sum_{j \geq N+1} \int_0^T \lambda_j^2 (M \operatorname{Im} \phi_j, \operatorname{Im} \phi_j) |\operatorname{Im} (e^{-ti\sqrt{\lambda_j}} \overline{b_j})|^2\, dt \end{aligned} \tag{6.42}$$

$$
\begin{aligned}
Z_3 = \; & 4 \sum_{\substack{k \geq N+1 \\ j \neq k}} \int_0^T ((\Lambda \operatorname{Re}\phi_j, \operatorname{Re}\phi_k) + (\Lambda \operatorname{Re}\phi_k, \operatorname{Re}\phi_j)) \\
& \qquad \cdot \operatorname{Re}(e^{-ti\sqrt{\lambda_j}}\overline{b_j}) \operatorname{Re}(e^{-ti\sqrt{\lambda_k}}\overline{b_k})\, dt \\
& + 4 \sum_{\substack{k \geq N+1 \\ j \neq k}} \int_0^T \lambda_j \lambda_k (M \operatorname{Im}\phi_j, \operatorname{Im}\phi_k) \\
& \qquad \cdot \operatorname{Im}(e^{-ti\sqrt{\lambda_j}}\overline{b_j}) \operatorname{Im}(e^{-ti\sqrt{\lambda_k}}\overline{b_k})\, dt.
\end{aligned}
\tag{6.43}
$$

We can write

$$
\begin{aligned}
& \int_0^T ((\Lambda \operatorname{Re}\phi_j, \operatorname{Re}\phi_k) + (\Lambda \operatorname{Re}\phi_k, \operatorname{Re}\phi_j)) \\
& \quad \cdot \operatorname{Re}(e^{-ti\sqrt{\lambda_j}}\overline{b_j}) \operatorname{Re}(e^{-ti\sqrt{\lambda_k}}\overline{b_k})\, dt \\
= & -(\Lambda \operatorname{Re}\phi_j, \operatorname{Re}\frac{\phi_k}{\sqrt{\lambda_k}})(\operatorname{Re}(e^{-Ti\sqrt{\lambda_j}}\overline{b_j}) \\
& \quad \cdot \operatorname{Im}(e^{-Ti\sqrt{\lambda_k}}\overline{b_k}) - \operatorname{Re}\overline{b_j} \operatorname{Im}\overline{b_k}) \\
& -(\Lambda \operatorname{Re}\phi_k, \operatorname{Re}\frac{\phi_j}{\sqrt{\lambda_j}})(\operatorname{Re}(e^{-Ti\sqrt{\lambda_k}}\overline{b_k}) \\
& \quad \cdot \operatorname{Im}(e^{-Ti\sqrt{\lambda_j}}\overline{b_j}) - \operatorname{Re}\overline{b_k} \operatorname{Im}\overline{b_j}) \\
& + \int_0^T (\frac{(\Lambda \operatorname{Re}\phi_j, \operatorname{Re}\phi_k)\sqrt{\lambda_j}}{\sqrt{\lambda_k}} + \frac{(\Lambda \operatorname{Re}\phi_k, \operatorname{Re}\phi_j)\sqrt{\lambda_k}}{\sqrt{\lambda_j}}) \\
& \quad \cdot \operatorname{Im}(e^{-ti\sqrt{\lambda_j}}\overline{b_j}) \operatorname{Im}(e^{-ti\sqrt{\lambda_k}}\overline{b_k})\, dt
\end{aligned}
\tag{6.44}
$$

and thus from the assumption (6.35)

$$
\begin{aligned}
Z_3 = & -8(\Lambda[\sum_{j \geq N+1} \operatorname{Re}\phi_j \operatorname{Re}(e^{-Ti\sqrt{\lambda_j}}\overline{b_j})], \\
& \sum_{k \geq N+1} \operatorname{Re}\frac{\phi_k}{\sqrt{\lambda_k}} \operatorname{Im}(e^{-Ti\sqrt{\lambda_k}}\overline{b_k})) \\
& + 8(\Lambda[\sum_{j \geq N+1} \operatorname{Re}\phi_j \operatorname{Re}(\overline{b_j})], \sum_{k \geq N+1} \operatorname{Re}\frac{\phi_k}{\sqrt{\lambda_k}} \operatorname{Im}(\overline{b_k}))
\end{aligned}
$$

$$+8\sum_{j\geq N+1}(\Lambda\,\mathrm{Re}\,\phi_j,\mathrm{Re}\,\frac{\phi_j}{\sqrt{\lambda_j}})(\mathrm{Re}\,(e^{-Ti\sqrt{\lambda_j}}\overline{b_j})\,\mathrm{Im}\,(e^{-Ti\sqrt{\lambda_j}}\overline{b_j})$$

$$-\mathrm{Re}\,(\overline{b_j})\,\mathrm{Im}\,(\overline{b_j}))$$

$$+4\int_0^T(Q\sum_{j\geq N+1}\sqrt{\lambda_j}\mathrm{Re}\,\phi_j\,\mathrm{Im}\,(e^{-Ti\sqrt{\lambda_j}}\overline{b_j}),$$

$$\sum_{k\geq N+1}\sqrt{\lambda_k}\mathrm{Re}\,\phi_k\,\mathrm{Im}\,(e^{-Ti\sqrt{\lambda_k}}\overline{b_k}))$$

$$-4\int_0^T\sum_{j\geq N+1}\lambda_j(Q\,\mathrm{Re}\,\phi_j,\mathrm{Re}\phi_j)|\mathrm{Im}\,(e^{-ti\sqrt{\lambda_j}}\overline{b_j})|^2\,dt. \qquad (6.45)$$

Since $Q\geq 0$, we can assert that

$$X_1+X_2+X_3\geq$$

$$2T\{\sum_{j=1}^{N}|B^*P_jh|^2$$
$$+\sum_{j\geq N+1}|b_j|^2[\lambda_j^2((M\,\mathrm{Re}\,\phi_j,\mathrm{Re}\,\phi_j)$$
$$+(M\,\mathrm{Im}\,\phi_j,\mathrm{Im}\,\phi_j))-\lambda_j(Q\,\mathrm{Re}\,\phi_j,\mathrm{Re}\,\phi_j)]\}$$

$$+Y_1+X_2+\Delta_1+\Delta_2 \qquad (6.46)$$

with

$$\Delta_1=$$

$$\sum_{j\geq N+1}\{\lambda_j^{\frac{3}{2}}[(M\,\mathrm{Re}\,\phi_j,\mathrm{Re}\,\phi_j)-(M\,\mathrm{Im}\,\phi_j,\mathrm{Im}\,\phi_j)]$$

$$+\lambda_j^{\frac{1}{2}}(Q\,\mathrm{Re}\,\phi_j,\mathrm{Re}\,\phi_j)-4\frac{(\Lambda\,\mathrm{Re}\,\phi_j,\mathrm{Re}\,\phi_j)}{\lambda_j^{\frac{1}{2}}}\}$$

$$\cdot\,\mathrm{Im}\,(b_j^2(e^{2Ti\sqrt{\lambda_j}}-1))$$

$\Delta_2 =$

$$-8(\Lambda \sum_{j\geq N+1} \operatorname{Re}\phi_j \operatorname{Re}(e^{-Ti\sqrt{\lambda_j}}\overline{b_j})), \sum_{k\geq N+1} \operatorname{Re}\frac{\phi_k}{\sqrt{\lambda_k}}\operatorname{Im}(e^{-Ti\sqrt{\lambda_k}}\overline{b_k}))$$

$$+8(\Lambda[\sum_{j\geq N+1} \operatorname{Re}\phi_j \operatorname{Re}(\overline{b_j})], \sum_{k\geq N+1} \operatorname{Re}\frac{\phi_k}{\sqrt{\lambda_k}}\operatorname{Im}(\overline{b_k})).$$

Then, according to the assumption (6.37)

$$|\Delta_1| \leq K||h||^2_{(D(A^2))'}$$

and

$$|\Delta_2| \leq 4||\Lambda||\,||h||^2_{(D(A^2))'}.$$

Then we deduce from the assumptions (6.21) and (6.36)

$$(\Gamma_T h, h) \geq c(T - T_0)||h||^2_{(D(A^2))'} \tag{6.47}$$

with $c = \min\{c_0, c_1\}$(c_0 if $N = 0$) and

$$T_0 = \frac{\beta'_N + K + 4||\Lambda||}{c}$$

and the desired result has been proved. ∎

6.3 Approximation

We shall study the properties of Γ_T, assuming only (6.7) to (6.10) and thus nothing on B except naturally $B \in \mathcal{L}(U; H)$.

We now keep only the eigenvalues with different values denoted $0 < \lambda_1 < \lambda_2 \ldots < \lambda_j < \ldots$ and call P_j the projector which has been defined in (6.21) (but now j is any integer). An element ϕ of H is represented as

$$\phi = \sum_j (P_j\phi + \overline{P_j}\phi) \tag{6.48}$$

and

$$e^{At}\phi = \sum_j (e^{i\sqrt{\lambda_j}t}P_j\phi + e^{-i\sqrt{\lambda_j}t}\overline{P_j}\phi). \tag{6.49}$$

Call

$$\Pi^N \phi = \sum_{j=1\ldots N} (P_j\phi + \overline{P_j}\phi). \tag{6.50}$$

We also define the operator Γ by

$$\Gamma = 2\,\mathrm{Re} \sum_j P_j BB^* P_j. \tag{6.51}$$

We can state the

Proposition 6.2 *Assume* (6.7) *to* (6.10) *then*

$$\frac{\Gamma_T}{T} \to \Gamma h \text{ in } H, \quad \forall\, h \in H, \text{ as } T \to \infty. \tag{6.52}$$

Proof. We can write, picking η in H

$$\begin{aligned}(\Gamma_T h, \eta) &= (\Gamma_T(h - \Pi^N h), \eta) \\ &\quad + 2T\,\mathrm{Re} \sum_{j=1}^{N} (P_j BB^* P_j h, \eta) \\ &\quad + X_{N,T} + Y_{N,T}\end{aligned} \tag{6.53}$$

with

$$\begin{aligned}X_{N,T} = \;& 2\,\mathrm{Re}\,(\sum_{j=1}^{N} BB^* P_j h, \\ & \sum_{k\neq j} \{P_k \eta \frac{1 - e^{-iT(\sqrt{\lambda_j}+\sqrt{\lambda_k})}}{i(\sqrt{\lambda_j}+\sqrt{\lambda_k})} \\ & \quad + \overline{P_k}\eta \frac{1 - e^{-iT(\sqrt{\lambda_j}-\sqrt{\lambda_k})}}{i(\sqrt{\lambda_j}-\sqrt{\lambda_k})}\})\end{aligned}$$

and

$$Y_{N,T} = \mathrm{Re} \sum_{j=1}^{N} (BB^* P_j h, P_j \eta)\, \frac{(1 - e^{2iT\sqrt{\lambda_j}})}{i\sqrt{\lambda_j}}.$$

We can check that

$$|X_{N,T}| \leq \beta_N ||h||_{(D(A))'} ||\eta||_{(D(A))'} \tag{6.54}$$

where β_N has been defined in the proof of Theorem 6.1.

Similarly

$$|Y_{N,T}| \leq \sqrt{2}\sum_{j=1}^{N} |BB^* P_j h| \, ||\eta||_{(D(A))'}. \tag{6.55}$$

Now

$$2T \operatorname{Re} \sum_{j=1}^{N} (P_j BB^* P_j h, \eta) = 2T(\Gamma h, \eta) - 2T(\Gamma(h - \Pi^N h), \eta).$$

Therefore collecting results, we deduce noticing that $||\Gamma_T|| \leq T||BB^*||$ and $||\Gamma|| \leq ||BB^*||$

$$|(\frac{\Gamma_T}{T} - \Gamma)h| \leq \quad 2||BB^*|| \, |h - \Pi^N h|$$

$$+ \frac{\beta_N ||h||_{(D(A))'} + \sqrt{2}\sum_{j=1}^{N} |BB^* P_j h|}{T\lambda_1}$$

and the desired result follows. ∎

We next prove the following estimate

Lemma 6.1

$$|(\Gamma_T h, \eta)| \leq ||BB^*||(2|h| + T||h||_{D(A)})||\eta||_{(D(A))'} \tag{6.56}$$

Proof. From the definition of Γ_T we have

$$(\Gamma_T h, \eta) =$$

$$4\int_0^T (BB^* \sum_j \operatorname{Re} (e^{it\sqrt{\lambda_j}} P_j h), \frac{d}{dt}\sum_k \operatorname{Im} (e^{it\sqrt{\lambda_k}} \frac{P_k \eta}{\sqrt{\lambda_k}})) \, dt$$

and by integrating by parts

$$= 4(BB^* \sum_j \mathrm{Re}\,(e^{iT\sqrt{\lambda_j}} P_j h), \sum_k \mathrm{Im}\,(e^{iT\sqrt{\lambda_k}} \frac{P_k\eta}{\sqrt{\lambda_k}}))$$

$$-4(BB^* \sum_j \mathrm{Re}\,(P_j h), \sum_k \mathrm{Im}\,(\frac{P_k\eta}{\sqrt{\lambda_k}}))$$

$$+4\int_0^T (BB^* \sum_j \mathrm{Im}\,(\sqrt{\lambda_j} e^{it\sqrt{\lambda_j}} P_j h), \sum_k \mathrm{Im}\,(e^{it\sqrt{\lambda_k}} \frac{P_k\eta}{\sqrt{\lambda_k}}))\, dt.$$

Thus the result (6.56) is easily deduced. ∎

One deduces from (6.56) that

$$\Gamma_T \in \mathcal{L}(D(A); D(A)). \tag{6.57}$$

Remark 6.3 The formula used in the proof of Lemma 6.1, shows also that that if $BB^* \in \mathcal{L}(D(A); D(A))$, then

$$\Gamma_T \in \mathcal{L}(D(A^2); D(A^2)).$$

∎

Similarly we check the estimate

$$|(\Gamma h, \eta)| \leq ||BB^*||\; ||h||_{D(A)} ||\eta||_{(D(A))'} \tag{6.58}$$

and then we can assert the following proposition.

Proposition 6.3 *One has*

$$\frac{\Gamma_T}{T} h \to \Gamma h \text{ in } D(A), \quad \forall\, h \in D(A), \text{ as } T \to \infty. \tag{6.59}$$

Proof. Using the formulas of Proposition 6.2 and Lemma 6.1 we obtain easily the estimate

$$\begin{aligned} &\left\| \left(\frac{\Gamma_T}{T} - \Gamma\right) h \right\|_{D(A)} \\ &\leq 2||BB^*|| \left\{ ||h - \Pi^N h||_{D(A)} + \frac{|h - \Pi^N h|}{T} \right\} \\ &+ \frac{\beta_N ||h||_{(D(A))'}}{T} + \sqrt{2} \sum_{j=1}^{N} |BB^* P_j h| \end{aligned} \tag{6.60}$$

and thus (6.60) obtains ∎

We now make the assumptions of Theorem 6.1. We first notice that $(\Gamma h, h)$ is also a norm on H, thanks to the

Proposition 6.4 *We make the assumptions of Theorem 6.1, then*

$$(\Gamma h, h) \geq c\|h\|^2_{(D(A))'} \quad \forall\, h \in H. \tag{6.61}$$

Proof. We have

$$(\Gamma h, h) = 2\sum_j (|B^* \operatorname{Re} P_j h|^2 + |B^* \operatorname{Im} P_j h|^2)$$

and from (6.15) and (6.21)

$$\begin{aligned} &\geq 2 \sum_{j \geq N+1} ((M \operatorname{Re} P_j h, \operatorname{Re} P_j h) + (M \operatorname{Im} P_j h, \operatorname{Im} P_j h)) \\ &+ 2c_1 \sum_{j=1}^{N} \frac{|P_j h|^2}{\lambda_j}. \end{aligned}$$

Now (6.16) to (6.18) imply that if

$$P_j h = \sum_{k=1}^{rj} \overline{c_{jk}} \phi_{jk}$$

then

$$\sum_{j \geq N+1} ((M \operatorname{Re} P_j h, \operatorname{Re} P_j h) + (M \operatorname{Im} P_j h, \operatorname{Im} P_j h))$$

$$= \sum_{j \geq N+1} \sum_{k=1}^{r_j} |\overline{c_{jk}}|^2 ((M \operatorname{Re} \phi_{jk}, \operatorname{Re} \phi_{jk}) + (M \operatorname{Im} \phi_{jk}, \operatorname{Im} \phi_{jk}))$$

and from (6.19)

$$\geq c_0 \sum_{j \geq N+1} \sum_{k=1}^{r_j} \frac{|\overline{c_{jk}}|^2}{\lambda_j} = c_0 \sum_{j \geq N+1} \frac{|P_j h|^2}{\lambda_j}.$$

Therefore collecting results, the desired result obtains. ∎

Let us define the space F' completing H with the norm (6.61). Clearly one has as for (6.32)

$$D(A) \subset F \subset H \subset F' \subset (D(A))'. \tag{6.62}$$

Since Γ_T is invertible from F'_T to F_T and Γ from F' to F, we can consider $(\Gamma_T)^{-1}$ from F_T to F'_T and Γ^{-1} from F to F', hence also from $D(A)$ to $(D(A))'$ for both.

In fact, it will be useful to notice that something more can be said for Γ^{-1}, namely

Lemma 6.2

$$\Gamma^{-1} \in \mathcal{L}(D(A^2); H)$$

Proof. Setting $\phi = \Gamma^{-1}h$, then from the definition of Γ

$$2\,\mathrm{Re}\sum_j P_j(BB^*P_j\phi) = 2\,\mathrm{Re}\sum_j P_jh$$

hence

$$P_j(BB^*P_j\phi) = P_jh, \quad \forall j.$$

Therefore

$$|B^*P_j\phi|^2 = (P_jh, \overline{P_j}\phi), \quad \forall j.$$

But from the proof of Proposition 6.4 it follows that

$$|B^*P_j\phi|^2 \geq c\frac{|P_j\phi|^2}{\lambda_j}$$

therefore

$$|P_j\phi| \leq \frac{\lambda_j}{c}|P_jh|$$

hence

$$\begin{aligned} |\phi|^2 = 2\sum_j |P_j\phi|^2 &\leq \frac{2}{c}\sum_j \lambda_j|P_jh|^2 \\ &\leq \frac{1}{c}||h||^2_{D(A^2)} \end{aligned}$$

which is the desired result. ∎

We can then state the

Theorem 6.3 *We make the assumptions of Theorem 6.1, then one has*

$$T(\Gamma_T)^{-1}h \to \Gamma^{-1}h \text{ in } (D(A))', \quad \forall\, h \in D(A) \tag{6.63}$$

Proof. Let us set $\rho_T = T(\Gamma_T)^{-1}h$, and $\rho = \Gamma^{-1}h$.

We begin by proving the weak convergence in $(D(A))'$. But

$$(h, \rho_T) = (\frac{\Gamma_T}{T}\rho_T, \rho_T) \geq c\,\frac{T - T_0}{T}\,||\rho_T||^2_{(D(A))'}$$

and thus

$$||\rho_T||_{(D(A))'} \leq \frac{2}{c}||h||_{D(A)}$$

as soon as $T > 2T_0$.

Therefore we can extract a subsequence converging weakly to some σ in $(D(A))'$. But, we can write for any η in $D(A)$

$$(\frac{\Gamma_T}{T}\eta, \rho_T) = (h, \eta)$$

and using Proposition 6.3 we get

$$(\Gamma\eta, \sigma) = (h, \eta).$$

Therefore

$$(\Gamma(\sigma - \rho), \eta) = 0$$

for any η in $D(A)$. Hence $\Gamma(\sigma - \rho) = 0$. But Γ is invertible from $(D(A))'$ to $D(A)$, hence $\rho = \sigma$. Note that by the uniqueness of the limit, we can assert that the full subsequence converges. To prove the strong convergence, we first assume that $h \in D(A^2)$, in which case we kow that ρ belongs to H. But

$$(\frac{\Gamma_T}{T}(\rho_T - \rho), \rho_T - \rho) = (h, \rho_T) - 2(h, \rho) + (\frac{\Gamma_T}{T}\rho, \rho)$$

and therefore the right hand side tends to 0. It easily follows that ρ_T tends to ρ in $(D(A))'$. In the general case, we pick a sequence h_μ which belongs to $D(A^2)$ and converges to h in $D(A)$. Let ρ_μ and $\rho_{T\mu}$ correspond to h_μ. We first notice that

$$||\rho_{T\mu} - \rho_T|| \leq \frac{2}{c}||h - h_\mu||_{D(A)}. \tag{6.64}$$

Now

$$||\rho_T - \rho|| \leq ||\rho_{T\mu} - \rho_T|| + ||\rho_{T\mu} - \rho_\mu|| + ||\rho_\mu - \rho||.$$

From the uniform estimate (6.64) and from the fact that $||\rho_{T\mu} - \rho_\mu||$ tends to 0 as T tends to infinity and μ is fixed, we deduce that ρ_T tends to ρ in $(D(A))'$ as T tends to infinity. ∎

6.4 Exact controllability for T arbitrarily small

In this section, we consider additional assumptions, which will guarantee that the pair (A, B) is not only exactly controllable for large T, but in fact for arbitrarily small T. We work in the framework of Theorem 6.1. We replace (6.20) by

$$\begin{aligned} &|(M \operatorname{Re}\phi_j, \operatorname{Re}\phi_j) - (M \operatorname{Im}\phi_j, \operatorname{Im}\phi_j) - 4\frac{(\Lambda \operatorname{Re}\phi_j, \operatorname{Re}\phi_j)}{\lambda_j}| \\ &\leq \frac{K}{\lambda_j}, \quad \forall j \geq N+1. \end{aligned} \tag{6.65}$$

In fact, we shall need (6.65) for j sufficiently large, but modifying the constant, we can always assume that it is satisfied for $j \geq N+1$, without loss of generality. We also assume that

$$\begin{aligned} &\text{there exists a Hilbert space } W \text{ such that } D(A) \subset W \subset H, \\ &\text{the injection of } D(A) \text{ in } W \text{ is compact, and } \Gamma \in \mathcal{L}(H; W'). \end{aligned} \tag{6.66}$$

Our objective is to prove the following.

Theorem 6.4 *We make the assumptions of Theorem 6.1, except* (6.20) *which is replaced by* (6.65) *and we assume* (6.66). *Then the pair A, B is exactly controllable for any time $T > 0$.*

Proof. We begin by proving that for any $T > 0$, then

$$(\Gamma_T h, h) = 0 \quad \Rightarrow \quad h = 0. \tag{6.67}$$

This will imply that $(\Gamma_T h, h)^{\frac{1}{2}}$ is a norm on H, for $T > 0$, arbitrary. Of course, we already know that it is true for $T > T_0$. Naturally (6.67) is equivalent to

$$\Gamma_T h = 0 \quad \Rightarrow \quad h = 0. \tag{6.68}$$

We denote by Π^J the projector on the subspace of H, generated by the eigenvalues $\pm i\sqrt{\lambda_j}$, with $j \leq J$. We shall need to consider only J large, so in particular $J > N$, hence for any element h of H, one has

$$\Pi^J h = \sum_{j=1}^{N}(P_j h + \overline{P_j} h) + \sum_{j=N+1}^{J} (\overline{c_j}\phi_j + c_j\overline{\phi_j}). \tag{6.69}$$

We shall prove that for any $T > 0$, there exists J depending on T, sufficiently large and a constant β^J, such that

$$(\Gamma_T h, h) + \beta^J |\Pi^J h|^2 \geq c_T ||h||^2_{(D(A))'}, \quad \forall\, h \in H. \tag{6.70}$$

Assume for a while that this is proved.

We shall then prove that

$$\Gamma_T h = 0 \text{ implies } h \in D(A). \tag{6.71}$$

Indeed, from the assumption we have

$$B^* e^{A^* t} h = 0, \quad t \in (0, T). \tag{6.72}$$

Let τ such that $T - 2\tau > 0$. Consider a sequence of C^∞ functions on R, with compact support on $(0, \tau)$, denoted by ρ_j, which converges to $\delta(0)$ in the distribution sense. Define successively $h_{1,j}$ and $h_{2,j}$ by the formulas

$$h_{1,j} = \int_0^\tau e^{A^* s} h \rho_j(s)\, ds \tag{6.73}$$

$$h_{2j} = \int_0^\tau e^{A^* s} h_{1,j} \rho_j(s)\, ds \tag{6.74}$$

then $h_{1,j} \in D(A)$, $h_{2,j} \in D(A^2)$.

Indeed

$$\begin{aligned} Ah_{1,j} &= \int_0^\tau e^{A^* s} h \rho_j'(s)\, ds \\ Ah_{2,j} &= \int_0^\tau e^{A^* s} h_{1,j} \rho_j'(s)\, ds. \end{aligned}$$

From (6.72), we deduce easily

$$\begin{aligned} &B^* e^{A^* t} h_{1,j} = 0, \quad t \in (0, T - \tau) \\ &B^* e^{A^* t} h_{2j} = 0, \quad t \in (0, T - 2\tau). \end{aligned} \tag{6.75}$$

Since $h_{2,j} \in D(A^2)$, we deduce by differentiating twice the 2nd relation that

$$B^* e^{A^* t} A^2 h_{2,j} = 0, \quad t \in (0, T - 2\tau).$$

Moreover $A^2 h_{2,j} \in H$. Therefore, we may apply (6.66) with $h = A^2 h_{2,j}$, and T changed into $T - 2\tau$. Since $\Gamma_{T-2\tau} A^2 h_{2j} = 0$, we deduce that $A^2 h_{2,j}$ remains in a bounded subset of $(D(A))'$, hence $h_{2,j}$ remains in a bounded subset of $D(A)$. On the other hand, h_{2j} converges to h in $(D(A))'$ weakly. This follows from the fact that if $\phi \in D(A)$, then

$$< \phi, h_{2,j} >= \\ ([A+I] \int_0^T e^{As} \phi \rho_j(s)\, ds, [A+I]^{-1} \int_0^T e^{A^* s} h \rho_j(s)\, ds)$$

and

$$[A+I]^{-1} \int_0^T e^{A^* s} h \rho_j(s)\, ds$$

tends to h in H strongly,

$$[A+I] \int_0^T e^{As} \phi \rho_j(s)\, ds$$

tends to ϕ in H weakly. Therefore $h \in D(A)$.

Let Y be the subspace of H of elements such that $h \in H$, $\Gamma_T h = 0$. Our objective is to prove that $Y = \{0\}$. We first prove that Y is finite dimensional. But if $h \in Y$, then one can write

$$\Gamma_T h + \beta^J \Pi^J h = \beta^J \Pi^J h$$

hence

$$h = \beta^J (\Gamma_T + \beta^J \Pi^J)^{-1} \Pi^J h$$

and

$$\Pi^J h = \beta^J \Pi^J (\Gamma_T + \beta^J \Pi^J)^{-1} \Pi^J h. \tag{6.76}$$

These formulas prove that Y is a subvectorspace of the span of $(\Gamma_T + \beta^J \Pi^J)^{-1} \phi_j$, $(\Gamma_T + \beta^J \Pi^J)^{-1} \overline{\phi_j}, j = 1, \ldots, J$, where we have written for simplicity Π^J in (6.76) as

$$\Pi^J h = \sum_{j=1}^{J} (\overline{c_j} \phi_j + c_j \overline{\phi_j})$$

which is always possible by a convenient renumbering.

Now, thanks to (6.71), $Ah \in Y$. Furthermore, if $h \in Y, A^2h \in Y$ and thus

$$\Gamma_T A^2 h = \beta^J \Pi^J A^2 h = \beta^J \Pi^J A^2 h$$

and also

$$(\Gamma_T A^2 h, A^2 h) + \beta^J (\Pi^J A^2 h, A^2 h) = \beta^J (\Pi^J A^2 h, A^2 h).$$

Therefore, by (6.70) we deduce

$$c_T ||A^2 h||^2_{(D(A))'} \leq \beta^J (\lambda^J)^2 |h|^2$$

or

$$c_T |Ah|^2_H \leq \beta^J (\lambda^J)^2 |h|^2$$

which proves that A is linear continuous on Y. But then Y can be spanned by the eigenvectors of A. If ϕ_j is an eigenvector of A belonging to Y, then

$$B^* \phi_j e^{it\sqrt{\lambda_j}} = 0, \quad \forall t \in (0, T).$$

By the analyticity property, it follows that

$$B^* \phi_j e^{it\sqrt{\lambda_j}} = 0, \quad \forall t \in (-\infty, +\infty).$$

In particular $(\Gamma_T \phi_j, \phi_j) = 0, \quad \forall T > 0$. This is impossible, because we know that for T sufficiently large $(\Gamma_T h, h)^{\frac{1}{2}}$ is a norm on H. Therefore $Y = \{0\}$, and (6.68) is proved. It is easy to check then, as a consequence of (6.68), (6.70) one has also

$$(\Gamma_T h, h) \geq c'_T ||h||^2_{(D(A))'}, \quad \forall h \in H, \forall T > 0. \tag{6.77}$$

Suppose that (6.77) is not true, then there exists a sequence h_n with $||h_n||_{(D(A))'} = 1$, and $(\Gamma_T h_n, h_n)$ tends to 0. Since $(\Gamma_T h, h)^{\frac{1}{2}}$ is a norm on H, we have $h_n \to 0$ in F_T (completed of H with the norm $(\Gamma_T h, h)^{\frac{1}{2}}$). Since h_n is bounded in $(D(A))'$, it converges also weakly to 0 in $(D(A))'$, hence $\Pi^J h_n \to 0$. But then from (6.70), $||h_n||_{(D(A))'} \to 0$, which is impossible.

It remains to prove (6.70). It is sufficient to prove that

$$\begin{array}{l}\forall T>0,\ \exists\, J=J_T \text{ and } c_T \text{ such that}\\ (\Gamma_T(h-\Pi^J h), h-\Pi^J h) \geq c_T \|h-\Pi^J h\|^2_{(D(A))'}\end{array} \tag{6.78}$$

Now

$$(\Gamma_T(h-\Pi^J h), h-\Pi^J h) = 4\int_0^T |B^*(\sum_{j\geq J+1} \mathrm{Re}\,(e^{-ti\sqrt{\lambda_j}}\overline{c_j}\phi_j))|^2\,dt$$

and with calculations already done

$$\geq 2T \sum_{j\geq J+1} \|c_j\|^2((M\,\mathrm{Re}\,\phi_j, \mathrm{Re}\,\phi_j) + (M\,\mathrm{Im}\,\phi_j, \mathrm{Im}\,\phi_j)) + Z_3 + Y_3',$$

with (recalling that $a_j = \frac{c_j}{\sqrt{\lambda_j}}$)

$$Z_3 =$$

$$-8(\Lambda[\sum_{j\geq J+1} \mathrm{Re}\,\phi_j\,\mathrm{Im}\,(e^{-iT\sqrt{\lambda_j}}\,\overline{a_j})], \sum_{k\geq J+1} \mathrm{Re}\,\phi_k \frac{\mathrm{Re}\,(e^{iT\sqrt{\lambda_k}}\overline{a_k})}{\sqrt{\lambda_k}})$$

$$+8(\Lambda[\sum_{j\geq J+1} \mathrm{Re}\,\phi_j\,\mathrm{Im}\,(\overline{a_j})], \sum_{k\geq J+1} \mathrm{Re}\,\phi_k \frac{\mathrm{Re}\,(\overline{a_k})}{\sqrt{\lambda_k}})$$

$$+8 \sum_{j\geq J+1} (\Lambda \mathrm{Re}\,\phi_j, \mathrm{Re}\,\phi_j)\,\mathrm{Im}\,(e^{-iT\sqrt{\lambda_j}}\,\overline{a_j})\,\frac{\mathrm{Re}\,(e^{-iT\sqrt{\lambda_j}}\,\overline{a_j})}{\sqrt{\lambda_j}}$$

$$-8 \sum_{j\geq J+1} (\Lambda\,\mathrm{Re}\,\phi_j, \mathrm{Re}\,\phi_j)\,\mathrm{Im}\,(\overline{a_j})\,\frac{\mathrm{Re}\,(\overline{a_j})}{\sqrt{\lambda_j}}$$

and

$$Y_3' =$$

$$\sum_{j\geq J+1} [(M\,\mathrm{Re}\,\phi_j, \mathrm{Re}\,\phi_j) - (M\,\mathrm{Im}\,\phi_j, \mathrm{Im}\,\phi_j)]\,\frac{\mathrm{Im}\,(c_j^2(e^{2iT\sqrt{\lambda_j}}-1))}{\sqrt{\lambda_j}}.$$

Again we can write

$$Z_3 + Y_3' = T_3 + T_3'$$

where

$$T_3 =$$

$$-8(\Lambda[\sum_{j\geq N+1} \operatorname{Re}\phi_j \operatorname{Im}(e^{-iT\sqrt{\lambda_j}}\overline{a_j})\,],\ \sum_{k\geq N+1} \operatorname{Re}\phi_k \frac{\operatorname{Re}(e^{-iT\sqrt{\lambda_k}}\overline{a_k})}{\sqrt{\lambda_k}})$$

$$+8(\Lambda[\sum_{j\geq N+1} \operatorname{Re}\phi_j \operatorname{Im}(\overline{a_j})],\ \sum_{k\geq N+1} \operatorname{Re}\phi_k \frac{\operatorname{Re}(\overline{a_k})}{\sqrt{\lambda_k}})$$

and

$$T_3' = \sum_{j\geq N+1} [(M\operatorname{Re}\phi_j, \operatorname{Re}\phi_j) - (M\operatorname{Im}\phi_j, \operatorname{Im}\phi_j)$$
$$- 4\frac{(\Lambda\operatorname{Re}\phi_j, \operatorname{Re}\phi_j)}{\lambda_j}]\,\frac{\operatorname{Im}(c_j^2(e^{2iT\sqrt{\lambda_j}}-1))}{\sqrt{\lambda_j}}.$$

We use (6.65) to assert that

$$|T_3'| \leq 2K \sum_{j\geq J+1} \frac{|c_j|^2}{\lambda_j^{\frac{3}{2}}} = 2K \sum_{j\geq J+1} \frac{|a_j|^2}{\lambda_j^{\frac{1}{2}}}.$$

Next, from (6.66) we can assert that for any ε there exists $C(\varepsilon)$ such that

$$||h||_W \leq \varepsilon||h||_{D(A)} + C(\varepsilon)|h|$$

(see J.-L. LIONS [1]). Therefore, as easily seen

$$|T_3| \leq 8||\Lambda||_{\mathcal{L}(H;W')}$$
$$\cdot [\varepsilon \sum_{j\geq J+1} |\overline{a}_j|^2 + C(\varepsilon)(\sum_{j\geq J+1} |\overline{a}_j|^2)^{\frac{1}{2}}(\sum_{j\geq J+1} \frac{|\overline{a}_j|^2}{\lambda_j})^{\frac{1}{2}}].$$

Therefore, collecting results we have

$$(\Gamma_T(h-\Pi^J h), h-\Pi^J h)$$
$$\geq ||h-\Pi^J h||^2_{(D(A))'}\left\{c_0T - \frac{K}{\sqrt{\lambda_J}} - 4||\Lambda||\varepsilon - \frac{C(\varepsilon)}{\sqrt{\lambda_j}}\right\}. \tag{6.79}$$

Therefore for any $T > 0$, choose

$$\varepsilon = \frac{c_0 T}{8||\Lambda||},$$

and J sufficiently large so that

$$\frac{c_0 T}{2} > \frac{K + C(\varepsilon)}{\sqrt{\lambda_J}}.$$

This concludes (6.78). ∎

Remark 6.4 The idea of the regularizing functions ρ_j has been given to us by J.-L. LIONS [5]. ∎

7 General framework: skew-symmetric operators

7.1 Operator A

Let L be a self adjoint operator in a Hilbert space H_L (which is identified with its dual). Let $D(L)$ be the domain of L. We assume that there exists a Hilbert space V_L such that

$$V_L \subset H_L \subset V_L', \tag{7.1}$$

continuously and densily embedded.

We assume that

$$< Lz_1, z_2 >= ((z_1, z_2)), \quad \forall\, z_1, z_2 \in V_L \tag{7.2}$$

where $((\ ,\ ))$ denotes the scalar product in V_L.

We furthermore assume that

$$\text{the injection of } V_L \text{ into } H_L \text{ is compact.} \tag{7.3}$$

In that case $L^{-1} \in \mathcal{L}(H_L; H_L)$ is compact and thus there exists an orthonormal base w_j of H_L, of eigenvectors of L, i.e.,

$$Lw_j = \lambda_j w_j, \quad |w_j|_{H_L} = 1$$

with $0 < \lambda_1 \leq \lambda_2 \ldots \leq \lambda_j \ldots \uparrow \infty$.

Besides $D(L)$ we shall need

$$\Delta(L) = \{z \in V_L \; : \; Lz \in V_L\}$$

and we have the sequence of spaces, each of them being densily and continuously embedded in the next one

$$\Delta_L \subset D(L) \subset V_L \subset H_L \subset V_L' \subset (D(L))' \subset \Delta_L'.$$

We can associate to L a skew symmetric operator A, satisfying the assumptions (6.7) to (6.10). Take indeed

$$H = H_L \times V_L', \quad A = \begin{bmatrix} 0 & I \\ -L & 0 \end{bmatrix}, \quad D(A) = V_L \times H_L. \tag{7.4}$$

We have to pay attention to the fact, that since H is identified with its dual, V_L' is in the 2nd component the pivot space, identified with its dual. In that framework the dual of H_L is $(D(L))'$, with the duality

$$< z, \zeta >= (z, L^{-1}\zeta)_{H_L}, \quad \forall z \in H_L, \zeta \in (D(L))'.$$

Note that the scalar product in H is expressed as follows

$$(h, h')_H = (h_1, h_1') + < L^{-1}h_2, h_2' > .$$

Moreover $(D(A))' = V_L' \times (D(L))'$.

The eigenvectors of A are given by

$$\phi_j = \frac{1}{\sqrt{2}} \begin{bmatrix} w_j \\ i\sqrt{\lambda_j} w_j \end{bmatrix}$$

7.2 Operator B

We next define the operator B. We shall consider a control space $U = U_1 \times U_2$ where U_1, U_2 are Hilbert spaces, identified with their duals. Let q_1, q_2 be operators such that

$$\begin{aligned} q_1 &\in \mathcal{L}(D(L); U_1) \\ q_2 &\in \mathcal{L}(V_L; U_2). \end{aligned} \tag{7.5}$$

We set

$$Bv = -\begin{bmatrix} L^{-1}q_1^* v_1 \\ q_2^* v_2 \end{bmatrix} \tag{7.6}$$

and $B \in \mathcal{L}(U; H)$. Note that

$$B^* h = -\begin{bmatrix} q_1 L^{-1} h_1 \\ q_2 L^{-1} h_2 \end{bmatrix} \tag{7.7}$$

The presence of L^{-1} in the second component of B^* arises from the fact that the pivot space is V_L'.

7.3 Dynamical system

Consider the dynamic system

$$\begin{aligned} &z' = Az + Bv, \quad z(0) = 0 \\ &z \in C([0,T]; H), \quad z' \in L^2(0,T; (D(A))') \end{aligned} \tag{7.8}$$

we deduce

$$\begin{aligned} z_1' &= z_2 - L^{-1} q_1^* v_1, \quad z_1(0) = 0, \\ z_2' &= -Lz_1 - q_2^* v_2, \quad z_2(0) = 0. \end{aligned} \tag{7.9}$$

Hence $z_1 \in C([0,T]; H_L), z_2 \in C([0,T]; V_L'); z_1' \in L^2(0,T; V_L')$, $z_2' \in L^2(0,T; (D(L))')$.

We can associate to (7.9) a 2nd order (in time) equation as follows. Set $\eta = z_2$, then we also write

$$\begin{aligned} &\eta'' + L\eta = q_1^* v_1 - q_2^* v_2' \\ &\eta(0) = 0, \quad \eta'(0) + q_2^* v_2(0) = 0, \\ &\eta \in C([0,T]; V_L'), \quad \eta' + q_2^* v_2 \in C([0,T]; (D(L))'), \\ &\eta' \in L^2(0,T; (D(L))'), \\ &\eta'' \in C([0,T]; \Delta_L') \oplus L^2(0,T; (D(L))') \oplus H^{-1}(0,T; V_L'). \end{aligned} \tag{7.10}$$

Note that all terms in (7.10) make perfect sense. However, it is important to notice that the value of η' is not defined at each point, and is not 0 in any sense in general. It is useful to reinterpret (7.10) in terms of the transposition method of J.-L. LIONS and E. MAGENES [1]. Note that for any $\psi_0 \in D(L), \psi_1 \in V_L$ there exists one and only one solution ψ such that

$$\begin{aligned} &\psi'' + L\psi = f \\ &\psi(T) = \psi_0, \quad \psi'(T) = \psi_1 \\ &\psi \in C([0,T]; D(L)), \quad \psi' \in C([0,T]; V_L) \end{aligned} \tag{7.11}$$

By making appropriate integration by parts between (7.10) and (7.11) we deduce the formula

$$\begin{aligned} &< \psi_1, \eta(T) > - < \psi_0, \eta'(T) + q_2^* v_2(T) > \\ &= \int_0^T [(v_1, q_1\psi) + (v_2, q_2\psi')]\, dt \end{aligned} \tag{7.12}$$

which can be considered as the definition of $\eta(T)$ and $\eta'(T)+q_2^* v_2(T)$.

There is a formal but mnemonic way of writing (7.10). Introduce the operator

$$J_0 \in \mathcal{L}(H^1(0,T;U_2); L^2(0,T;U_2))$$

defined by

$$J_0 u = u'.$$

Its transpose $J_0^* \in \mathcal{L}(L^2(0,T;U_2); (H^1(0,T;U_2))')$. One should not mix up $J_0^* u$ which belongs to $(H^1(0,T;U_2))'$ with $-u'$ which belongs to $H^{-1}(0,T;U_2)$. They coincide only, when applied to $H_0^1(0,T;U_2)$. Next, consider q_2 as a linear bounded operator from $H^1(0,T;V_L)$ to $H^1(0,T;U_2)$, and thus q_2^* as an element of $\mathcal{L}((H^1(0,T;U_2))'; (H^1(0,T;V_L))')$. Therefore $q_2^* J_0^* \in \mathcal{L}(L^2(0,T;U_2); (H^1(0,T;V_L))')$. We write (7.10) as

$$\begin{aligned} &\eta'' + L\eta = q_1^* v_1 + q_2^* J_0^* v_2 \\ &\eta(0) = 0, \quad \eta'(0) = 0. \end{aligned} \tag{7.13}$$

We know that

$$\eta \in C([0,T]; V_L') \cap L^2(0,T;H_L), \quad \eta' \in L^2(0,T;(D(L))').$$

From (7.13) we read

$$\eta'' \in C([0,T];\Delta_L') \oplus L^2(0,T;(D(L))') \oplus (H^1(0,T;V_L))'.$$

The writing is formal as far $\eta'(0)$ (which is not defined) and η'' are concerned. What it means is that for any

$$\psi \in H^1(0,T;D(L)) \cap L^2(0,T;\Delta_L)$$

with $\psi(T)=0$, then one has

$$-\int_0^T (\eta',\psi')\,dt + \int_0^T (\eta, L\psi)\,dt = \int_0^T [(v_1,q_1\psi)+(v_2,q_2\psi')]\,dt \quad (7.14)$$

where all terms make sense. This equation defines a unique η since it implies in particular (7.12). The value of η' at any given time t (in particular 0) is not the value of the derivative, but a short for $\eta' + q_2^* v_2$.

7.4 Exact controllability

Let us first express the operator $\Gamma_T \in \mathcal{L}(H;H)$. By the definition

$$(\Gamma_T h, k) = \int_0^T (B^* e^{A^* t} h, B^* e^{A^* t} k)\,dt. \quad (7.15)$$

If we write

$$h = \begin{bmatrix} h_1 \\ h_2 \end{bmatrix}, \quad k = \begin{bmatrix} k_1 \\ k_2 \end{bmatrix}$$

and define ϕ, ψ by

$$\begin{array}{ll} \phi'' + L\phi = 0 & \psi'' + L\psi = 0 \\ \phi(0) = L^{-1}h_1 & \psi(0) = L^{-1}k_1 \\ \phi'(0) = -L^{-1}h_2 & \psi'(0) = -L^{-1}k_2 \end{array} \quad (7.16)$$

we deduce easily from (7.7) that

$$B^* e^{A^* t} h = \begin{bmatrix} -q_1\phi(t) \\ q_2\phi'(t) \end{bmatrix} \quad (7.17)$$

$$(\Gamma_T h, k) = \int_0^T [(q_1\phi, q_1\psi) + (q_2\phi', q_2\psi')]\,dt. \quad (7.18)$$

Therefore the estimate

$$(\Gamma_T h, k) \geq c(T - T_0)||h||^2_{(D(A))'}$$

is clearly equivalent to

$$\int_0^T (|q_1\phi|^2 + |q_2\phi'|^2)\, dt \geq c(T - T_0)(||\phi(0)||^2_{V_L} + |\phi'(0)|^2_{H_L}).$$

Moreover

$$\Gamma_T h = \begin{bmatrix} -L^{-1}(\eta'(T) + q_2^* q_2 \phi'(T)) \\ \eta(T) \end{bmatrix} \tag{7.19}$$

where η is the solution of (7.10) with

$$v_1(t) = -q_1\phi(t), \quad v_2(t) = q_2\phi'(t).$$

If we use the formulation (7.13) for the η equation, then we may write

$$\Gamma_T h = \begin{bmatrix} -L^{-1}\eta'(T) \\ \eta(T) \end{bmatrix} \tag{7.20}$$

but $\eta'(T)$ is not the value of η' at T, which is not meaningful since η' is only an L^2 function, but a mnemonic to denote an element of $(D(L))'$ defined by the relation

$$\begin{aligned} &(\eta'(T), \psi(T)) - \int_0^T (\eta', \psi')\, dt + \int_0^T (\eta, L\psi)\, dt \\ &= \int_0^T [(v_1, q_1\psi) + (v_2, q_2\psi')]\, dt \end{aligned} \tag{7.21}$$

for any $\psi \in H^1(0, T; D(L)) \cap L^2(0, T; \Delta_L)$.

Let us now try to check how we can satisfy the sufficient assumptions of exact controllability as stated in Theorem 6.1 and find the operators M, Λ satisfying (6.15), (6.20). We shall suppose that there exist operators π_1, π_2 and a form $b(\xi, \xi')$ such that

$$\begin{aligned} &\pi_1 \in \mathcal{L}(D(L); D(L)'), \quad \pi_2 \in \mathcal{L}(V_L; V_L'), \\ &b \text{ is bilinear continuous on } H_L \times V_L \\ &|q_1\zeta|^2 \geq (\pi_1\zeta, \zeta), \quad \forall \zeta \in D(L) \\ &|q_2\zeta|^2 \geq (\pi_2\zeta, \zeta), \quad \forall \zeta \in V_L \end{aligned} \tag{7.22}$$

$$(\pi_1 w_j, w_k) = \lambda_j b(w_j, w_k) + \lambda_k b(w_k, w_j), \quad \forall j \neq k \geq N+1 \quad (7.23)$$

$$b(w_j, w_k) + b(w_k, w_j) + (\pi_2 w_j, w_k) = 0, \quad \forall j \neq k \geq N+1 \quad (7.24)$$

$$\frac{1}{\lambda_j}(\pi_1 w_j, w_j) + (\pi_2 w_j, w_j) \geq 2c_0, \quad \forall j \geq N+1 \quad (7.25)$$

$$\left| \frac{(\pi_1 w_j, w_j)}{\lambda_j^{\frac{3}{2}}} - \frac{(\pi_2 w_j, w_j)}{\lambda_j^{\frac{1}{2}}} - 4\frac{b(w_j, w_j)}{\lambda_j^{\frac{1}{2}}} \right| \leq 2K, \quad \forall j \geq N+1. \quad (7.26)$$

We then define

$$(M\xi, \eta) = (\pi_1 L^{-1}\xi_1, L^{-1}\eta_1) + (\pi_2 L^{-1}\xi_2, L^{-1}\eta_2), \quad \forall \xi, \eta \in H. \quad (7.27)$$

$$(\Lambda\xi, \eta) = b(\xi_1, \eta_1), \quad \forall \xi \in H, \eta \in D(A). \quad (7.28)$$

It is easy to check that all assumptions (6.15) ... (6.20) are satisfied.

As a consequence of Theorem 6.1 we can state the following.

Theorem 7.1 *Consider the pair* (A, B) *defined in* §*7.1 and* §*7.2. Assume* (7.22) *to* (7.26) *and* (6.21). *Then the pair* (A, B) *is exactly controllable for* T *sufficiently large.*

Let us check what must be added to obtain exact controllability at any $T > 0$. We replace (7.26) by

$$\left| \frac{(\pi_1 w_j, w_j)}{\lambda_j} - (\pi_2 w_j, w_j) - 4b(w_j, w_j) \right| \leq 2K, \quad \forall j \geq N+1. \quad (7.29)$$

We also assume that there exists a Hilbert space W_L such that

$$V_L \subset W_L \subset H_L, \text{ the injection of } V_L \text{ in } W_L \text{ is compact}$$
$$b \text{ is bilinear continuous on } H_L \times W_L. \quad (7.30)$$

Then (7.29) implies (6.65) and (7.30) implies (6.66).

We can then state the

Theorem 7.2 *We make the assumptions of Theorem 7.1, except* (7.26) *which is replaced by* (7.29) *and we assume* (7.30), *then the pair* (A, B) *is exactly controllable for any* $T > 0$.

8 Exact controllability of hyperbolic equations

We shall apply in this section the results of §6 and §7 to the exact controllability of the wave equation, Maxwell equations, the plate equation with boundary control. The domain Ω considered in the sequel is smooth and bounded.

8.1 Wave equation with Dirichlet boundary control

We take $L = -\Delta$, with

$$\begin{aligned} H_L &= L^2(\Omega), \\ V_L &= H_0^1(\Omega), \\ D(L) &= H^2(\Omega) \cap H_0^1(\Omega), \\ \Delta_L &= \{z \in D(L) \,:\, \Delta z \in H_0^1(\Omega)\}. \end{aligned}$$

We take $U_1 = U_2 = L^2(\Gamma)$ and $q_1 = \frac{\partial}{\partial \nu}, q_2 = 0$. Let

$$m(x) = x - x_0, \quad R(x_0) = \sup_{x \in \Gamma} |m(x)|.$$

We define

$$(\pi_1 \zeta, \zeta') = \frac{1}{R(x_0)} \int_\Gamma m \cdot \nu \frac{\partial \zeta}{\partial \nu} \frac{\partial \zeta'}{\partial \nu}\, d\Gamma, \quad \pi_2 = 0. \tag{8.1}$$

$$b(z, \zeta) = -\frac{1}{R(x_0)} \int_\Omega \sum_\alpha m_\alpha z \frac{\partial \zeta}{\partial x_\alpha}\, dx. \tag{8.2}$$

Consider the eigenvectors w_j defined by

$$-\Delta w_j = \lambda_j w_j, \quad w_j|_\Gamma = 0 \tag{8.3}$$

then we have the relation

$$\begin{aligned} &\int_\Gamma m \cdot \nu \frac{\partial w_j}{\partial \nu} \frac{\partial w_k}{\partial \nu}\, d\Gamma = \\ &(2-n)\sqrt{\lambda_j \lambda_k}\, \delta_{jk} - \int_\Omega \sum_\alpha m_\alpha \Big(\lambda_j w_j \frac{\partial w_k}{\partial x_\alpha} + \lambda_k w_k \frac{\partial w_j}{\partial x_\alpha}\Big)\, dx \end{aligned} \tag{8.4}$$

for any j, k. Then taking $N = 0$, all assumptions of Theorem 7.1 are satisfied. The system to be controlled is, according to (7.12)

$$< \psi_1, \eta(T) > - < \psi_0, \eta'(T) >= \int_0^T (v_1, \frac{\partial \psi}{\partial \nu})\, dt \tag{8.5}$$

where ψ satisfies

$$\begin{aligned} &\psi'' - \Delta\psi = 0 \\ &\psi_{|\Gamma} = 0 \\ &\psi(T) = \psi_0, \quad \psi'(T) = \psi_1, \quad \psi_0 \in H^2 \cap H_0^1, \quad \psi_1 \in H_0^1 \end{aligned} \tag{8.6}$$

The relation (8.6) is the "transposition method" definition of

$$\begin{aligned} &\eta'' - \Delta\eta = 0 \\ &\eta_{|\Gamma} = -v_1 \\ &\eta(0) = \eta'(0) = 0. \end{aligned} \tag{8.7}$$

The exact controllability property is expressed as follows: given $y_0 \in L^2$, $y_1 \in H^{-1}$, for T sufficiently large the system

$$\begin{aligned} &\eta'' - \Delta\eta = 0 \qquad \phi'' - \Delta\phi = 0 \\ &\eta_{|\Gamma} = \frac{\partial \phi}{\partial \nu} \qquad \phi_\Gamma = 0 \\ &\eta(0) = \eta'(0) = 0, \quad \eta(T) = y_0, \quad \eta'(T) = y_1 \end{aligned} \tag{8.8}$$

has a solution with $\phi(0) \in H_0^1$, and $\phi'(0) \in L^2$.

8.2 Wave equation with Neumann boundary control

We take now

$$\begin{aligned} &L = -\Delta + I, \quad H_L = L^2(\Omega), \quad V_L = H^1(\Omega), \\ &D(L) = \{z \in H^2(\Omega) : \frac{\partial z}{\partial \nu} = 0\}, \\ &\Delta_L = \{z \in H^3(\Omega) : \frac{\partial z}{\partial \nu} = 0\}. \end{aligned}$$

The eigenvectors are defined by

$$\begin{aligned} &-\Delta w_j + w_j = \lambda_j w_j \\ &\frac{\partial w_j}{\partial \nu} = 0. \end{aligned} \tag{8.9}$$

Note that $\lambda_1 = 1$, and $w_1 = \frac{1}{|\Omega|}$.

The eigenspace corresponding to λ_1 is one dimensional. We next take $U_1 = (L^2(\Gamma_1))^{n+1}$, $U_2 = L^2(\Gamma_0)$ (see (5.21) for the definition of Γ_0 and Γ_1), and

$$q_1 z = \begin{bmatrix} \gamma_1 z \\ \gamma_1 D z \end{bmatrix} \text{ for } z \in D(L), \quad q_2 z = \gamma_0 z, \text{ for } z \in H^1. \tag{8.10}$$

Note that for

$$\forall z \in D(L), \quad \gamma_1 D z = D_{\sigma,1} \gamma_1 z.$$

See §5.3 for the notation. Define then

$$(\pi_1 \zeta, \zeta') = -\frac{1}{R(x_0)} \int_\Gamma m \cdot \nu \, (\zeta\zeta' + D\zeta \cdot D\zeta') \, d\Gamma, \quad \forall \zeta, \zeta' \in D(L) \tag{8.11}$$

$$(\pi_2 \zeta, \zeta') = \frac{1}{R(x_0)} \int_\Gamma m \cdot \nu \, \zeta\zeta' \, d\Gamma, \quad \forall \zeta, \zeta' \in H^1. \tag{8.12}$$

Similarly define

$$b(z, \zeta) = -\frac{1}{R(x_0)} \int_\Omega \sum_\alpha m_\alpha z \frac{\partial \zeta}{\partial x_\alpha} \, dx. \tag{8.13}$$

We use the following relation between eigenvalues

$$\begin{aligned} &\int_\Gamma m \cdot \nu \, (w_j w_k + D w_j D w_k) \, d\Gamma \\ &= (n-2) \sqrt{\lambda_j \lambda_k} \, \delta_{jk} + 2\delta_{jk} \\ &\quad + \int_\Omega \sum_\alpha m_\alpha (\lambda_j w_j \frac{\partial w_k}{\partial x_\alpha} + \lambda_k w_k \frac{\partial w_j}{\partial x_\alpha}) \, dx. \end{aligned} \tag{8.14}$$

Pick $N = 1$. Consider $j \neq k \geq 2$. It is easily checked that the properties (7.22) to (7.26) are satisfied. Let us check (6.21). It follows from the fact that

$$|B^* \phi_1|^2 = \frac{|\Gamma|}{2|\Omega|}.$$

Therefore all assumptions of Theorem 7.1 are satisfied.

The system to be controlled is, according to (7.12),

$$< \psi_1, \eta(T) > - < \psi_0, \eta'(T) + q_2^* v_2(T) >$$

$$= \int_0^T \int_{\Gamma_1} (v_{10}\psi + \sum_{i=1}^n v_{1i} \frac{\partial \psi}{\partial x_i})\, d\Gamma + \int_0^T \int_{\Gamma_0} v_2 \psi' \, d\Gamma \tag{8.15}$$

where

$$v_1 = \begin{bmatrix} v_{10} \\ v_{12} \\ \dots \\ \dots \\ v_{1n} \end{bmatrix}$$

and ψ is the solution of

$$\psi'' - \Delta\psi + \psi = f$$

$$\frac{\partial \psi}{\partial \nu} = 0 \tag{8.16}$$

$$\psi(T) = \psi'(T) = 0.$$

We can consider v_1 as an element of $(H^1(\Gamma_1))'$, by the following duality formula (see also (5.29))

$$< v_1, \theta >= \int_{\Gamma_1} (v_{10}\gamma_1\theta + \sum_{i=1}^n v_{1i}\gamma_1 \frac{\partial \theta}{\partial x_i})\, d\Gamma. \tag{8.17}$$

Consider the special case

$$v_{10} = \gamma_1 \chi, \quad v_{1i} = \gamma_1 \frac{\partial \chi}{\partial x_i} \text{ with } \frac{\partial \chi}{\partial \nu} = 0. \tag{8.18}$$

Then

$$\begin{aligned} < v_1, \theta > &= \int_{\Gamma_1} \left\{ \gamma_1\chi\gamma_1\theta + \sum_{i=1}^n \gamma_1 \frac{\partial \chi}{\partial x_i} \gamma_1 \frac{\partial \theta}{\partial x_i} \right\} d\Gamma \\ &= \int_{\Gamma_1} \left\{ \gamma_1\chi\gamma_1\theta + \sum_{i=1}^n \sigma_{i,1}\gamma_1\chi\sigma_{i,1}\gamma_1\theta \right\} d\Gamma \qquad (8.19) \\ &= \int_{\Gamma_1} \{\gamma_1\chi - \Delta_{\sigma 1}\gamma_1\chi\}\, \gamma_1\theta \, d\Gamma \end{aligned}$$

where

$$-\Delta_{\sigma 1} = \sum_i \sigma_{i,1}^* \sigma_{i,1},$$

represents the tangential Laplacian on Γ_1, which is an element of $\mathcal{L}(H^1(\Gamma_1); (H^1(\Gamma_1))')$.

So we can write (8.15) as

$$\begin{aligned}
&\eta'' - \Delta\eta + \eta = 0 \\
&\frac{\partial\eta}{\partial\nu}|_{\Gamma 0} = -J_0^* v_2 \\
&\frac{\partial\eta}{\partial\nu}|_{\Gamma 1} = v_1 \\
&\eta(0) = \eta'(0) = 0
\end{aligned} \tag{8.20}$$

with the interpretation of the operator J_0 already discussed in §7.3.

Let us express the property of exact controllability. Let $y_0 \in L^2$, $y_1 \in (H^1(\Omega))'$, then for T sufficiently large the system

$$\begin{aligned}
&\eta'' - \Delta\eta + \eta = 0 && \phi'' - \Delta\phi + \phi = 0 \\
&\frac{\partial\eta}{\partial\nu}|\Gamma_0 = -J_0^*\phi' && \frac{\partial\phi}{\partial v} = 0 \\
&\frac{\partial\eta}{\partial\nu}|_{\Gamma_1} = \Delta_{\sigma 1}\phi - \phi && \eta(T) = y_0, \quad \eta'(T) = y_1 \\
&\eta(0) = 0, \quad \eta'(0) = 0
\end{aligned} \tag{8.21}$$

has a solution with $\phi(0) \in H^1, \phi'(0) \in L^2$. We recall that $J_0^*\phi'$ belongs to $(H^1(0,T; L^2(\Gamma_0))'$ and $-\Delta_{\sigma 1}\phi + \phi$ to $L^2(0,T; (H^1(\Gamma_1))')$.

8.3 Maxwell equations

Let us introduce some notations. Let $\mathbf{L}^2(\Omega) = (L^2(\Omega))^3$, $\mathbf{L}^2(\Gamma) = (L^2(\Gamma))^3$ and similarly, $\mathbf{H}^k(\Omega) = (H^k(\Omega))^3$. If $v \in \mathbf{H}^1(\Omega)$, we set curl $v = D \times v$, where $\times$ symbolizes the external product. We recall the relations

$$\text{curl grad } v = 0, \qquad \text{div curl } v = 0.$$

Furthermore, the set Ω being assumed smooth and singly connected, we have

$$\begin{aligned} \text{curl } u = 0 \quad &\Rightarrow \quad u = Dp \\ \text{div } u = 0 \quad &\Rightarrow \quad u = \text{curl } v. \end{aligned}$$

Define

$$\begin{aligned} J &= \text{closure in } \mathbf{L^2}(\Omega) \text{ of functions belonging to} \\ &\quad (C^\infty(\overline{\Omega}))^3 \text{with divergence 0.} \\ K &= \text{closure in } \mathbf{L}^2(\Omega) \text{ of functions belonging to} \\ &\quad (C_0^\infty(\Omega))^3 \text{ with divergence 0.} \\ J_1 &= \{v \in \mathbf{H}^1(\Omega) \,:\, \text{div } v = 0, \quad v \times \nu = 0\} \\ K_1 &= \{v \in \mathbf{H}^1(\Omega) \,:\, \text{div } v = 0, \quad v \cdot \nu = 0\} \end{aligned}$$

where ν represents the external unit normal. We provide J_1 and K_1 with the scalar product

$$((u, v)) = (\text{curl } u, \ \text{curl } v) \tag{8.22}$$

for which they become Hilbert spaces. Indeed if $v \in J_1$, curl $v = 0$, then

$$v = \text{curl } \chi, \quad \text{curl curl } \chi = 0.$$

Using the integration by parts formula

$$(\text{curl } \phi, \psi) = (\phi, \ \text{curl } \psi) + \int_\Gamma \phi \times \nu \cdot \psi \, d\Gamma \tag{8.23}$$

we deduce easily that curl $\chi = 0$, hence $v = 0$. Similarly if $v \in K_1$, curl $v = 0$, then $v = Dp$ and $\Delta p = 0$, with $\frac{\partial p}{\partial \nu} = 0$, hence p is a constant, which implies again $v = 0$. It can be proved (see (8)) that

$$J_1 \subset J, \quad K_1 \subset K \tag{8.24}$$

with continuous and dense embedding. Moreover curl is an isometry from J_1 to K, and K_1 to J. The norm thus defined on J_1 and K_1 is equivalent to that induced by , $\mathbf{H}^1(\Omega)$. We shall further need the spaces

$$\begin{aligned} J_2 &= \{v \in \mathbf{H}^2(\Omega) \,:\, \text{div } v = 0, v \times \nu = 0, \nu \cdot \text{curl } v = 0\} \\ J_3 &= \{v \in \mathbf{H}^3(\Omega) \,:\, \text{div } v = 0, v \times \nu = 0, \nu \cdot \text{curl } v = 0, \\ &\qquad \nu \times \text{ curl curl } v = 0\} \end{aligned} \tag{8.25}$$

and

$$\begin{aligned} K_2 &= \{v \in \mathbf{H}^2(\Omega) \, : \, \text{div } v = 0, \, v \cdot \nu = 0, \, \nu \times \text{curl } v = 0\} \\ K_3 &= \{v \in \mathbf{H}^3(\Omega) \, : \, \text{div } v = 0, \, \nu \cdot v = 0, \\ &\qquad \nu \times \text{curl } v = 0, \, \nu \cdot \text{curl curl } v = 0\}. \end{aligned} \tag{8.26}$$

The spaces J_2 and K_2 are provided with the norms

$$||v|| = |\text{curl curl } v|.$$

and J_3, K_3 with the norms

$$||v|| = |\text{curl curl curl } v|.$$

Then curl is an isometry from J_2 to K_1, J_3 to K_2, K_2 to J_1, K_3 to J_2.

Note that from (8.23) we have

$$(\text{curl } \phi, \psi) = (\text{curl } \psi, \phi) \tag{8.27}$$

with $\phi \in J_1, \psi \in K_1$, or $\phi \in K_1, \psi \in J_1$. It follows that curl can be extended as an isometry from J to $(K_1)'$ and K to $(J_1)'$. Similarly it extends as an isometry from $(K_1)'$ to $(J_2)', (J_1)'$ to $(K_2)', (J_2)'$ to $(K_3)', (K_2)'$ to $(J_3)'$.

Summarizing we have the sequences

$$\begin{array}{l} J_3 \subset J_2 \subset J_1 \subset J \subset (J_1)' \subset (J_2)' \subset (J_3)' \\ K_3 \subset K_2 \subset K_1 \subset K \subset (K_1)' \subset (K_2)' \subset (K_3)' \end{array} \tag{8.28}$$

each space being continuously and densely embedded in the next one. Moreover curl curl is an isometry from J_3 to J_1, J_2 to J, J_1 to $(J_1)'$, J to $(J_2)'$, $(J_1)'$ to $(J_3)'$ and similar result with J changed into K.

Define

$$\begin{array}{l} H_L = K, \quad V_L = K_1, \quad D(L) = K_2, \quad \Delta(L) = K_3, \\ L = \text{curl curl} \end{array} \tag{8.29}$$

The eigenvectors w_j are given by the relations

$$\begin{aligned}&\text{curl curl } w_j = \lambda_j w_j, && \text{div } w_j = 0\\ &\nu \cdot w_j = 0, && \nu \times \text{curl } w_j = 0 \text{ on } \Gamma\\ &|w_j| = 1.\end{aligned} \tag{8.30}$$

We shall use the following relations between eigenvectors

$$\begin{aligned}&\int_\Gamma m \cdot \nu \,\text{curl } w_j \cdot \text{curl } w_k \, d\Gamma\\ &= (n-2)\sqrt{\lambda_j\lambda_k}\,\delta_{jk}\\ &\quad + \int_\Omega \sum_\alpha m_\alpha \Big(\lambda_j w_j \cdot \frac{\partial w_k}{\partial x_\alpha} + \lambda_k w_k \cdot \frac{\partial w_j}{\partial x_\alpha}\Big)\, dx\end{aligned} \tag{8.31}$$

with again $m(x) = x - x_0$.

Define next

$$\begin{aligned}&U_2 = \mathbf{L}^2(\Gamma_0), && U_1 = \{v \in \mathbf{L}^2(\Gamma_1) \,:\, \nu \times v = 0\}\\ &q_1 = -\gamma_1 \text{ curl}, && q_2 = -\gamma_0.\end{aligned}$$

We can take

$$(\pi_1\zeta, \zeta') = -\frac{1}{R(x_0)}\int_\Gamma m \cdot \nu \text{ curl}\zeta \cdot \text{curl}\zeta' \, d\Gamma, \quad \forall \zeta, \zeta' \in K_2 \tag{8.32}$$

$$(\pi_2\zeta, \zeta') = \frac{1}{R(x_0)}\int_\Gamma m \cdot \nu \, \zeta \cdot \zeta' \, d\Gamma, \quad \forall \zeta, \zeta' \in K_1 \tag{8.33}$$

$$b(z, \zeta) = -\frac{1}{R(x_0)}\int_\Omega \sum_\alpha m_\alpha \, z \cdot \frac{\partial \zeta}{\partial x_\alpha}\, dx, \quad \forall z \in K, \zeta \in K_1. \tag{8.34}$$

Then all assumptions of Theorem 7.1 are satisfied, with $N = 0$.

We next interpret the system to be controlled. Consider (7.11) and (7.12), we get for $\psi_0 \in K_2$, $\psi_1 \in K_j$

$$\begin{aligned}&\psi'' + \text{ curl curl } \psi = 0\\ &\text{div } \psi = 0\\ &\nu \cdot \psi = 0, \quad \nu \times \text{curl } \psi = 0 \text{ on } \Gamma\\ &\psi(T) = \psi_0, \quad \psi'(T) = \psi_1\end{aligned} \tag{8.35}$$

$$< \psi_1, \eta(T) > - < \psi_0, \eta'(T) + q_2^* v_2(T) >$$
$$= - \int_0^T [\int_{\Gamma_1} v_1 \cdot \gamma_1 \operatorname{curl} \psi \, d\Gamma + \int_{\Gamma_0} v_2 \cdot \gamma_0 \, \psi' \, d\Gamma] \, dt. \quad (8.36)$$

Now, we use the formula (see (5.22))

$$\gamma \operatorname{curl} \psi = \nu \times \frac{\partial \psi}{\partial \nu} + D_\sigma \times \gamma \psi \text{ on } \Gamma \quad (8.37)$$

and similar relations on Γ_0 and Γ_1. Using the fact that v_1 belongs to U_1, we have

$$\int_{\Gamma_1} v_1 \cdot \gamma_1 \operatorname{curl} \psi \, d\Gamma = - < D_{\sigma 1}^* \times v_1, \gamma_1 \psi >$$
$$\int_{\Gamma_0} v_2 \cdot \gamma_0 \, \psi' \, d\Gamma =< J_0^* v_2, \gamma_0 \psi >$$

We then can write (8.36) as follows (using the transposition method)

$$\begin{aligned} &\eta'' + \operatorname{curl} \operatorname{curl} \eta = 0 \\ &\operatorname{div} \eta = 0 \\ &\eta(0) = \eta'(0) = 0 \\ &\nu \times \operatorname{curl} \eta|_{\Gamma_1} = D_{\sigma 1}^* \times v_1 \\ &\nu \times \operatorname{curl} \eta|_{\Gamma_0} = -J_0^* v_2 \end{aligned} \quad (8.38)$$

We see that $\nu \times \eta|_{\Gamma_1}$ belongs to $L^2(0, T; H^1(\Gamma_1)')$ and $\nu \times \eta|_{\Gamma_0}$ belongs to $(H^1(0, T; L^2(\Gamma_0)))'$. Let us then express the condition of exact controllability. Given y_0 in K and y_1 in $(K_1)'$ we consider the system

$$\begin{aligned} &\eta'' + \operatorname{curl} \operatorname{curl} \eta = 0 && \phi'' + \operatorname{curl} \operatorname{curl} \phi = 0 \\ &\operatorname{div} \eta = 0 && \operatorname{div} \phi = 0 \\ &\eta(0) = \eta'(0) = 0 && \eta(T) = y_0, \quad \eta'(T) = y_1 \\ &\nu \times \operatorname{curl} \eta|_{\Gamma_1} = D_{\sigma 1}^* \times \gamma_1 \operatorname{curl} \phi && \phi \cdot \nu = 0, \\ &\nu \times \operatorname{curl} \eta|_{\Gamma_0} = J_0^* \phi' && \nu \times \operatorname{curl} \phi = 0, \end{aligned} \quad (8.39)$$

which has a solution with $\phi(0) \in K_1$ and $\phi'(0) \in K$, at least for T sufficiently large. Furthermore we know that $\eta \in C([0, T]; K_1')$ and $\eta' \in L^2([0, T]; K_2')$. Reducing (8.39) to Maxwell equations by setting

$$H = \operatorname{curl} \eta, \quad E = -\eta'$$
$$\psi = - \operatorname{curl}^{-1} \phi'$$

we deduce the relations

$$
\begin{array}{ll}
E' = \text{curl } H, & \phi' + \text{ curl } \psi = 0, \\
H' + \text{ curl } E = 0 & \psi' = \text{ curl } \phi \\
\text{div } E = \text{div } H = 0 & \text{div } \phi = \text{div } \psi = 0 \\
\nu \times H|_{\Gamma_1} = D^*_{\sigma 1} \times \gamma_1 \psi' & E(T) = E_T, \quad H(T) = H_T \\
\nu \times H|_{\Gamma_0} = J_0^* \phi' & \phi \cdot \nu = 0, \quad \nu \times \psi = 0 \\
E(0) = H(0) = 0 &
\end{array}
\tag{8.40}
$$

with

$$
\begin{array}{ll}
H \in C([0,T]; J_2'), & E \in L^2([0,T]; K_2'), \\
H' \in L^2([0,T]; J_3'), & E' \in L^2([0,T]; K_3'), \\
\phi \in C([0,T]; K_1), & \psi \in C([0,T]; J_1), \\
\phi' \in C([0,T]; K), & \psi' \in C([0,T]; J).
\end{array}
$$

The values E_T and H_T are any elements of K_1' and J_1' respectively. The system (8.40) has a solution for T sufficiently large.

We now turn to another situation where we shall be able to use Theorem 6.2 instead of Theorem 6.1 (which permitted to state Theorem 7.1). However we assume the geometric condition

$$
m \cdot \nu \geq 0. \tag{8.41}
$$

Consider again (8.29). Take $U_1 = U_2 = \mathbf{L}^2(\Gamma)$ and $q_1 = -\gamma$, $q_2 = 0$. We define A, B as in §7.1 and 7.2. We note that $B \in D(A)$ and thus (6.32) is satisfied. We then take

$$
(M\xi, \eta) = \frac{1}{R(x_0)} \int_\Gamma m \cdot \nu \, L^{-1}\xi_1 \cdot L^{-1}\eta_1 \, d\Gamma \tag{8.42}
$$

$$
(\Lambda\xi, \eta) = \frac{1}{R(x_0)} \int_\Omega \sum_\alpha m_\alpha \, \xi_1 \cdot \frac{\partial \eta_1}{\partial x_\alpha} \, dx \tag{8.43}
$$

$$
(Q\xi, \eta) = \frac{1}{R(x_0)} \int_\Gamma m \cdot \nu \text{ curl } L^{-1}\xi_1 \cdot \text{curl } L^{-1}\eta_1 \, d\Gamma. \tag{8.44}
$$

Let us check the assumptions of Theorem 6.2, with $N = 0$. Thanks to (8.41) the assumption (6.15) is satisfied and (6.16) is trivial. The property (6.34) is clear from the definition of ϕ_j and that of

M, Λ. From the definition of Q and the relation (8.31) the property (6.35) is easily checked. It remains to check (6.36) and (6.37). But

$$\begin{aligned}(M\,\mathrm{Re}\,\phi_j, \mathrm{Re}\,\phi_j) &= \frac{1}{2R(x_0)\lambda_j^2}\int_\Gamma m\cdot\nu\,|w_j|^2\,d\Gamma \\ (Q\,\mathrm{Re}\,\phi_j, \mathrm{Re}\,\phi_j) &= \frac{1}{2R(x_0)\lambda_j^2}\int_\Gamma m\cdot\nu\,|\mathrm{curl}\;w_j|^2\,d\Gamma\end{aligned}$$

and from (8.31)

$$(M\,\mathrm{Re}\,\phi_j, \mathrm{Re}\,\phi_j) - \frac{1}{\lambda_j}(Q\,\mathrm{Re}\,\phi_j, \mathrm{Re}\,\phi_j) = \frac{1}{R(x_0)\lambda_j^2}$$

therefore (6.36) is satisfied with

$$c_0 = \frac{1}{R(x_0)}.$$

Now

$$\begin{aligned}&(M\,\mathrm{Re}\,\phi_j, \mathrm{Re}\,\phi_j) + \frac{1}{\lambda_j}(Q\,\mathrm{Re}\,\phi_j, \mathrm{Re}\,\phi_j) \\ &= \frac{1}{R(x_0)\lambda_j^2}\left\{\int_\Gamma m\cdot\nu\,|w_j|^2\,d\Gamma - 1\right\}\end{aligned}$$

and

$$\begin{aligned}\int_\Gamma m\cdot\nu\,|w_j|^2\,d\Gamma &= 2\int_\Omega \sum_\alpha m_\alpha\, w_j\cdot\frac{\partial w_j}{\partial x_\alpha}\,dx + n \\ &= 4R(x_0)\,(\Lambda\,\mathrm{Re}\,\phi_j, \mathrm{Re}\,\phi_j) + n\end{aligned}$$

and in particular the property (6.37) follows. The dynamic system is given by

$$\begin{aligned}&\eta'' + \mathrm{curl\;curl}\;\eta = 0 \\ &\mathrm{div}\;\eta = 0 \\ &\eta(0) = \eta'(0) = 0 \\ &\nu\times\mathrm{curl}\;\eta|_\Gamma = -v\end{aligned} \tag{8.45}$$

We express the property of exact controllability as follows.

The system

$$\begin{array}{ll} \eta'' + \text{curl curl } \eta = 0 & \phi'' + \text{curl curl } \phi = 0 \\ \text{div } \eta = 0 & \text{div } \phi = 0 \\ \eta(0) = \eta'(0) = 0 & \eta(T) = y_0, \quad \eta'(T) = y_1 \\ \nu \times \text{ curl } \eta|_\Gamma = \gamma\phi & \phi \cdot \nu = 0, \quad \nu \times \text{ curl } \phi = 0 \end{array} \tag{8.46}$$

has a solution for T sufficiently large, with $y_0 \in K_1$, $y_1 \in K$. Moreover $\phi(0) \in K, \phi'(0) \in (K_1)'$. We have $\eta \in C([0,T];K)$, $\eta' \in C([0,T];(K_1)')$. Reducing to Maxwell equations as above we obtain the system

$$\begin{array}{ll} E' = \text{curl } H, & \phi' + \text{curl } \psi = 0, \\ H' + \text{curl } E = 0 & \psi' = \text{curl } \phi \\ \text{div } E = \text{div } H = 0 & \text{div } \phi = \text{ div } \psi = 0 \\ \nu \times H|_\Gamma = \gamma\phi & E(T) = E_T, \quad H(T) = H_T \\ E(0) = H(0) = 0 & \phi \cdot \nu = 0, \quad \nu \times \psi = 0 \end{array} \tag{8.47}$$

For any given E_T and H_T in K and J respectively, and for sufficiently large T, the system (8.47) has a solution such that

$$\begin{array}{l} H \in C([0,T];J_1'), \quad E \in C([0,T];K_1'), \\ H' \in C([0,T];J_2'), \quad E' \in C([0,T];K_2'), \\ \phi \in C([0,T];K), \quad \psi \in C([0,T];J), \\ \phi' \in C([0,T];(K_1)'), \quad \psi' \in C([0,T];(J_1)') \end{array}$$

and $\gamma\phi$ belongs to $L^2(0,T;\mathbf{L}^2(\Gamma))$.

8.4 Plate equation

We begin with Neumann control. We take $H_L = L^2(\Omega)$, $V_L = H_0^2(\Omega)$, $D(L) = H^4(\Omega) \cap H_0^2(\Omega)$, and

$$L = \Delta^2, \quad \Delta(L) = \{z \in H_0^2(\Omega) \,:\, \Delta^2 z \in H_0^2(\Omega)\}.$$

The eigenvalues related to L are defined by

$$\begin{array}{l} \Delta^2 w_j = \lambda_j w_j \\ w_{j|\Gamma} = \left.\dfrac{\partial w_j}{\partial \nu}\right|_\Gamma = 0, \quad |w_j| = 1. \end{array} \tag{8.48}$$

Considering again the multiplier $m(x) = x - x_0$, we can check the relation

$$\int_\Gamma m \cdot \nu \, \Delta w_j \Delta w_k \, d\Gamma = \tag{8.49}$$
$$(4-n)\sqrt{\lambda_j \lambda_k}\delta_{jk} - \int_\Omega \sum_\alpha m_\alpha (\lambda_j w_j \frac{\partial w_k}{\partial x_\alpha} + \lambda_k w_k \frac{\partial w_j}{\partial x_\alpha}) \, dx.$$

We take $U_1 = U_2 = L^2(\Gamma_0)$, where we recall that Γ_0 is the part of the boundary on which $m \cdot \nu \geq 0$. We pick $q_1 = -\gamma_0 \Delta$ and $q_2 = 0$.

It is useful to introduce the space $H^3(\Omega) \cap H_0^2(\Omega)$, intermediary between V_L and $D(L)$, equipped with the norm $||z|| = |D(\Delta z)|$. Note that L is an isomorphism between $H^3 \cap H_0^2$ and H^{-1}. We shall check the property of approximate controllability for any $T > 0$, by checking the assumptions of Theorem 7.2. We take $N = 0$, and we define

$$(\pi_1 \zeta, \zeta') = \frac{1}{R(x_0)} \int_\Gamma m \cdot \nu \Delta\zeta \Delta\zeta' d\Gamma, \quad \pi_2 = 0, \tag{8.50}$$

$$b(z, \zeta) = -\frac{1}{R(x_0)} \int_\Omega \sum_\alpha m_\alpha z \frac{\partial \zeta}{\partial x_\alpha} \, dx. \tag{8.51}$$

The properties (7.22), (7.23), (7.24) are easily verified. Since

$$\begin{aligned} (\pi_1 w_j, w_j) &= \frac{4\lambda_j}{R(x_0)} \\ b(w_j, w_j) &= \frac{n}{2R(x_0)} \end{aligned} \tag{8.52}$$

the properties (7.25) and (7.29) are clear. It remains to check (7.30). We define $W_L = H_0^1$ and the assumption (7.30) is satisifed. Therefore, all the assumptions of Theorem 7.2 are now satisfied. Thus the pair (A, B) is approximately controllable for arbitrary positive T. Let us interpret this result. The dynamical system is described by

$$\begin{aligned} &< \psi_1, \eta(T) > - < \psi_0, \eta'(T) + q_2^* v_2(T) > \\ &= -\int_0^T \int_{\Gamma_0} v \, \Delta\psi \, dx \, dt \end{aligned} \tag{8.53}$$

for $\psi_0 \in H^4 \cap H_0^2$ and $\psi_1 \in H_0^2$ and ψ being the solution of

$$\begin{aligned}
&\psi'' + \Delta^2\psi = f \\
&\psi(T) = \psi'(T) = 0 \\
&\psi|_\Gamma = \frac{\partial\psi}{\partial\nu}|_\Gamma = 0 \\
&\psi \in C([0,T]; H^4 \cap H_0^2), \quad \psi' \in C([0,T]; H_0^2)
\end{aligned} \tag{8.54}$$

We interpret (8.53) as follows

$$\begin{aligned}
&\eta'' + \Delta^2\eta = 0 \\
&\eta(0) = \eta'(0) = 0 \\
&\eta|_\Gamma = 0, \quad \frac{\partial\eta}{\partial\nu}|_{\Gamma_1} = 0 \\
&\frac{\partial\eta}{\partial\nu}|_{\Gamma_0} = v \\
&\eta \in C([0,T]; V_L'), \quad \eta' \in C([0,T]; (D(L))').
\end{aligned} \tag{8.55}$$

Let us express the controllability property. Given $y_0 \in L^2(\Omega)$, $y_1 \in H^{-2}(\Omega)$, the system

$$\begin{aligned}
&\eta'' + \Delta^2\eta = 0 & &\phi'' + \Delta^2\phi = 0 \\
&\eta(0) = \eta'(0) = 0 & &\phi|_\Gamma = \frac{\partial\phi}{\partial\nu}|_\Gamma = 0 \\
&\eta|_\Gamma = 0, \quad \frac{\partial\eta}{\partial\nu}|_{\Gamma_1} = 0 & &\eta(T) = y_0, \quad \eta'(T) = y_1 \\
&\frac{\partial\eta}{\partial\nu} = \Delta\phi|_{\Gamma_0} & & \\
&\phi \in C([0,T]; H_0^2), \quad \phi' \in C([0,T]; L^2), & & \\
&\Delta\phi|_{\Gamma_0} \in L^2(0,T; L^2(\Gamma_0)) & &
\end{aligned} \tag{8.56}$$

has a solution for any $T > 0$.

We next consider a more elaborate case, where we control y and Δy on the boundary. We need some notation. Let J_n be such that

$$J_n = \{z \in H^n \,:\, z|_\Gamma = 0, \quad \Delta z|_\Gamma = 0, \ldots, \Delta^{\frac{n-1}{2}} z\Big|_\Gamma = 0\}$$

Hence $J_1 = H_0^1$, $J_2 = H^2 \cap H_0^1 \ldots$ We provide J_n with the norm

$$||z||_{J_n}^2 = \begin{cases} |\Delta^{\frac{n}{2}} z|^2, & \text{if } n \text{ is even,} \\ |D\Delta^{\frac{n-1}{2}} z|^2, & \text{if } n \text{ is odd.} \end{cases}$$

We set $J_0 = L^2(\Omega)$. The operator $-\Delta$ is an isometry from J_n to J_{n-2}, $n \geq 2$.

Denote J_n the dual of J_n, when J_0 is the pivot space. Clearly $-\Delta$ extends as an isometry between J_n and J_{-n-2}, and we have the sequence

$$J_n \subset J_{n-1} \subset J_{n-2} \subset \ldots \subset J_1 \subset J_0 \subset J_{-1} \ldots \subset J_{-n} \ldots$$

In the above sequence, suppose that we pick J_1 to be the pivot space, identified with its dual. Then the dual of J_2 is J_0, the dual of J_3 is J_{-1}, the dual of J_n is J_{-n+2}.

The duality pairing is in the 1st case (J_0 pivot)

$$\begin{cases} ((-\Delta)^{\frac{n}{2}} u, (-\Delta)^{-\frac{n}{2}} v), & \text{if } n \text{ is even} \\ (D(-\Delta)^{\frac{n-1}{2}} u, D(-\Delta)^{-\frac{n+1}{2}} v), & \text{if } n \text{ is odd} \end{cases} \tag{8.57}$$

with $u \in J_n$, $v \in J_{-n}$.

Now in the case when J_1 is chosen as the pivot space, then the duality pairing is given by

$$\begin{cases} ((-\Delta)^{\frac{n}{2}} u, (-\Delta)^{-\frac{n}{2}-1} v), & \text{if } n \text{ is even} \\ (D(-\Delta)^{\frac{n-1}{2}} u, D(-\Delta)^{-\frac{n-1}{2}} v), & \text{if } n \text{ is odd} \end{cases} \tag{8.58}$$

with $u \in J_n, v \in J_{-n+2}$.

We choose in the sequel J_1 to be the pivot space. Let $H_L = J_1$, $V_L - J_3$, $D(L) = J_5$, $\Delta(L) = J_7$. Then $V_L' = J_{-1}$, $(D(L))' = J_{-3}$.

We pick $L = \Delta^2$. The eigenvectors are defined by

$$\begin{aligned} &\Delta^2 w_j = \lambda_j w_j \\ &w_j|_\Gamma = \Delta w_j|_\Gamma = 0, \qquad ||w_j|| = 1 \end{aligned} \tag{8.59}$$

and the w_j form an orthornormal system in J_1 the pivot space. We shall use the relations

$$\int_\Gamma m \cdot \nu \, \frac{\partial}{\partial \nu} \Delta w_j \frac{\partial}{\partial \nu} \Delta w_k \, d\Gamma =$$

$$(2+n)\sqrt{\lambda_j \lambda_k}\, \delta_{jk} - \int_\Omega \sum_\alpha m_\alpha (\lambda_j \Delta w_k \frac{\partial w_j}{\partial x_\alpha} + \lambda_k \Delta w_j \frac{\partial w_k}{\partial x_\alpha}) \, dx \tag{8.60}$$

$$\int_\Gamma m\cdot\nu\,(\frac{\partial}{\partial\nu}\Delta w_j)^2 d\Gamma + \lambda_j\int_\Gamma m\cdot\nu\,(\frac{\partial}{\partial\nu}w_j)^2 d\Gamma = 4\lambda_j. \qquad (8.61)$$

We take $U_1 = U_2 = L^2(\Gamma_0)$. We recall that

$$n_0 = \frac{\partial}{\partial\nu}$$

on Γ_0. We define $q_1 = n_0\Delta$, $q_2 = n_0$. We then take

$$(\pi_1\zeta,\zeta') = \frac{1}{R(x_0)}\int_\Gamma m\cdot\nu\,\frac{\partial}{\partial\nu}\Delta\zeta\frac{\partial}{\partial\nu}\Delta\zeta' d\Gamma \qquad (8.62)$$

$$(\pi_2\zeta,\zeta') = \frac{1}{R(x_0)}\int_\gamma m\cdot\nu\,\frac{\partial}{\partial\nu}\zeta\frac{\partial}{\partial\nu}\zeta' d\Gamma \qquad (8.63)$$

$$b(z,\zeta) = -\frac{1}{R(x_0)}\int_\Omega\sum_\alpha m_\alpha\Delta\zeta\frac{\partial z}{\partial x_\alpha}\,dx. \qquad (8.64)$$

Let us check that all the assumptions of Theorem 7.2 are satisifed, with $N = 0$. The properties (7.22), (7.23) are clearly satisfied. The relation (7.24) can also be checked, taking into account the fact that the w_j are orthogonal in H_0^1. Let us check (7.25). We have

$$\frac{1}{\lambda_j}(\pi_1 w_j, w_j) + (\pi_2 w_j, w_j)$$

$$= \frac{1}{R(x_0)}\left[\frac{1}{\lambda_j}\int_\Gamma m\cdot\nu\left|\frac{\partial}{\partial\nu}\Delta w_j\right|^2 d\Gamma + \int_\Gamma m\cdot\nu\left|\frac{\partial}{\partial\nu}w_j\right|^2 d\Gamma\right]$$

$$= \frac{4}{R(x_0)}$$

from (8.61). Hence (7.25) is verified. Let us check (7.29). We have

$$b(w_j,w_j) = -\frac{(\pi_2 w_j, w_j)}{2} + \frac{2-n}{2R(x_0)}$$

hence

$$\frac{1}{\lambda_j}(\pi_1 w_j, w_j) - (\pi_2 w_j, w_j) - 4b(w_j,w_j) = \frac{2n}{R(x_0)}$$

and (7.29) is satisfied. Let us finally check (7.30). We take $W_L = J_2$ and note that

$$|b(z,\zeta)| \leq \frac{\mu(x_0)}{R(x_0)} ||z||_{J_1} ||\zeta||_{J_2}.$$

Since the injection of J_3 into J_2 is compact the assumption (7.30) is also satisfied. Therefore all the assumptions of Theorem 7.2 are now satisfied. Thus the pair (A, B) is exactly controllable for arbitrary positive T.

The dynamic system is given by the relation

$$\begin{aligned} &< \psi_1, \eta(T) > - < \psi_0, \eta'(T) + q_2^* v_2(T) > \\ &= \int_0^T \int_{\Gamma_0} (v_1 \frac{\partial}{\partial \nu} \Delta\psi + v_2 \frac{\partial}{\partial \nu} \psi') \, d\Gamma \, dt \end{aligned} \tag{8.65}$$

where $\psi_0 \in J_5$, $\psi_1 \in J_3$ and ψ is the solution of

$$\begin{aligned} &\psi'' + \Delta^2 \psi = 0 \\ &\psi(T) = \psi_0, \quad \psi'(T) = \psi_1 \\ &\psi|_\Gamma = \Delta\psi_{|\Gamma} = \Delta^2 \psi_{|\Gamma} = 0 \\ &\psi \in C([0,T]; J_5), \quad \psi' \in C([0,T]; J_3). \end{aligned} \tag{8.66}$$

Writing $\zeta = -\Delta\eta$, we interpret (8.65) as follows

$$\begin{aligned} &\zeta'' + \Delta^2 \zeta = 0 \\ &\zeta(0) = \zeta'(0) = 0 \\ &\zeta|_{\Gamma_1} = 0, \quad \Delta\zeta|_{\Gamma_1} = 0 \\ &\zeta|_{\Gamma_0} = v_1, \quad \Delta\zeta|_{\Gamma_0} = J_0^* v_2 \\ &\zeta \in L^2(0,T; J_{-1}) \cap C([0,T]; J_{-3}), \quad \zeta' \in L^2([0,T]; J_{-5}). \end{aligned} \tag{8.67}$$

Let us express the controllability property. Given y_0 in J_1 and y_1 in J_{-3}, the system

$$\begin{aligned} &\zeta'' + \Delta^2 \zeta = 0 && \phi'' + \Delta^2 \phi = 0 \\ &\zeta(0) = \zeta'(0) = 0 && \phi|_\Gamma = \Delta\phi|_\Gamma = 0 \\ &\zeta|_{\Gamma_1} = 0, \quad \Delta\zeta|_{\Gamma_1} = 0 && \zeta(T) = y_0, \\ &\zeta|_{\Gamma_0} = -\frac{\partial}{\partial \nu}\Delta\phi, \quad \Delta\zeta|_{\Gamma_0} = J_0^* \frac{\partial}{\partial \nu}\phi' && \zeta'(T) = y_1 \\ &\zeta \in L^2(0,T; J_{-1}) \cap C([0,T]; J_{-3}), && \phi \in C([0,T]; J_3) \\ &\zeta' \in L^2([0,T]; J_{-5}) && \phi' \in C([0,T]; J_1) \end{aligned} \tag{8.68}$$

has a solution for $T > 0$ arbitrary.

Remark 8.1 For other examples and results of exact controllability see J.-L. LIONS [1], [2] where the HUM method is presented in full generality, I. LASIECKA and R. TRIGGIANI [1] [2] [3] [4], R. TRIGGIANI [2], J. LAGNESE [1] and J.E. LAGNESE and J.-L. LIONS [1]. The Maxwell equation is treated in J. LAGNESE [2]. The result of exact controllability for the plate equation is given in E. ZUAZUA [1] and I. LASIECKA and R. TRIGGIANI [3] . For the use of microlocal analysis in the context of exact controllability, we refer to C. BARDOS, G. LEBEAU and J. RAUCH [1].

References

C. BARDOS, G. LEBEAU and J. RAUCH [1], *Contrôle et stabilisation daus les problèmes hyperboliques*, in Appendix II to J.-L. LIONS, "Contrôlabilité exacte, perturbations et stabilisation des systèmes distribués", Vol. 1, Masson, Paris 1988.

T. BASAR and P. BERNHARD [1], "H^∞-Optimal Control and Related Minimax Design Problems: A Dynamic Game Approach", Birkhäuser, Boston, Basel, Berlin 1991.

A. BEL FEKIH [1], "Une nouvelle approche pour la contrôlabilité et l'observabilité des systèmes distribués", thèse, Université de Kénitra, June 1990.

R.E. BELLMAN [1], "Dynamic programming", Princeton University Press, Princeton 1957.

R.W. BROCKETT [1], "Finite dimensional linear systems", Wiley, New York 1970.

M.C. DELFOUR and S.K. MITTER [1], *Controllability and observability for infinite dimensional systems*, SIAM J. Control 10 (1972), 329–333.

YU. V. EGOROV [1], *Some problems in the theory of optimal control*, Soviet Math. 3 (1962), 1080–1084.

[2], Ž. Vyčisl. Mat. i Mat. Fiz. 5 (1963), 887-904.

A. EL JAI and A.J. PRITCHARD [1], "Sensors and controls in the analysis of distributed systems", Wiley, New York 1988.

[2], "Capteurs et actionneurs daus l'analyse des systèmes distribués", RMA 3, Masson, Paris 1986.

H.O. FATTORINI [1], *Some remarks on complete controllability*, SIAM J. Control 4 (1966), 686-694.

[2], *On Complete Controllability of Linear Systems*, J. Differential Equations 3 (1967), 391–402.

[3], *Boundary control systems*, SIAM J. Control, 6 (1968), 349-388.

H.O. FATTORINI and D.L. RUSSELL [1], *Exact controllability theorems for Linear parabolic equations in one space dimension*, Arch. Rational Mech. Anal. 43 (1971), 272–292.

B.A. FRANCIS [1], "A Course in H_∞ Control Theory", Lecture, Notes in Control and Information Sciences, Volume 88, Springer-Verlag, New York 1987.

E. GILBERT [1], *Controllability and observability in multivariable control systems*, SIAM J. Control 1 (1963), 128–151.

M.L.J. HAUTUS [1], *Controllability and observability conditions for linear autonomous systems*, Ned. Akad. Wetenschappen, Proc. Ser. A 72 (1969), 443–448.

L.F. HO [1], *Observabilité frontière de l'équation des ondes*, C.R. Acad. Sc. Paris sér A 302 (1986), 443–446.

R.E. KALMAN [1], *On the general theory of control systems*, in "Proc. First IFAC Congress Automatic Control (Moscow)", Vol. I, pp. 481–492, Butterworths, London 1960.

[2], *Mathematical description of linear dynamical systems*, SIAM J. Control 1 (1963), 152–192.

R.E. KALMAN, Y.C. HO and K.S. NARENDRA [1], *Controllability of linear dynamical systems*, Contrib. Diff. Eqs 1 (1963), 189–213.

T. KATO [1], "Perturbation theory for linear operators", Springer Verlag, Berlin, New York 1966.

J.E. LAGNESE [1], "Boundary stabilization of thin plates", SIAM Studies in Applied Mathematics, Vol. 10, SIAM, Philadelphia 1989.

[2], *Exact controllability of Maxwell's equations in a general region*, SIAM J. Control and Optim. 27 (1989), 374–388.

J.E. LAGNESE and J.-L. LIONS [1], "Modelling, analysis and control of thin plates", RMA, Vol. 6, Masson, Paris, Milan, Barcelone, Mexico 1988

J.P. LASALLE [1], *The Time Optimal Control Problem*, in "Theory of Nonlinear Oscillations", Vol. 5, pp. 1-24, Princeton University Press, Princeton, N.J. 1959.

I. LASIECKA and R. TRIGGIANI [1], *Exact controllability for the wave equation with Neumann boundary control*, Appl. Math. Optim. 19 (1989), 263–290.

[2], *Exact controllability of the Euler-Bernoulli equation with controls in the Dirichlet and Neumann boundary conditions: a non conservative code*, SIAM J. Control and Optim. 27 (1989), 330–373.

[3], *Regularity, exact controllability and uniform stabilization of Kirchoff plates via only the bending moment*, J. Differential Equations, to appear.

[4], "Differential and Algebraic Riccati Equations with applications to boundary/point control problems: continuous theory and approximation theory", Lecture Notes in Control and Information Sciences, Vol. 164, Springer-Verlag, Berlin, Heidelberg, New York 1991.

J.-L. LIONS [1], "Equations différentielles opérationnelles et problèmes aux limites", Springer Verlag, Berlin, New York 1962.

[2], "Contrôlabilité exacte, perturbations et stabilisation des systèmes distribués", Vol. 1 and 2, Masson, Paris 1988.

[3], *Exact controllability, stabilization and perturbations*, SIAM Rev. 30 (1988), 1–68.

[4], "Optimal control of systems governed by partial differential equations", Springer-Verlag, Berlin, Heidelberg, New York 1971 (translated by S.K. Mitter).

[5], Personal communication.

J.-L. LIONS and E. MAGENES [1], "Problèmes aux limites non homogènes", Dunod, Paris 1968.

W. LITTMAN [1], *Boundary control theory for hyperbolic and parabolic equations with constant coefficients*, Annali Scuola Normale Superiore, Pisa IV (1978), 567–580.

D.G. LUENBERGER [1], *An introduction to observers*, IEEE Trans. Aut. Control 16 (1971), 596–602.

R.C. MACCAMY, V.J. MIZEL, T.I. SEIDMAN [1], *Approximate boundary controllability of the heat equation*, J. Math. Anal. Appl. 23 (1968), 699–703.

L. MARKUS [1], "Introduction to the theory of distributed control systems", University of Minnesota, Minneapolis, Nov. 1989.

S.K. MITTER [1], *Successive approximation methods for the solution of optimal control problems*, Automatica 3 (1966), 135–149.

S.K. MITTER and J.S. SHAMMA [1], *An elementary proof of an $\mathcal{H}^\infty$-optimal state feedback result*, unpublished manuscript 1989.

V.J. MIZEL and T.I. SEIDMAN [1], *Observation and Prediction for the heat equation*, J. Math. Anal. Appl. 28 (1969), 303–312.

[2], *Observation and Prediction for the heat equation*, II, J. Math. Anal. Appl. 38 (1972), 149–166.

S. MIZOHATA [1], *Unicité du prolongement des solutions pour quelques opérateurs différentiels paraboliques*, Mém. Coll. Sc. Univ. Kyoto, Sér. A 31 (1958), 219–239.

D.L. RUSSELL [1], *Controllability and stabilizability theory for linear partial differential equations: recent progress and open questions*, SIAM Review 20 (1978), 639–739.

M. SCHECHTER [1], "Principles of Functional Analysis", Academic Press, New York, London 1971.

E.J.P.G. SCHMIDT and N. WECK [1], *On the boundary behaviour of solutions to elliptic and parabolic equations, with applications to boundary control for parabolic equations*, SIAM J. Control and Optim. 16 (1978), 593–598.

T.I. SEIDMAN [1], *Observation and prediction for the heat equation, II*, J. Math. Anal. Appl. 38 (1972), 149-166.

[2], *Observation and prediction for the heat equation, III*, J. Differential Equations 20 (1976), 18–27 (the result was announced in *A well-posed problem for the heat equation*, Bull. Amer. Math. Soc. 80 (1974), 901–902).

[3], *Time invariance of the reachable set for linear control problems*, J. Math. Anal. Appl. 72 (1979), 17–20.

[4], *Observation and prediction for the heat equation, IV: patch observability and controllability*, SIAM J. Control 15 (1977), 412–427.

J.D. SIMON and S.K. MITTER [1], *A theory of modal control*, Information and Control 13, no. 4 (1968), 316–353.

R. TRIGGIANI [1], *On a lack of exact controllability for mild solutions in Banach spaces*, J. Math. Anal. and Appl. 50 (1975), 438-446.

[2], *Exact boundary controllability on $L^2(\Omega) \times H^{-1}(\Omega)$ for the wave equation with Dirichlet control acting on a portion of the boundary, and related problems*, Appl. Math. and Optim. 18 (1988), 241–277.

J.C. WILLEMS [1], *Least squares stationary optimal control and the algebraic Riccati equation*, IEEE Trans. Autom. Control AC-16 (1971), 621–634.

W.M. WONHAM [1], *On a matrix Riccati equation of stochastic control*, SIAM J. Control 6, no. 4 (1968), 681–697.

[2], *On pole assignment in multi-input controllable linear systems*, IEEE Trans. Aut. Control AC-12, no. 6 (1967), 660–665.

[3], "Linear multivariable control", Springer-Verlag, New York, Berlin, Heidelberg, Tokyo 1979, 1985.

J. ZABCZYK [1], *Remark on the algebraic Riccati equation in Hilbert space*, J. Appl. Math. and Optimization 2 (1976), 351–358.

E. ZUAZUA [1], *Controlabilité exacte d'un modèle de plaques vibrantes en un temps arbitrairement petit*, C.R. Acad. Sc. Paris sér. A. 304 (1987), 173–176.

Part II

Quadratic Optimal Control: Finite Time Horizon

Chapter 1

Systems with Bounded Control Operators: Control Inside the Domain

1 Introduction and setting of the problem

In this chapter we consider the dynamical system governed by the equation

$$\begin{cases} x'(t) = Ax(t) + Bu(t), \quad t \geq 0 \\ \\ x(0) = x_0 \in H, \end{cases} \tag{1.1}$$

where $A : D(A) \subset H \to H$, $B : U \to H$ are linear operators defined on the Hilbert spaces H (*state space*) and U (*control space*) respectively. x is the *state* and u the *control* of the system. We shall also consider another Hilbert space Y, the space of *observations*. The inner product and norm in H, U and Y will be denoted by $(\cdot, \cdot)$ and $|\cdot|$. Whenever confusion is possible a subscript H, U or Y will be added.

Given $T > 0$, we want to minimize the *cost function*

$$J(u) = \int_0^T \{|Cx(s)|^2 + |u(s)|^2\}\, ds + (P_0 x(T), x(T)) \tag{1.2}$$

over all controls $u \in L^2(0,T;U)$ subject to the differential equation constraint (1.1). Concerning the operators A, B, C and P_0 we shall assume that

$$(\mathcal{H}) \quad \begin{cases} (i) \;\; A \text{ generates a strongly continuous semigroup } e^{tA} \\ \qquad \text{on } H, \\ (ii) \;\; B \in \mathcal{L}(U;H), \\ (iii) \; P_0 \in \mathcal{L}(H) \;\text{ is hermitian and non negative,} \\ (iv) \;\; C \in \mathcal{L}(H;Y). \end{cases}$$

We shall also say that (A, B, C, P_0) verifies assumptions $(\mathcal{H})$. Assumption $(\mathcal{H})$-(iv) means that the observation operator is bounded. However in §4 below we shall also consider the more general situation when C is unbounded. Under assumptions $(\mathcal{H})$-(i)-(ii) problem (1.1) has a unique mild solution $x \in C([0,T];H)$ for any $x_0 \in H$ and any $u \in L^2(0,T;U)$ (see Volume I, Chapter 1, §3.1). Moreover x is given by the expression

$$x(t) = e^{tA}x_0 + \int_0^t e^{(t-s)A}Bu(s)\,ds.$$

A function $u^\star \in L^2(0,T;U)$ is called an *optimal control* if

$$J(u^\star) \le J(u), \quad \forall\, u \in L^2(0,T;U).$$

In this case the corresponding solution of (1.1) is called an *optimal state* and the pair $(u^\star, x^\star)$ an *optimal pair*. Under hypotheses $(\mathcal{H})$ it is easy to see that there exists a unique optimal control (since the quadratic form $J(u)$ is coercive on $L^2(0,T;U)$). However we are interested in showing that the optimal control can be obtained as a *feedback control* (*synthesis problem*). For this purpose we shall describe the *Dynamic Programming* approach which consists, essentially, in the following two steps:

Step 1. We solve the Riccati equation

$$\begin{cases} P' = A^\star P + PA - PBB^\star P + C^\star C, \\ \\ P(0) = P_0. \end{cases}$$

Step 2. We prove that the optimal control $u^\star$ is related to the optimal state $x^\star$ by the feedback formula

$$u^\star(t) = -B^\star P(T-t)x^\star(t), \quad t \in [0,T],$$

and moreover that $x^\star$ is the solution of the *closed loop equation*

$$\begin{cases} x'(t) = [A - BB^\star P(T-t)]x(t), & t \in [0,T], \\ x(0) = x_0. \end{cases}$$

2 Solution of the Riccati equation

2.1 Notation and preliminaries

We first introduce some notation. Let H be a complex Hilbert space, define

$$\Sigma(H) = \{T \in \mathcal{L}(H) : \ T \text{ is hermitian}\},$$

$$\Sigma^+(H) = \{T \in \Sigma(H) : \ (Tx, x) \geq 0, \ \forall \, x \in H\}.$$

The space $\Sigma(H)$, endowed with the norm of $\mathcal{L}(H)$, is a real Banach space and $\Sigma^+(H)$ is a cone in $\Sigma(H)$. For any interval I in $\mathbf{R}$, we shall denote by $C(I; \Sigma(H))$ the set of all continuous mappings from I to $\Sigma(H)$. We next consider the set of all strongly continuous mappings $F : I \to \Sigma(H)$ (that is, such that $F(\cdot)x$ is continuous for any $x \in H$). We shall mainly use the following concept of convergence, called *strong convergence*. A sequence $\{F_n\}$ is strongly convergent to F if

$$\forall x \in H, \quad \lim_{n\to\infty} F_n(\cdot)x = F(\cdot)x \text{ in } C(I; H),$$

where $C(I; H)$ is endowed with the topology of uniform convergence on compact subsets of I. This topological space will be denoted by $C_s(I; \Sigma(H))$. If I is compact we can also consider a stronger topology. Indeed, if F is strongly continuous, the number

$$\| \, F \, \| = \sup_{t \in I} \|F(t)\| \tag{2.1}$$

is finite by virtue of the Uniform Boundedness Theorem. The space of all strongly continuous mappings $F : I \to \Sigma(H)$, endowed with the norm (2.1) is a Banach space and will be referred to as $C_u(I; \Sigma(H))$, where the subscript u stands for uniform convergence. Note that the spaces $C_u(I; \Sigma(H))$ and $C_s(I; \Sigma(H))$ are equal as sets, but their topologies are different. In particular $C(I; \Sigma(H))$ is a proper closed subspace of $C_u(I; \Sigma(H))$, but it is a dense subspace of $C_s(I; \Sigma(H))$

$$\gamma_n^1 = \gamma_n, \quad \gamma_n^{m+1}(P) = \gamma_n(\gamma_n^m(P)),$$

for $m = 2, 3, \ldots$, and $P \in C_s([0,T]; \Sigma(H))$. It is easy to check that

$$\|\gamma(P) - \gamma(Q)\| \leq \alpha \|P - Q\|, \quad \forall P, Q \in C_u([0,T]; \Sigma(H)).$$

Then, by the Contraction Mapping Principle, there exist unique P_n and P in $C_u([0,T]; \Sigma(H))$ such that

$$\gamma_n(P_n) = P_n \text{ and } \gamma(P) = P.$$

However, since we do not assume that

$$\gamma_n(P) \to \gamma(P) \text{ in } C_u([0,T]; \Sigma(H))$$

we cannot conclude that $P_n \to P$ in $C_u([0,T]; \Sigma(H))$, but a weaker result holds.

Lemma 2.1 *Under the previous hypotheses on the sequence of mappings* $\{\gamma_n\}$,

$$P_n \to P \text{ in } C_s([0,T]; \Sigma(H)).$$

Proof. Set

$$P^0 = 0, \quad P_n^0 = 0,$$

and define

$$P^m = \gamma^m(P^0), \quad P_n^m = \gamma_n^m(P^0), \quad m = 1, 2, \ldots$$

By the Contraction Mapping Principle, we have

$$\lim_{m \to \infty} P^m = P, \lim_{m \to \infty} P_n^m = P_n \text{ in } C_u([0,T]; \Sigma(H)), \ n = 1, 2, \ldots$$

Moreover

$$\|P - P^m\| \leq \sum_{k=m}^{\infty} \alpha^k \|\gamma(P^0)\|, \ \|P_n - P_n^m\| \leq \sum_{k=m}^{\infty} \alpha^k \|\gamma_n(P^0)\|.$$

Now fix x in H, then for all t in $[0,T]$

$$\begin{aligned} &|P(t)x - P_n(t)x| \\ &\quad \leq |P(t)x - P^m(t)x| + |P^m(t)x - P_n^m(t)x| \\ &\quad + |P_n^m(t)x - P_n(t)x|. \end{aligned} \tag{2.4}$$

and moreover that $x^\star$ is the solution of the *closed loop equation*

$$\begin{cases} x'(t) = [A - BB^\star P(T-t)]x(t), & t \in [0,T], \\ x(0) = x_0. \end{cases}$$

2 Solution of the Riccati equation

2.1 Notation and preliminaries

We first introduce some notation. Let H be a complex Hilbert space, define

$$\Sigma(H) = \{T \in \mathcal{L}(H) : \ T \ \text{ is hermitian}\},$$

$$\Sigma^+(H) = \{T \in \Sigma(H) : \ (Tx, x) \geq 0, \ \forall \, x \in H\}.$$

The space $\Sigma(H)$, endowed with the norm of $\mathcal{L}(H)$, is a real Banach space and $\Sigma^+(H)$ is a cone in $\Sigma(H)$. For any interval I in $\mathbf{R}$, we shall denote by $C(I; \Sigma(H))$ the set of all continuous mappings from I to $\Sigma(H)$. We next consider the set of all strongly continuous mappings $F : I \to \Sigma(H)$ (that is, such that $F(\cdot)x$ is continuous for any $x \in H$). We shall mainly use the following concept of convergence, called *strong convergence.* A sequence $\{F_n\}$ is strongly convergent to F if

$$\forall \, x \in H, \quad \lim_{n \to \infty} F_n(\cdot)x = F(\cdot)x \text{ in } C(I; H),$$

where $C(I; H)$ is endowed with the topology of uniform convergence on compact subsets of I. This topological space will be denoted by $C_s(I; \Sigma(H))$. If I is compact we can also consider a stronger topology. Indeed, if F is strongly continuous, the number

$$\| \ F \ \| = \sup_{t \in I} \|F(t)\| \tag{2.1}$$

is finite by virtue of the Uniform Boundedness Theorem. The space of all strongly continuous mappings $F : I \to \Sigma(H)$, endowed with the norm (2.1) is a Banach space and will be referred to as $C_u(I; \Sigma(H))$, where the subscript u stands for uniform convergence. Note that the spaces $C_u(I; \Sigma(H))$ and $C_s(I; \Sigma(H))$ are equal as sets, but their topologies are different. In particular $C(I; \Sigma(H))$ is a proper closed subspace of $C_u(I; \Sigma(H))$, but it is a dense subspace of $C_s(I; \Sigma(H))$

$$\gamma_n^1 = \gamma_n, \quad \gamma_n^{m+1}(P) = \gamma_n(\gamma_n^m(P)),$$

for $m = 2, 3, \ldots$, and $P \in C_s([0,T]; \Sigma(H))$. It is easy to check that

$$\|\gamma(P) - \gamma(Q)\| \le \alpha \|P - Q\|, \quad \forall\, P, Q \in C_u([0,T]; \Sigma(H)).$$

Then, by the Contraction Mapping Principle, there exist unique P_n and P in $C_u([0,T]; \Sigma(H))$ such that

$$\gamma_n(P_n) = P_n \ \text{ and } \ \gamma(P) = P.$$

However, since we do not assume that

$$\gamma_n(P) \to \gamma(P) \ \text{ in } \ C_u([0,T]; \Sigma(H))$$

we cannot conclude that $P_n \to P$ in $C_u([0,T]; \Sigma(H))$, but a weaker result holds.

Lemma 2.1 *Under the previous hypotheses on the sequence of mappings* $\{\gamma_n\}$,

$$P_n \to P \ \text{ in } \ C_s([0,T]; \Sigma(H)).$$

Proof. Set

$$P^0 = 0, \quad P_n^0 = 0,$$

and define

$$P^m = \gamma^m(P^0), \quad P_n^m = \gamma_n^m(P^0), \quad m = 1, 2, \ldots$$

By the Contraction Mapping Principle, we have

$$\lim_{m \to \infty} P^m = P, \ \lim_{m \to \infty} P_n^m = P_n \ \text{ in } \ C_u([0,T]; \Sigma(H)), \ n = 1, 2, \ldots$$

Moreover

$$\|P - P^m\| \le \sum_{k=m}^{\infty} \alpha^k \|\gamma(P^0)\|, \ \|P_n - P_n^m\| \le \sum_{k=m}^{\infty} \alpha^k \|\gamma_n(P^0)\|.$$

Now fix x in H, then for all t in $[0,T]$

$$\begin{aligned} &|P(t)x - P_n(t)x| \\ &\quad \le |P(t)x - P^m(t)x| + |P^m(t)x - P_n^m(t)x| \\ &\qquad + |P_n^m(t)x - P_n(t)x|. \end{aligned} \tag{2.4}$$

and moreover that $x^\star$ is the solution of the *closed loop equation*

$$\begin{cases} x'(t) = [A - BB^\star P(T-t)]x(t), \quad t \in [0,T], \\ x(0) = x_0. \end{cases}$$

2 Solution of the Riccati equation

2.1 Notation and preliminaries

We first introduce some notation. Let H be a complex Hilbert space, define

$$\Sigma(H) = \{T \in \mathcal{L}(H) : \ T \ \text{ is hermitian}\},$$

$$\Sigma^+(H) = \{T \in \Sigma(H) : \ (Tx, x) \geq 0, \ \forall \, x \in H\}.$$

The space $\Sigma(H)$, endowed with the norm of $\mathcal{L}(H)$, is a real Banach space and $\Sigma^+(H)$ is a cone in $\Sigma(H)$. For any interval I in $\mathbf{R}$, we shall denote by $C(I; \Sigma(H))$ the set of all continuous mappings from I to $\Sigma(H)$. We next consider the set of all strongly continuous mappings $F : I \to \Sigma(H)$ (that is, such that $F(\cdot)x$ is continuous for any $x \in H$). We shall mainly use the following concept of convergence, called *strong convergence*. A sequence $\{F_n\}$ is strongly convergent to F if

$$\forall x \in H, \quad \lim_{n\to\infty} F_n(\cdot)x = F(\cdot)x \text{ in } C(I;H),$$

where $C(I;H)$ is endowed with the topology of uniform convergence on compact subsets of I. This topological space will be denoted by $C_s(I;\Sigma(H))$. If I is compact we can also consider a stronger topology. Indeed, if F is strongly continuous, the number

$$\| \ F \ \| = \sup_{t\in I} \|F(t)\| \tag{2.1}$$

is finite by virtue of the Uniform Boundedness Theorem. The space of all strongly continuous mappings $F : I \to \Sigma(H)$, endowed with the norm (2.1) is a Banach space and will be referred to as $C_u(I;\Sigma(H))$, where the subscript u stands for uniform convergence. Note that the spaces $C_u(I;\Sigma(H))$ and $C_s(I;\Sigma(H))$ are equal as sets, but their topologies are different. In particular $C(I;\Sigma(H))$ is a proper closed subspace of $C_u(I;\Sigma(H))$, but it is a dense subspace of $C_s(I;\Sigma(H))$

if H is separable. (The fact that H is separable will not be used in the sequel). If $F \in C_s(I; \Sigma(H))$, then it is easy to check that $\|F(\cdot)\|$ is Lebesgue measurable (see N. DUNFORD and J.T. SCHWARTZ [1, Lemma 3, page 616].

Remark 2.1 The above terminology is both natural and convenient, but it may deserve some extra general comments. Given a compact interval I in $\mathbf{R}$, there are two underlying spaces $C(I; \mathcal{L}(H))$ and $\mathcal{L}(H; C(I; H))$ and the space $\mathcal{L}(H; C(I; H))$ has two different topologies. The notation $C_u(I; \mathcal{L}(H))$ applies to the space $\mathcal{L}(H; C(I; H))$ endowed with its natural Banach space topology; $C_s(I; \mathcal{L}(H))$ denotes the same space endowed with the weaker non-Banach topology $\mathcal{T}_s$ of *pointwise convergence*:

$$a_n \to a \Leftrightarrow \forall\, x \in H,\ a_n(x) \to a(x) \ \text{ in } \ C(I; H).$$

The space $\mathcal{L}(H; C(I; H))$ coincides with the space of all strongly continuous mappings: $A : I \to \mathcal{L}(H)$. The mapping A induces a map $a : H \to C(I; H)$ defined by $a(x)(t) = A(t)x$ and since

$$\forall\, x \in H, \quad \sup_{t \in I} |A(t)x| < \infty,$$

we conclude from the Uniform Boundedness Theorem that

$$\sup_{t \in I} \|A(t)\|_{\mathcal{L}(H)} < \infty.$$

By interchanging the order of the sup's

$$\begin{aligned} \sup_{|x|=1} \sup_{t \in I} |A(t)x| &= \sup_{t \in I} \sup_{|x|=1} |A(t)x| \\ &= \sup_{t \in I} \|A(t)\|_{\mathcal{L}(H)}, \end{aligned}$$

we conclude that a belongs to $\mathcal{L}(H; C(I; H))$. Using this construction the space $C(I; \mathcal{L}(H))$ of all uniformly continuous mappings can be identified with a closed but smaller subspace of $\mathcal{L}(H; C(I; H))$ since the norm on $\mathcal{L}(H; C(I; H))$ coincides with the norm on $C(I; \mathcal{L}(H))$ by interchanging the sup's. In particular $\mathcal{L}(H; C(I; H))$ contains all strongly continuous semigroups of class C_0, while $C(I; \mathcal{L}(H))$ only contains uniformly continuous semigroups. The reader can easily check that the space $\mathcal{L}(H; C(I; H))$, endowed with the weaker topology $\mathcal{T}_s$ remains closed, but the space $C(I; \mathcal{L}(H))$ is now dense when H is separable. ∎

We define $C_s^1(I; \Sigma(H))$ as the set of all the mappings $F \in C_s(I; \Sigma(H))$ which are strongly differentiable (that is, such that $F(\cdot)x$ is differentiable for any $x \in H$) and such that their derivative belongs to $C_s(I; \Sigma(H))$. We set

$$C_s^1(I; \Sigma^+(H)) = \{F \in C_s^1(I; \Sigma(H)) : \ F(t) \in \Sigma^+(H), \quad \forall t \in I\}.$$

Finally we denote by $C^\alpha(I; \Sigma(H))$ the set of all Hölder continuous mappings from I to $\Sigma(H)$ with exponent $\alpha \in]0, 1[$.

We are given a linear operator A which is the infinitesimal generator of a strongly continuous semigroup e^{tA} in H. By the Hille-Yosida theorem there exist $M \geq 1$ and $\omega \in \mathbf{R}$ such that

$$\|e^{tA}\| \leq Me^{\omega t} \text{ and } \|e^{tA_n}\| \leq Me^{\omega t}, \quad t > 0, \tag{2.2}$$

where A_n is the Yosida approximation of A. For any $T > 0$ we set

$$M_T = \sup\{\|e^{tA_n}\| : t \in [0, T], \, n \in \mathbf{N}\}.$$

Finally we denote by $A^\star$ the adjoint operator of A.

In the sequel, we shall need a generalization of the Contraction Mapping Principle. Let $T > 0$, and let $\{\gamma_n\}$ be a sequence of mappings from $C_u([0, T]; \Sigma(H))$ into itself such that

$$\|\gamma_n(P) - \gamma_n(Q)\| \leq \alpha \|P - Q\|,$$

for all $P, Q \in C_u([0, T]; \Sigma(H))$ and all $n \in \mathbf{N}$, where $\alpha \in [0, 1[$.

Moreover assume that there exists a mapping γ from the space $C_u([0, T]; \Sigma(H))$ into itself such that

$$\lim_{n \to \infty} \gamma_n^m(P) = \gamma^m(P) \ \text{ in } \ C_s([0, T]; \Sigma(H)), \tag{2.3}$$

for all $P \in C_u([0, T]; \Sigma(H))$ and all $m \in \mathbf{N}$, where γ^m and γ_n^m are defined by recurrence as

$$\gamma^1 = \gamma, \quad \gamma^{m+1}(P) = \gamma(\gamma^m(P)),$$

$$\gamma_n^1 = \gamma_n, \quad \gamma_n^{m+1}(P) = \gamma_n(\gamma_n^m(P)),$$

for $m = 2, 3, \ldots$, and $P \in C_s([0,T];\Sigma(H))$. It is easy to check that

$$\|\gamma(P) - \gamma(Q)\| \leq \alpha\|P - Q\|, \quad \forall\, P, Q \in C_u([0,T];\Sigma(H)).$$

Then, by the Contraction Mapping Principle, there exist unique P_n and P in $C_u([0,T];\Sigma(H))$ such that

$$\gamma_n(P_n) = P_n \ \text{ and } \ \gamma(P) = P.$$

However, since we do not assume that

$$\gamma_n(P) \to \gamma(P) \ \text{ in } \ C_u([0,T];\Sigma(H))$$

we cannot conclude that $P_n \to P$ in $C_u([0,T];\Sigma(H))$, but a weaker result holds.

Lemma 2.1 *Under the previous hypotheses on the sequence of mappings* $\{\gamma_n\}$,

$$P_n \to P \ \text{ in } \ C_s([0,T];\Sigma(H)).$$

Proof. Set

$$P^0 = 0, \quad P_n^0 = 0,$$

and define

$$P^m = \gamma^m(P^0), \quad P_n^m = \gamma_n^m(P^0), \quad m = 1, 2, \ldots$$

By the Contraction Mapping Principle, we have

$$\lim_{m\to\infty} P^m = P, \ \lim_{m\to\infty} P_n^m = P_n \ \text{ in } \ C_u([0,T];\Sigma(H)), \ n = 1, 2, \ldots$$

Moreover

$$\|P - P^m\| \leq \sum_{k=m}^{\infty} \alpha^k \|\gamma(P^0)\|, \ \|P_n - P_n^m\| \leq \sum_{k=m}^{\infty} \alpha^k \|\gamma_n(P^0)\|.$$

Now fix x in H, then for all t in $[0,T]$

$$\begin{aligned} &|P(t)x - P_n(t)x| \\ &\quad \leq |P(t)x - P^m(t)x| + |P^m(t)x - P_n^m(t)x| \\ &\quad\quad + |P_n^m(t)x - P_n(t)x|. \end{aligned} \tag{2.4}$$

Given $\varepsilon > 0$, there exists $m_\varepsilon \in \mathbf{N}$ such that

$$\sum_{k=m}^{\infty} \alpha^k \{\|\gamma(P^0)\| + \|\gamma_n(P^0)\|\} \le \frac{\varepsilon}{2}, \tag{2.5}$$

for all $m \ge m_\varepsilon$ and all $n \in \mathbf{N}$. By (2.4) and (2.5) it follows that

$$|P(t)x - P_n(t)x| \le \frac{\varepsilon}{2} + |P^{m_\varepsilon}(t)x - P_n^{m_\varepsilon}(t)x|, \ \forall\, t \in [0,T]. \tag{2.6}$$

Now (2.3) yields the conclusion. ∎

2.2 Riccati equation

Let A, B, C and P_0 be given linear operators such that assumptions $(\mathcal{H})$ be verified. Consider the following Riccati equation

$$\begin{cases} P' = A^\star P + PA - PBB^\star P + C^\star C, \\ \\ P(0) = P_0. \end{cases} \tag{2.7}$$

Problem (2.7) was studied by several authors when H, U and Y are finite dimensional. We recall the pioneering work in the early papers by R.E. Kalman [1] and W.M. Wonham [1]. The first systematic approach in the infinite dimensional case is due to J.-L. Lions [1] who solved (2.7) by variational methods. Here we follow the direct approach by G. Da Prato [1]. For other results see D.L. Lukes and D.L. Russel [1], R. Temam [1], L. Tartar [1], R. Curtain and A. Pritchard [1].

Since the operator A is unbounded it is not clear a priori what a solution of (2.7) means. We shall define now two kinds of solutions: *mild solutions* and *weak solutions*. It is also possible to define a notion of *strict solution* and *classical solution*. These definitions are more technical and will be given in §3. We first notice that if $A \in \mathcal{L}(H)$, then, as is easily checked, problem (2.7) is equivalent to the following integral equation

$$\begin{aligned} P(t)x &= e^{tA^\star} P_0 e^{tA} x + \int_0^t e^{sA^\star} C^\star C e^{sA} x\, ds \\ &\quad - \int_0^t e^{(t-s)A^\star} P(s) BB^\star P(s) e^{(t-s)A} x\, ds, \ x \in H. \end{aligned} \tag{2.8}$$

Now equation (2.8) is meaningful for any A which satisfies assumptions $(\mathcal{H})$. In particular the right-hand side of (2.8) belongs to the function space $C_s([0,T];\Sigma(H))$ and it is natural to seek solutions in that space.

Definition 2.1 *(i) A* mild solution *of problem* (2.7) *in the interval* $[0,T]$ *is a function* $P \in C_s([0,T];\Sigma(H))$ *that verifies the integral equation* (2.8).
(ii) A weak solution *of problem* (2.7) *in the interval* $[0,T]$ *is a function* $P \in C_s([0,T];\Sigma(H))$ *such that,* $P(0) = P_0$ *and, for any* $x, y \in D(A)$, $(P(\cdot)x, y)$ *is differentiable in* $[0,T]$ *and verifies the equation*

$$\begin{aligned} \frac{d}{dt}(P(t)x, y) &= (P(t)x, Ay) + (P(t)Ax, y) \\ &\quad -(B^\star P(t)x, B^\star P(t)y) + (Cx, Cy). \end{aligned} \tag{2.9}$$

Note that the choice of the space $C_s([0,T];\Sigma(H))$ is natural because the mapping $t \mapsto e^{tA}$ is in general strongly continuous but not uniformly continuous.

Proposition 2.1 *Let* $P \in C_s([0,T];\Sigma(H))$. *Then* P *is a mild solution of problem* (2.7) *if and only if* P *is a weak solution of* (2.7).

Proof. If P is a mild solution of (2.7), then for any $x, y \in D(A)$ we have

$$\begin{aligned} (P(t)x, y) &= (P_0 e^{tA}x, e^{tA}y) \\ &\quad + \int_0^t ([C^\star C - P(s)BB^\star P(s)]e^{(t-s)A}x, e^{(t-s)A}y)ds. \end{aligned}$$

It follows that $(P(t)x, y)$ is differentiable with respect to t and, by a simple computation, that (2.9) holds. Conversely if P is a weak solution, then it is easy to check that for all $x, y \in D(A)$

$$\frac{d}{ds}(P(s)e^{(t-s)A}x, e^{(t-s)A}y)$$

$$= (Ce^{(t-s)A}x, Ce^{(t-s)A}y) - (B^\star P(t)e^{(t-s)A}x, B^\star P(t)e^{(t-s)A}y).$$

Integrating from 0 to t we obtain

$$(P(t)x,y) = (e^{tA^\star}P_0e^{tA}x,y) + \int_0^t (e^{(t-s)A^\star}[C^\star C - P(s)BB^\star P(s)]e^{(t-s)A}x,y)\,ds$$

for all $x,y \in D(A)$. Since $D(A)$ is dense in H equation (2.8) follows. ■

It is useful to introduce the approximating problem:

$$\begin{cases} P_n' = A_n^\star P_n + P_nA_n - P_nBB^\star P_n + C^\star C \\ P_n(0) = P_0 \end{cases} \tag{2.10}$$

where $A_n = n^2R(n,A) - nI$ is the Yosida approximation of A and $R(n,A)$ is the resolvent of A. Problem (2.10) is obviously equivalent to the following integral equation

$$P_n(t) = e^{tA_n^\star}P_0e^{tA_n} + \int_0^t e^{sA_n^\star}C^\star Ce^{sA_n}\,ds - \int_0^t e^{(t-s)A_n^\star}P_n(s)BB^\star P_n(s)e^{(t-s)A_n}\,ds.$$

We now solve problem (2.7). We first prove the local existence of a solution.

Lemma 2.2 *Assume that* $(\mathcal{H})$ *is verified, fix* $T > 0$, *set*

$$r = 2\,M_T^2\,\|P_0\| \tag{2.11}$$

and let τ *be such that*

$$\tau \in]0,T], \quad \tau\left(\|C\|^2 + r^2\|B\|^2\right) \le \|P_0\|, \quad 2rM_T^2\tau \le \frac{1}{2}, \tag{2.12}$$

where M_T *has been previously defined just below equation* (2.2). *Then problems* (2.7) *and* (2.10) *have unique mild solutions* P *and* P_n *in the ball*

$$B_{r,\tau} = \{F \in C_u([0,\tau];\Sigma(H)) : \|F\| \le r\}.$$

Moreover

$$\lim_{n\to\infty} P_n = P \ \text{ in } \ C_s([0,\tau];\Sigma(H)). \tag{2.13}$$

Proof. Equation (2.8) (resp. the integral version of (2.10)) can be written in the form

$$P = \gamma(P) \quad (resp.\ P_n = \gamma_n(P_n)),$$

where for $x \in H$

$$\gamma(P)(t)x$$

$$= e^{tA^\star} P_0 e^{tA} x + \int_0^t e^{(t-s)A^\star}[C^\star C - P(s)BB^\star P(s)]e^{(t-s)A} x\, ds$$

and

$$\gamma_n(P)(t)x = e^{tA_n^\star} P_0 e^{tA_n} x$$

$$+ \int_0^t e^{(t-s)A_n^\star}[C^\star C - P(s)BB^\star P(s)]e^{(t-s)A_n} x\, ds.$$

Choose now r and τ such that (2.11) and (2.12) hold. We show that γ and γ_n are $\frac{1}{2}$-contractions on the ball $B_{r,\tau}$ of $C_u([0,\tau];\Sigma(H))$. Let in fact $P \in B_{r,\tau}$. Then, recalling (2.2), we have

$$\begin{aligned}|\gamma(P)(t)x| &\leq M_T^2 \{\|P_0\| + \tau[\|C\|^2 + r^2\|B\|^2]\}\, |x| \\ &\leq 2M_T^2\|P_0\||x|,\end{aligned}$$

and analogously

$$|\gamma_n(P)(t)x| \leq 2\, M_T^2\, \|P_0\||x|.$$

It follows that

$$\|\gamma(P)(t)\| \leq r,\ \|\gamma_n(P)(t)\| \leq r, \quad \forall\, t \in [0,\tau],\ n \in \mathbf{N}, P \in B_{r,\tau}$$

so that γ and γ_n map $B_{r,\tau}$ into $B_{r,\tau}$.

For $P, Q \in B_{r,\tau}$ we have

$$\gamma(P)(t)x - \gamma(Q)(t)x$$

$$= \int_0^t e^{(t-s)A^\star}[PBB^\star(Q-P) + (Q-P)BB^\star Q](s)e^{(t-s)A} x\, ds,$$

and a similar formula holds for $\gamma_n(P)(t)x - \gamma_n(Q)(t)x$. It follows that

$$\|\gamma(P)(t) - \gamma(Q)(t)\| \le 2rM_T^2\tau\|B\|^2\|P-Q\| \le \frac{1}{2}\|P-Q\|,$$

$$\|\gamma_n(P)(t) - \gamma_n(Q)(t)\| \le 2rM_T^2\tau\|B\|^2\|P-Q\| \le \frac{1}{2}\|P-Q\|.$$

Thus γ and γ_n are $\frac{1}{2}$-contractions in $B_{r,\tau}$ and there exist unique mild solutions P and P_n in $B_{r,\tau}$. Finally (2.13) follows from Lemma 2.1. ■

We now prove global uniqueness.

Lemma 2.3 *Assume that* $(\mathcal{H})$ *is verified, let* $T > 0$ *and let* P, Q *be two mild solutions of problem* (2.7) *in* $[0,T]$. *Then* $P = Q$.

Proof. Set

$$\alpha = \sup_{t\in[0,T]} \max\{\|P(t)\|, \|Q(t)\|\};$$

α is finite by the Uniform Boundedness Theorem. Choose $r > 0$ and $\tau \in [0,T]$ such that

$$r = 2M_T^2\alpha, \quad \tau\left(\|C\|^2 + r^2\|B\|^2\right) \le \alpha, \quad 2rM_T^2\tau \le \frac{1}{2}.$$

By Lemma 2.2 it follows that $P(t) = Q(t)$ for any $t \in [0,\tau]$. It is now sufficient to repeat this argument in the interval $[\tau, 2\tau]$ and so on. ■

The main result of this section is the following theorem.

Theorem 2.1 *Assume that* $(\mathcal{H})$ *is verified. Then problem* (2.7) *has a unique mild solution* $P \in C_s([0,\infty[;\Sigma^+(H))$. *Moreover for each* n, *problem* (2.10) *has a unique solution* $P_n \in C([0,\infty[;\Sigma^+(H))$ *and*

$$\lim_{n\to\infty} P_n = P \ \text{ in } \ C_s([0,T];\Sigma(H)),$$

for any $T > 0$.

Proof. Fix $T > 0$, set $\beta = M_T^2(\|P_0\| + T\|C\|^2)$, and choose $r > 0$ and $\tau \in]0,T]$ such that

$$r = 2\beta M_T^2, \quad \tau\left(\|C\|^2 + r^2\|B\|^2\right) \le \beta, \quad 2r\tau M_T^2 \le \frac{1}{2}.$$

By Lemma 2.3 there exists a unique solution P (resp. P_n) of (2.7) (resp. (2.10)) in $[0, \tau]$, and $P_n \to P$ in $C_s([0,\tau];\Sigma(H))$. We now prove that

$$P_n(t) \geq 0, \quad \forall t \in [0,\tau]. \tag{2.14}$$

This will imply that

$$P(t) \geq 0, \quad \forall t \in [0,\tau]. \tag{2.15}$$

To this end we notice that P_n is the solution of the following linear problem in $[0,\tau]$

$$P_n' = L_n^\star P_n + P_n L_n + C^\star C, \quad P_n(0) = P_0,$$

where $L_n = A_n - \frac{1}{2} BB^\star P_n$. Denote by $U_n(t,s)$, $0 \leq s \leq t \leq \tau$, the evolution operator associated to $L_n^\star$ (see Volume I, Chapter 1, §3.5). Then we can write the solution $P_n(t)$ as

$$P_n(t) = U_n(t,0)\, P_0\, U_n^\star(t,0) + \int_0^t U_n(t,s)\, C^\star C\, U_n^\star(t,s)\, ds.$$

Thus (2.14) and (2.15) follow immediately.

We now prove that, for n large enough, we have

$$P(t) \leq \beta I, \; P_n(t) \leq \beta I, \quad \forall t \in [0,\tau]. \tag{2.16}$$

These inequalities will allow us to repeat the previous argument in the interval $[\tau, 2\tau]$ and so on. In this way the theorem will be proved. We have in fact

$$\begin{aligned}(P_n(t)x, x) &= (P_0 e^{tA_n} x, e^{tA_n} x) + \int_0^t |Ce^{sA_n} x|^2 ds \\ &\quad - \int_0^t |B^\star P_n(s) e^{(t-s)A_n} x|^2 ds \leq \beta |x|^2.\end{aligned}$$

Since $P_n(t) \geq 0$ this implies (2.16). The proof is complete. ■

We now prove *continuous dependence with respect to the data.* Consider a sequence of Riccati equations

$$\begin{cases} (P^k)' = A_k^\star P^k + P^k A_k - P^k B_k B_k^\star P^k + C_k^\star C_k, \\ P^k(0) = P_0^k, \end{cases} \tag{2.17}$$

under the following hypotheses:

$$\begin{cases} (i) \ \textit{for any } k \in \mathbf{N}, \ (A_k, B_k, C_k, P_0^k) \ \textit{fulfill } (\mathcal{H}), \\ (ii) \ \textit{for all } T > 0 \textit{ and all } x \in H, \\ \qquad \lim_{k\to\infty} e^{tA_k}x = e^{tA}x \ \textit{uniformly in } [0,T], \\ (iii) \ \textit{for all } T > 0 \textit{ and all } x \in H, \\ \qquad \lim_{k\to\infty} e^{tA_k^\star}x = e^{tA^\star}x \ \textit{uniformly in } [0,T], \\ (iv) \ \textit{the sequences } \{B_k\}, \{B_k^\star\}, \{C_k\}, \{C_k^\star\}, \{P_0^k\} \textit{ are} \\ \qquad \textit{strongly convergent to } B, B^\star, C, C^\star, P_0, \\ \qquad \textit{respectively.} \end{cases} \tag{2.18}$$

Theorem 2.2 *Assume that* $(\mathcal{H})$ *and* (2.18) *hold. Let* P *and* P_k *be the respective mild solutions to* (2.7) *and* (2.17). *Then, for any* $T > 0$ *we have*

$$\lim_{k\to\infty} P_k = P \ \text{ in } \ C_s([0,T];H). \tag{2.19}$$

Proof. Fix $T > 0$. By the Uniform Boundedness Theorem there exist positive numbers p, b and c such that

$$\|P_0^k\| \le p, \quad \|C_k^\star C_k\| \le c, \quad \|B_k B_k^\star\| \le b, \quad \forall\, k \in \mathbf{N}.$$

Set $\beta = M_T^2(p + cT)$ and choose r and $\tau \in]0,T]$ such that

$$r = 2M_T^2\beta, \quad \tau(c + r^2 b) \le \beta, \quad 2M_T^2\tau \le \frac{1}{2}.$$

Then, arguing as we did in the proof of Lemma 2.2 we can show that $P^k(\cdot)x \to P(\cdot)x$ in $C([0,\tau];H)$ for any x in H. Finally, proceeding as in the proof of Theorem 2.1, we prove that this argument can be iterated in the interval $[\tau, 2\tau]$ and so on. ■

We conclude this section by proving an important *monotonicity property* of the solutions of the Riccati equation (2.7).

Proposition 2.2 *Consider the Riccati equations:*

$$\begin{cases} P_i' = A^\star P_i + P_i A - P_i B_i B_i^\star P_i + C_i^\star C_i \\ \\ P_i(0) = P_{i,0}, \quad i = 1,2. \end{cases} \tag{2.20}$$

Assume that $(A_i, B_i, C_i, P_{i,0})$ *verify* $(\mathcal{H})$ *for* $i = 1,2$, *and, in addition, that*

$$P_{1,0} \le P_{2,0}, \quad C_1^\star C_1 \le C_2^\star C_2, \quad B_2 B_2^\star \le B_1 B_1^\star.$$

Then we have

$$P_1(t) \le P_2(t), \quad t \ge 0. \tag{2.21}$$

Proof. Due to Theorem 2.1 it is sufficient to prove (2.21) when A is bounded (since we can approximate A with the sequence of bounded operators A_n). Set $Z = P_2 - P_1$, then, as easily checked, Z is the solution to the linear problem:

$$\begin{cases} Z' = X^\star Z + ZX - P_2[B_2B_2^\star - B_1B_1^\star]P_2 + C_2^\star C_2 - C_1^\star C_1, \\ \\ Z(0) = P_{2,0} - P_{1,0}, \end{cases}$$

where

$$X = A - \frac{1}{2}B_1B_1^\star(P_1 + P_2).$$

Let $V(t,s)$ be the evolution operator associated with $X^\star$. Then we have

$$\begin{aligned} Z(t) &= V(t,0)(P_{2,0} - P_{1,0})V^\star(t,0) \\ &+ \int_0^t V(t,s)\{C_2^\star C_2 - C_1^\star C_1\}V^\star(t,s)\,ds \\ &+ \int_0^t V(t,s)P_1(s)\left[B_1B_1^\star - B_2B_2^\star\right]P_1(s)V^\star(t,s)\,ds \end{aligned}$$

so that $Z(t) \ge 0$ and the conclusion follows. ∎

2.3 Representation formulas for the solution of the Riccati equation

We want here to give an explicit formula for the solution of the Riccati equation (2.7). In fact several variants are possible, see A.V. Balakrishnan [1], and I. Lasiecka and R. Triggiani [1], [3].

We assume that $(\mathcal{H})$ is verified and introduce the operator

$$K_t \in \mathcal{L}(L^2(0,t;H); L^2(0,t;H))$$

defined by

$$(K_t x)(s) = \int_s^t e^{(\rho - s)A}x(\rho)\,d\rho, \quad x \in L^2(0,t;H),\ s \in [0,t].$$

Its adjoint $K_t^\star \in \mathcal{L}(L^2(0,t;H);L^2(0,t;H))$ is given by

$$(K_t^\star x)(s) = \int_0^s e^{(s-\sigma)A^\star} x(\sigma)\,d\sigma, \quad s \in [0,t].$$

In fact $K_t, K_t^\star \in \mathcal{L}(L^2(0,t;H);C([0,t];H))$, (with $K_t^\star$ not representing the dual for these functional spaces). Therefore we can define $K_t^0 \in \mathcal{L}(L^2(0,t;H);H)$ as the operator

$$K_t^0 x = (K_t x)(0) = \int_0^t e^{\rho A} x(\rho)\,d\rho, \quad x \in L^2(0,t;H)$$

and its dual $(K_t^0)^\star \in \mathcal{L}(H;L^2(0,t;H))$ is given by

$$[(K_t^0)^\star h](s) = e^{sA^\star} h, \ h \in H.$$

We can assert the following proposition.

Proposition 2.3 *Assume that* $(\mathcal{H})$ *is verified and let* P *be the mild solution of* (2.7). *Then one has the formula*

$$\begin{aligned} P(t) = {} & [I + (K_t^\star C^\star C K_t + (K_t^0)^\star P_0 K_t)\, BB^\star]^{-1} \\ & \cdot [K_t^\star C^\star C e^{(t-\cdot)A} + (K_t^0)^\star P_0 e^{tA}](t). \end{aligned} \tag{2.22}$$

Proof. We first clarify the meaning of the right-hand side of (2.22). For fixed t, $e^{(t-\cdot)A}$ is the operator in $\mathcal{L}(H;L^2(0,t;H))$ defined by $h \mapsto e^{(t-s)A}h, \ s \in]0,t[$. Next $C \in \mathcal{L}(H;Y)$ is identified with an element of $\mathcal{L}(L^2(0,t;H);L^2(0,t;Y))$ defined by

$$Cx(\cdot)(s) = Cx(s).$$

Similarly $B \in \mathcal{L}(L^2(0,t;U);L^2(0,t;H))$ and analogous considerations hold for $C^\star, B^\star$. Then the operator

$$\Lambda_t = I + \Big(K_t^\star C^\star C K_t + (K_t^0)^\star P_0 K_t\Big)\, BB^\star$$

is an element of $\mathcal{L}(L^2(0,t;H);L^2(0,t;H))$ which even belongs to $\mathcal{L}(L^2(0,t;H);C([0,t];H))$. According to Proposition 1.1 in Appendix A to this volume, it is invertible and also

$$\Lambda_t^{-1} \in \mathcal{L}(L^2(0,t;H);C([0,t];H))$$

and finally

$$\Lambda_t^{-1}\left[K_t^\star C^\star C e^{(t-\cdot)A} + (K_t^0)^\star P_0 e^{tA}\right] \in \mathcal{L}(H; C([0,t]; H)),$$

and it makes sense to take its value at time t, defining in this way an element of $\mathcal{L}(H)$, and therefore (2.22) has a meaning.

We shall proceed with the formal derivation of (2.22). The rigorous derivation is first done for the approximation (2.10), where the formal computations are valid, since A_n is bounded, and then the final result is obtained by going to the limit. Define $X(t,s)$ to be the solution of the problem

$$\begin{cases} -\dfrac{\partial X(t,s)}{\partial s} = (A - BB^\star P(s))X(t,s), \quad 0 < t < s \\ X(t,t) = I. \end{cases} \tag{2.23}$$

Define next

$$Z(t,s) = P(s)X(t,s).$$

An easy computation (from the Riccati equation and (2.23)) shows that the pair $X(t,s), Z(t,s)$ is the solution of the Hamiltonian system

$$\begin{cases} -\dfrac{\partial X(t,s)}{\partial s} = AX(t,s) - BB^\star Z(t,s), \\ \dfrac{\partial Z(t,s)}{\partial s} = A^\star Z(t,s) + C^\star C X(t,s) \\ Z(t,0) = P_0 X(t,0), \quad X(t,t) = I. \end{cases} \tag{2.24}$$

Therefore we have

$$\begin{cases} Z(t,s) = e^{sA^\star} P_0 X(t,0) + \displaystyle\int_0^s e^{(s-\sigma)A^\star} C^\star C X(t,\sigma)\, d\sigma, \\ X(t,s) = e^{(t-s)A} - \displaystyle\int_s^t e^{(\rho-s)A} BB^\star Z(t,\rho)\, d\rho, \end{cases}$$

and combining the two relations we obtain

$$Z(t,s) =$$

$$e^{sA^\star}P_0\left\{e^{tA} - \int_0^t e^{\rho A}BB^\star Z(t,\rho)\,d\rho\right\}$$

$$+\int_0^s e^{(s-\sigma)A^\star}C^\star C\left\{e^{(t-\sigma)A} - \int_\sigma^t e^{(\rho-\sigma)A}BB^\star Z(t,\rho)\,d\rho\right\}\,d\sigma.$$

Setting $Z_t(s) = Z(t,s) \in \mathcal{L}(L^2(0,t;H);L^2(0,t;H))$, this relation reads

$$Z_t + (K_t^0)^\star P_0 K_t^0 BB^\star Z_t + K_t^\star C^\star C K_t BB^\star Z_t$$

$$= K_t^\star C^\star C e^{(t-\cdot)A} + (K_t^0)^\star P_0 e^{tA}.$$

Since $P(t) = Z_t(t) = Z(t,t)$, we obtain formula (2.22). ∎

3 Strict and classical solutions of the Riccati equation

In this section we are interested in strict and classical solutions of the Riccati equation. The section is divided into three subsections, namely the general case, the case when e^{tA} is an analytic semigroup and finally the case when A is a variational operator. In §3.1 and §3.2 we follow G. Da Prato [1].

3.1 The general case

In order to define strict solutions to equation (2.7) we need to give a precise meaning to the linear mapping $\mathcal{A} : T \mapsto \mathcal{A}(T) = A^\star T + TA$. For any $T \in \Sigma(H)$ we set

$$\varphi_T(x,y) = (Tx, Ay) + (Ax, Ty), \quad x,y \in D(A) \tag{3.1}$$

and define

$$D(\mathcal{A}) = \{T \in \Sigma(H) : \ \varphi_T \text{ is continuous in } H \times H\}. \tag{3.2}$$

If $T \in D(\mathcal{A})$ then φ_T has a unique extension (which we still denote by φ_T) as a continuous sesquilinear form in $H \times H$. In this case there exists a linear operator which we denote by $\mathcal{A}(T) \in \Sigma(H)$ such that

$$(\mathcal{A}(T)x, y) = \varphi_T(x, y), \quad x, y \in D(A),\ T \in D(\mathcal{A}). \tag{3.3}$$

Thus we have defined a linear operator from $D(\mathcal{A}) \subset \Sigma(H) \to \Sigma(H)$. It is easy to check that $\mathcal{A}$ is closed in $\Sigma(H)$. However $D(\mathcal{A})$ is not dense in $\Sigma(H)$ in general. The following proposition shows the relationship between $\mathcal{A}(T)$ and $A^\star T + TA$.

Proposition 3.1 *Let $T \in D(\mathcal{A})$. Then for any $x \in D(A)$ we have $Tx \in D(A^\star)$ and*

$$\mathcal{A}(T)x = A^\star Tx + TAx. \tag{3.4}$$

Proof. For any $x, y \in D(A)$ we have

$$(Tx, Ay) = \varphi_T(x, y) - (Ax, Ty).$$

It follows that, for $x \in D(A)$, the linear mapping

$$y \mapsto (Tx, Ay) : \ D(A) \to \mathbf{C},$$

is continuous for H. This implies that $Tx \in D(A^\star)$ and for all $y \in H$

$$(\mathcal{A}(T)x, y) = \varphi_T(x, y) = (A^\star Tx, y) + (TAx, y)$$

which proves (3.4). ∎

Remark 3.1 We define a semigroup on $\Sigma(H)$

$$e^{t\mathcal{A}}(T) = e^{tA^\star}Te^{tA}, \quad t \geq 0,\ T \in \Sigma(H). \tag{3.5}$$

Notice that $e^{t\mathcal{A}}$ is not strongly continuous in general (this should imply that $e^{t\mathcal{A}}(T) \to T$ as $t \to 0$ in $\mathcal{L}(H)$). However it is easy to see that

$$e^{t\mathcal{A}}(T)x \to Tx \ \text{ as } \ t \to 0, \quad \forall\, x \in H, \quad \forall\, T \in \Sigma(H).$$

∎

We now return to the Riccati equation (2.7) which we write in the form

$$P' = \mathcal{A}(P) - PBB^\star P + C^\star C, \quad P(0) = P_0. \tag{3.6}$$

We say that P is a *strict solution* to the Riccati equation (2.7) (or (3.6)) if

$$\begin{cases} (i) & P \in C_s^1([0,\infty[;\Sigma(H)), \\ (ii) & P(t) \in D(\mathcal{A}),\ \forall\, t \geq 0, \\ (iii) & \mathcal{A}(P) \in C_s([0,\infty[\,;\Sigma(H)) \text{ and (3.6) holds.} \end{cases} \tag{3.7}$$

Remark 3.2 Assume that P is a strict solution of problem (2.7). By Proposition 3.1 it follows that for any $x \in D(A)$, $P(t)x \in D(A^\star)$ and the following equation holds

$$P'(t)x = A^\star P(t)x + P(t)Ax - P(t)BB^\star P(t)x + C^\star Cx. \tag{3.8}$$

By (3.8) one can easily check that P is a weak solution of problem (2.7). Thus, by Proposition 2.2, a strict solution is a mild solution of (2.7). ■

Proposition 3.2 *Assume that* $(\mathcal{H})$ *is verified and that* $P_0 \in D(\mathcal{A})$. *Then the Riccati equation* (2.7) *has a unique strict solution.*

Proof. Uniqueness follows from Theorem 2.1. Let us prove existence. Fix $T > 0$ and let P (resp. P_n) be the mild solution of (2.7) (resp. (2.10)). Set

$$\mathcal{A}_n(S) = A_n^\star S + SA_n, \quad n > \omega,\ S \in \Sigma(H),$$

where A_n is the Yosida approximation of A and ω is given in (2.2) (note that the $\mathcal{A}_n$'s are not the Yosida approximations of $\mathcal{A}$). We have

$$\begin{cases} P_n' = \mathcal{A}_n(P_n) - P_nBB^\star P_n + C^\star C \\ \\ P_n(0) = P_0. \end{cases}$$

Moreover $V_n = P_n'$ is the solution of the problem

$$\begin{cases} V_n' = \mathcal{A}_n(V_n) - P_nBB^\star V_n - V_nBB^\star P_n \\ \\ V_n(0) = \mathcal{A}_n(P_0) - P_0BB^\star P_0 + C^\star C. \end{cases}$$

We want to show that $V_n(\cdot)x$ is convergent, as $n \to \infty$, to $V(\cdot)x$ in $C([0,T];H)$ where V is the solution to the following linear integral equation: for all x in H

$$\begin{aligned}
&V(t)x \\
&= e^{tA^\star} V(0) e^{tA} x \\
&\quad - \int_0^t e^{(t-s)A^\star} \left[P(s)BB^\star V(s) + V(s)BB^\star P(s)\right] e^{(t-s)A} x \, ds,
\end{aligned}$$

which can be easily solved by successive approximations. In fact, by using similar arguments as in the proof of Lemma 2.2, it follows that

$$P_n \to P, \; V_n \to V \;\text{ in }\; C_s([0,T];H).$$

Since $V_n = P'_n$ we have that $P \in C^1_s([0,T];H)$. It remains to show that $P(t) \in D(\mathcal{A})$ and that (3.6) holds. Fix $t \geq 0$ and let $x, y \in D(A)$. We have

$$\begin{aligned}
\varphi_{P(t)}(x,y) &= (P(t)x, Ay) + (Ax, P(t)y) \\
&= \lim_{n\to\infty} \left[(P_n(t)x, A_n y) + (A_n x, P_n(t)y)\right] \\
&= \lim_{n\to\infty} (\mathcal{A}_n(P_n(t))x, y) \\
&= \lim_{n\to\infty} \left[(V_n(t)x, y) + (P_n(t)BB^\star P_n(t)x, y) - (C^\star C x, y)\right] \\
&= (V(t)x, y) + (P(t)BB^\star P(t)x, y) - (C^\star C x, y).
\end{aligned}$$

It follows that $P(t) \in D(\mathcal{A})$ and

$$\mathcal{A}(P(t)) = P'(t) + P(t)BB^\star P(t) - C^\star C.$$

∎

3.2 The analytic case

We assume here that the semigroup e^{tA} is analytic. Then there exists $c > 0$ and $\gamma > 0$ such that

$$\|e^{tA_n}\| \leq c e^{t\gamma}, \; \|A_n e^{tA_n}\| \leq \frac{c}{t} e^{t\gamma}, \quad 0 < t < 1. \tag{3.9}$$

Since $e^{t\mathcal{A}_n}(T) = e^{tA_n^\star}Te^{tA_n}$, $\forall T \in \mathcal{L}(H)$, it follows that

$$\|e^{t\mathcal{A}_n}\| \le c^2 e^{2t\gamma}, \ \|\mathcal{A}_n e^{t\mathcal{A}_n}\| \le \frac{2c^2}{t} e^{2t\gamma}, \quad 0 < t < 1. \tag{3.10}$$

Moreover the semigroup $e^{t\mathcal{A}}$, defined by (3.5), is clearly analytic for $t > 0$.

We now introduce the notion of classical solution. We say that $P \in C_s([0,\infty[;\Sigma(H))$ is a *classical solution* of the Riccati equation (2.7) if

$$\begin{cases} (i) \ \ \text{for any} \ \ t>0, \ P(t) \ \text{is differentiable}, \ P(t) \in D(\mathcal{A}) \ \text{and} \\ \qquad\qquad P'(t) = \mathcal{A}(P(t)) - P(t)BB^\star P(t) + C^\star C, \quad t > 0, \\ (ii) \ \ P(0) = P_0. \end{cases}$$

Proposition 3.3 *Assume that* $(\mathcal{H})$ *is verified and that* e^{tA} *is an analytic semigroup. If* $P \in C_s([0,\infty[;\Sigma(H))$ *is a classical solution of* (2.7)*, then* P *is a weak solution and a mild solution of* (2.7).

Proof. Let P be a classical solution, let $x, y \in D(A)$ and $t > 0$. By Proposition 3.1 we have $P(t)x \in D(A^\star)$ and

$$\mathcal{A}(P(t))x = A^\star P(t)x + P(t)Ax.$$

It follows that

$$\begin{aligned}\frac{d}{dt}(P(t)x, y) &= (P(t)x, y) + (P(t)Ax, y) \\ &\quad - (B^\star P(t)x, B^\star P(t)y) + (Cx, Cy)\end{aligned}$$

for all $t > 0$. By integrating this identity between $\varepsilon > 0$ and t, and letting ε tend to 0 we find that P is a weak solution and then a mild solution of (2.7). ■

Remark 3.3 By Proposition 3.3 it follows that there exists at most one classical solution of (2.7). ■

Theorem 3.1 *Assume* $(\mathcal{H})$ *and that* e^{tA} *is an analytic semigroup. If* $P \in C_s([0,\infty[;\Sigma(H))$ *is the mild solution of* (2.7)*, then* P *is a classical solution and belongs to* $C^\infty([\varepsilon,\infty[;\Sigma(H))$, $\forall \varepsilon > 0$.

In order to prove Theorem 3.1 we need the following two lemmas.

Lemma 3.1 *Assume that e^{tA} is an analytic semigroup. Let $E \in C_s([0,T];\Sigma(H))$ for some $T > 0$ and let F be defined by*

$$F(t)x = \int_0^t e^{(t-s)A^\star} E(s) e^{(t-s)A} x\, ds, \quad x \in H.$$

Then for any $\alpha \in]0,1[$, we have $F \in C^\alpha([0,T];\Sigma(H))$ and there exists a constant $C_T > 0$ such that

$$|F(t)x - F(r)x| \le C_T |t-r|^\alpha \, \|E\|_{C([0,T];\Sigma(H))} \, |x|,$$

for all $t, r > 0$.

Proof. Let $x \in H$ and $0 < r < t \le T$. We have

$$F(t)x - F(r)x = \int_r^t e^{(t-s)\mathcal{A}}(E(s))x\, ds + \int_0^r \int_{r-s}^{t-s} (\mathcal{A}e^{\sigma\mathcal{A}})(E(s))x\, d\sigma.$$

Taking into account (3.10) we have

$$\|F(t) - F(r)\| \le c^2 e^{2t\gamma} \left\{ |t-r| + \int_0^r ds \int_{r-s}^{t-s} \frac{d\sigma}{\sigma} \right\} \|E\|.$$

Since

$$\int_0^r ds \int_{r-s}^{t-s} \frac{d\sigma}{\sigma} \le \int_0^r (r-s)^{-\alpha} ds \int_{r-s}^{t-s} \sigma^{\alpha-1}\, d\sigma$$

$$\le \frac{1}{\alpha}|t-r|^\alpha \int_0^r (r-s)^{-\alpha}\, ds,$$

the conclusion follows. ∎

Lemma 3.2 *Assume that e^{tA} is an analytic semigroup. Let $\alpha \in]0,1[$, $M \in C^\alpha([0,T];\Sigma(H))$, and let G be defined by*

$$G(t)x = \int_0^t e^{(t-s)A^\star} M(s) e^{(t-s)A}\, x\, ds, \quad x \in H. \tag{3.11}$$

Then $G \in C^1([0,T];\Sigma(H)) \cap C([0,T];D(\mathcal{A}))$ and

$$G'(t) = \mathcal{A}(G(t)) + M(t).$$

Proof. Set

$$G_n(t)x = \int_0^t e^{(t-s)A_n}(M(s))x\,ds.$$

Then we have

$$G_n'(t) = \int_0^t A_n e^{(t-s)A_n}(M(s) - M(t))\,ds + e^{tA_n}(M(t)). \tag{3.12}$$

By using (3.10) we have

$$\begin{aligned} &\|A_n e^{(t-s)A_n}(M(s) - M(t))\| \\ &\le 2c^2 e^{2t\gamma}\,|t-s|^{\alpha-1}\,\|M\|_\alpha, \end{aligned} \tag{3.13}$$

where

$$\|M\|_\alpha = \sup_{0 \le t < s \le T} \frac{\|M(t) - M(s)\|}{|t-s|^\alpha}.$$

By (3.12) and (3.13) it follows that $G \in C^1([0,T];\Sigma(H))$ and

$$G'(t)x = \int_0^t A e^{(t-s)A}(M(s) - M(t))x\,ds + e^{tA}(M(t))x.$$

Finally, arguing as we did in the proof of Proposition 3.2, we find that $G \in C([0,T];D(A))$ and that (3.11) holds. ∎

Proof of Theorem 3.1. Let P be the mild solution of equation (2.7), which we write in the form

$$P(t)x = e^{tA^\star} P_0 e^{tA} x + \int_0^t e^{(t-s)A^\star} E(s) e^{(t-s)A} x\,ds, \tag{3.14}$$

where $x \in H$ and $E = C^\star C - PBB^\star P$. Since $e^{tA^\star}P_0 e^{tA}$ is analytic in t for $t > 0$, from Lemma 3.1 we have $P \in C^\alpha([\varepsilon,T];\Sigma(H))$ for $0 < \varepsilon < T$, $\alpha \in \,]0,1]$. Moreover for all x in H and $t \ge \varepsilon$

$$P(t)x = e^{(t-\varepsilon)A^\star} P(\varepsilon) e^{(t-\varepsilon)A} x + \int_\varepsilon^t e^{(t-s)A^\star} E(s) e^{(t-s)A} x\,ds.$$

From Lemma 3.2, it follows that

$$P \in C^1([2\varepsilon,T];\Sigma(H)) \cap C([2\varepsilon,T];D(A)),\ 0 < 2\varepsilon < T,$$

which implies $E \in C^1([2\varepsilon, T]; \Sigma(H))$. Moreover, by the identity

$$P(t)x = e^{(t-2\varepsilon)A^\star} P(2\varepsilon) e^{(t-2\varepsilon)A} x + \int_{2\varepsilon}^{t} e^{\sigma A^\star} E(t-\sigma) e^{\sigma A} x \, d\sigma, \; x \in H,$$

with $t \geq 2\varepsilon$, it follows that $P \in C^2([3\varepsilon, T]; \Sigma(H))$, $0 < 3\varepsilon < T$. By repeating this argument several times we find $P \in C^\infty([\varepsilon, T]; \Sigma(H))$, for any $\varepsilon > 0$, as required. ∎

The next corollary follows easily from (3.9).

Corollary 3.1 *Assume that* $(\mathcal{H})$ *is verified and let* e^{tA} *be an analytic semigroup. Let* $P \in C_s([0, \infty[\,; H)$ *be the classical solution to* (2.7). *Then* P *belongs to* $C([\varepsilon, \infty[\,; \mathcal{L}(H; D((-A^\star)^{1-\varepsilon})))$ *for all* $\varepsilon \in]0, 1[$.

Theorem 3.2 *Assume that* $(\mathcal{H})$ *is verified and that the semigroup* e^{tA} *is analytic in the sector* $S_\theta = \{\lambda \in \mathbf{C} : |\arg \lambda| < \theta\}$ *for some* $\theta \in]0, \frac{\pi}{2}[$. *Let* P *be the classical solution of problem* (2.7). *Then* P *has an analytical extension, as a function with values in* $\Sigma(H)$, *on the sector* S_θ.

Proof. Define r and τ as in (2.12). Then by Lemma 2.2 we have

$$P(\cdot)x = \lim_{m \to \infty} P^{(m)}(\cdot)x \text{ in } C([0, \tau]; H), \quad x \in H$$

where $P^{(m)}$ are defined by recurrence as

$$P^{(0)}(t)x = e^{tA^\star} P_0 e^{tA} x$$

$$P^{(m+1)}(t)x = e^{tA^\star} P_0 e^{tA} x + \int_0^t e^{(t-s)A^\star} C^\star C e^{(t-s)A} x \, ds - \int_0^t e^{(t-s)A^\star} P^{(m)}(s) B B^\star P^{(m)}(s) e^{(t-s)A} x \, ds$$

Clearly $P^{(m)}(\cdot)x$ are analytic in S_θ. Thus by a classical result, $P(\cdot)x$ is analytic in S_θ for any $x \in H$. This yields the conclusion (see for instance A.E. Taylor [1, Thm 4.4.F]. ∎

3.3 The variational case

We consider here the situation described in Volume I (Theorem 2.12, Chapter 1 §2.7 pp. 45-46, or Chapter 2 §1.3). We assume that there exists a Hilbert space V such that $V \subset H$, algebrically and topologically and V is dense in H. Moreover, H is identified with its dual, and $V \subset H \subset V'$, where V' is the dual of V. We denote by $< \cdot, \cdot >$ the duality pairing between V and V' and by $\| \cdot \|$ the norm on V, (recall that $| \cdot |$ and $(\cdot, \cdot)$ represent the norm and the scalar product in H).

Let $a : V \times V \to \mathbf{R}$ be a V-H coercive continuous bilinear form (cf. Volume I, Chapter 1, end of §2.7 or Chapter 2 §§ 1.1.1 and 1.3). In particular

$$\exists\, \alpha > 0,\ \exists\, \lambda \in \mathbf{R},\ \forall\, x \in V, \quad a(v, v) + \lambda |v|^2 \geq \alpha \|v\|^2.$$

Let $-A : D(A) \subset H \to H$ be the operator generated by a

$$-(Av, v) = a(v, v), \quad \forall\, v \in D(A),$$

and denote by A_λ the operator $A - \lambda I$. The operator A_λ, as well as A, is the generator of an analytic semigroup on H. Moreover we have seen that when

$$D_A(1/2, 2) = D_{A^\star}(1/2, 2)$$

then

$$D_A(1/2, 2) = D((-A_\lambda)^{1/2}) = V = D((-A_\lambda^\star)^{1/2} = D_{A^\star}(1/2, 2)$$

(cf. Volume I, Chapter 2, §1.5 Proposition 1.1).

We shall need the following result due to F. FLANDOLI [5], which is true for arbitrary variational operators.

Lemma 3.3 *Let A be a variational operator in H. Then for any $T > 0$ there exists $K_T > 0$ such that*

$$\int_0^T \left|(-A)^{1/2} e^{tA} x\right|^2 dt \leq K_T |x|^2, \quad \forall\, x \in H. \tag{3.15}$$

Proof. For variational operators we know from the result of A. YAGI [1] (cf. Volume I, Chapter 1, Theorem 6.1), that $D_A(1/2, 2)$

and $D((-A)^{1/2})$ are isomorphic. It is sufficient to prove (3.15) for all $x \in D(A)$; if $x \in D(A)$, we have

$$\int_0^T \left|(-A)^{1/2}e^{tA}x\right|^2 dt = \int_0^T \left|Ae^{tA}(-A)^{-1/2}x\right|^2 dt$$

$$\leq |(-A)^{-1/2}x|^2_{D_A(1/2,2)}.$$

Since $D_A(1/2,2)$ is isomorphic to $D((-A)^{1/2})$ the conclusion follows. ∎

In this case we have the following additional regularity result.

Theorem 3.3 *Assume that* $(\mathcal{H})$ *is verified and let* A *be a variational operator. Let* $P \in C_s([0,\infty[\,;H)$ *be the classical solution to* (2.7). *Then for any* $t > 0$ *and* $x \in D((-A)^{1/2})$ *we have* $P(t)x \in D((-A^\star)^{1/2})$ *and the mapping* $(-A^\star)^{1/2}P(\cdot)(-A)^{1/2}x$ *is continuous for* $t > 0$.

Proof. Let $x \in D((-A)^{1/2})$ and let P be given by (3.14), then we have

$$(P(t)(-A)^{1/2}x, (-A)^{1/2}x)$$

$$= (P_0(-A)^{1/2}e^{tA}x, (-A)^{1/2}e^{tA}x)$$
$$+ \int_0^t (E(s)(-A)^{\frac{1}{2}}e^{(t-s)A}x, (-A)^{\frac{1}{2}}e^{(t-s)A}x)\, ds$$

$$\leq \|P_0\|\ \|(-A)^{1/2}\|^2\, |e^{tA}x|^2 + \|E\| \int_0^t |(-A)^{\frac{1}{2}}e^{sA}x|^2\, ds.$$

It follows, recalling (3.15), that

$$|(P(t)(-A)^{1/2}x, (-A)^{1/2}x)| \leq \frac{c^2}{t^2}e^{2\gamma T}\|(-A)^{1/2}\|^2|x|^2 + \|E\|\, K_T\, |x|^2$$

which yields the conclusion. ∎

So when A is a variational operator generated by a continuous and coercive bilinear form a such that

$$D((-A)^{1/2}) = D((-A^\star)^{1/2}),$$

Theorem 3.3 says that for all $t > 0$ the operator $P(t)$ belongs to $\mathcal{L}(V';V)$.

4 The case of the unbounded observation

In this section we consider the case when the observation C is unbounded. More precisely we make the following assumption:

$$\mathcal{H}\text{-}(v) \begin{cases} (i)\ C \in \mathcal{L}(D(A);Y) \\ (ii)\ \exists \text{ a real continuous function } K \text{such that} \\ \qquad \displaystyle\int_0^t |Ce^{sA}x|^2 ds \le K(t)|x|^2, \quad \forall x \in D(A). \end{cases}$$

Clearly, if $\mathcal{H}$-(v) holds, then for any $t \ge 0$ there exists a linear operator $F_C(t) \in \Sigma^+(H)$ such that

$$(F_C(t)x, y) = \int_0^t (Ce^{sA}x, Ce^{sA}y)\, ds, \quad \forall x, y \in D(A). \tag{4.1}$$

Lemma 4.1 *Assume that $\mathcal{H}$-(v) is verified and let $F_C(\cdot)$ be defined by (4.1). Then $F_C \in C_s([0,+\infty[\,;\Sigma^+(H))$. Moreover, setting*

$$F_{C,n}(t)x = \int_0^t e^{sA^\star}(CJ_n)^\star CJ_n e^{sA}x\, ds, \quad x \in H,$$

where $J_n = nR(n,A)$, one has for all $T > 0$

$$\lim_{n\to\infty} F_{C,n}(\cdot) = F_C(\cdot),$$

in $C_s([0,T];\Sigma(H))$.

Proof. Let $t_0 \in [0,+\infty[$, $t > t_0$ and $x \in D(A)$, then we have

$$|\sqrt{(F(t) - F(t_0))}\, x|^2 = (F(t)x - F(t_0)x, x)$$

$$= \int_{t_0}^t |Ce^{sA}x|^2 ds \to 0, \ \text{as } t \to t_0.$$

Since $D(A)$ is dense in H we have

$$\lim_{t\to t_0} \sqrt{(F(t) - F(t_0))}\, x = 0$$

for all $x \in H$. Consequently, the first part of the lemma follows easily. The last part follows immediately since

$$F_{C,n}(t) = J_n^\star F_C J_n.$$

■

We consider now the Riccati equation in the integral form

$$\begin{aligned} P(t)x &= e^{tA^\star}P_0e^{tA}x + F_C(t)x \\ &\quad -\int_0^t e^{(t-s)A^\star}P(s)BB^\star P(s)e^{(t-s)A}x\,ds. \end{aligned} \tag{4.2}$$

and the approximating equation

$$\begin{aligned} P_n(t)x &= e^{tA_n^\star}P_0e^{tA_n}x + F_{C,n}(t)x \\ &\quad -\int_0^t e^{(t-s)A_n^\star}P_n(s)BB^\star P_n(s)e^{(t-s)A_n}x\,ds,\ n\in\mathbf{N}. \end{aligned} \tag{4.3}$$

Equation (4.3) has clearly a unique solution $P_n \in C([0,\infty[\,;\Sigma^+(H))$ which is also the solution of the problem

$$\begin{cases} P_n' = A_n^\star P_n + P_nA_n - P_nBB^\star P_n + (CJ_n)^\star CJ_n. \\ P_n(0) = P_0. \end{cases}$$

We now prove the following theorem.

Theorem 4.1 *Assume that* $(\mathcal{H})$*-(i)-(ii)-(iii)-(v) are verified. Then equation* (4.2) *has a unique solution* $P \in C_s([0,\infty[\,;\Sigma^+(H))$ *and*

$$\lim_{n\to\infty} P_n(\cdot) = P(\cdot), \ \ \text{in } C_s([0,T];\Sigma(H)), \quad \forall\, T>0. \tag{4.4}$$

Proof. The proof is similar to the one of Theorem 2.1, with minor differences. Fix $T>0$, set

$$L_T = \sup_{t\in[0,T],\,n\in\mathbf{N}} \|F_{C,n}(t)\|, \quad r = 2(M_T^2\|P_0\| + L_T),$$

and choose $\tau\in[0,T]$ such that

$$\tau r^2 M_T^2\|B\|^2 \le M_T^2\|P_0\| + L_T, \quad 2\tau r M_T^2 \le \frac{1}{2}.$$

Now set

$$\begin{aligned} \gamma(P)(t)x &= e^{tA^\star}P_0e^{tA}x + F_C(t)x \\ &\quad -\int_0^t e^{(t-s)A^\star}P(s)BB^\star P(s)e^{(t-s)A}x\,ds, \end{aligned}$$

and let $\gamma_n(P)$ be defined in a similar way. Then γ and γ_n are $\frac{1}{2}$-contractions on the ball

$$\{P \in C_s([0,\tau];H) : \|P(t)\| \le r, \forall\, t \in [0,\tau]\}.$$

Thus there exists P (resp. P_n) such that $\gamma(P) = P$, (resp. $\gamma_n(P_n) = P_n$) and P, (resp. P_n) is the unique solution of (4.2) (resp. (4.3)) in $[0,\tau]$. Now, by Lemma 2.1 it follows that

$$\lim_{n\to\infty} P_n(\cdot) = P(\cdot), \text{ in } C_s([0,\tau];\Sigma(H)).$$

This yields $P(t) \ge 0, \forall\, t \in [0,\tau]$. From now on the proof is completely similar to the one of Theorem 2.1. ∎

4.1 The analytic case

We choose once and for all $\lambda_0 > \omega$ and set $A_{\lambda_0} = \lambda_0 - A$. We assume besides $(\mathcal{H})$-(i)-(ii)-(iii)

$(\mathcal{H})$-(vi) *A is the infinitesimal generator of an analytic semigroup,*

$(\mathcal{H})$-(vii) *there exists $\alpha \in [0, \frac{1}{2}[$ such that $C \in \mathcal{L}(D(A^{\alpha}_{\lambda_0});Y)$.*

We remark that $(\mathcal{H})$-(vi)-(vii) imply that there exists a constant $K_\alpha > 0$ such that (see Volume I, Chapter 1, §5, Thm 5.2)

$$\|Ce^{tA}\| \le K_\alpha e^{\omega t} t^{-\alpha}, \quad t \ge 0.$$

It follows that $(\mathcal{H})$-(v) holds true and so we have the following straightforward generalization of Theorem 2.1.

Proposition 4.1 *Assume that $(\mathcal{H})$-(i)-(ii)-(iii)-(vi)-(vii) are verified. Then the Riccati equation (2.7) has a unique mild solution $P \in C_s([0,\infty[\,;\Sigma^+(H))$.*

4.2 The variational case

As in §3.3. we are given a variational operator A which is V-H coercive for some $\alpha > 0$ and $\lambda \in \mathbf{R}$. We assume that V is isomorphic to $D((-A)^{\frac{1}{2}})$ and to $D((-A^\star)^{\frac{1}{2}})$. We assume in addition that

$(\mathcal{H})$-(viii) $C \in \mathcal{L}(V;Y)$.

In fact this framework corresponds to the limit case of the situation considered in §4.1, with $\alpha = \frac{1}{2}$. If $(\mathcal{H})$-(viii) holds, then by Lemma

3.3 it follows that $(\mathcal{H})$-(v) is also fulfilled. Thus we have the final result.

Proposition 4.2 *Assume that* $(\mathcal{H})$-*(i)-(ii)-(iii)-(viii) are verified. Then the Riccati equation* (2.7) *has a unique mild solution* $P \in C_s([0,\infty[\,;\Sigma^+(H))$.

5 The case when A generates a group

We shall denote by $\mathcal{L}_r(H)$ the set of elements in $\mathcal{L}(H)$ that have a continuous inverse. It is well known that $\mathcal{L}_r(H)$ is open in $\mathcal{L}(H)$.

In this section we assume $(\mathcal{H})$ and in addition that A is the infinitesimal generator of a strongly continuous group of operators. This is equivalent to say that A fulfils $(\mathcal{H})$-(i) and $-A$ generates a C_0 semigroup e^{-tA}. Besides Riccati equation (2.7) we shall consider the following

$$\begin{cases} Q' = -AQ - QA^\star - QC^\star CQ + BB^\star \\ \\ Q(0) = Q_0. \end{cases} \tag{5.1}$$

We shall show that, as proved in F. Flandoli [2], when $Q_0 = P_0^{-1}$, then we have $Q(t) = P^{-1}(t)$. We first consider the case when A is bounded.

Proposition 5.1 *Assume that* $(\mathcal{H})$ *is verified, that* $A \in \mathcal{L}(H)$ *and that* $P_0 \in \mathcal{L}_r(H)$. *Let* $P(\cdot)$ *be the mild solution of* (2.7) *and* $Q(\cdot)$ *the mild solution of* (5.1) *with* $Q_0 = P_0^{-1}$. *Then* $P(t) \in \mathcal{L}_r(H)$ *for all* $t > 0$ *and we have*

$$P(t)^{-1} = Q(t), \quad \forall\, t \geq 0.$$

Proof. Set $\Lambda = PQ - I$; as easily checked we have

$$\begin{cases} \Lambda' = (A^* - PBB^*)\Lambda - \Lambda(A^* - C^*CQ) =: L(\Lambda) \\ \\ \Lambda(0) = 0. \end{cases}$$

Since L is a linear bounded operator in $\mathcal{L}(H)$ we have $\Lambda(t) = 0$, $\forall\, t \geq 0$ and so $P(t)Q(t) = I$, $\forall\, t \geq 0$. In a similar way one shows that $Q(t)P(t) = I$, $\forall\, t \geq 0$. ∎

We prove now the main result of this section.

Theorem 5.1 *Assume that* $(\mathcal{H})$ *is verified, that* A *generates a strongly continuous group in* H *and that* $P_0 \in \mathcal{L}_r(H)$. *Let* $P(\cdot)$ *be the mild solution of* (2.7) *and* $Q(\cdot)$ *the mild solution of* (5.1) *with* $Q_0 = P_0^{-1}$. *Then* $P(t) \in \mathcal{L}_r(H)$ *for all* $t > 0$ *and we have*

$$P(t)^{-1} = Q(t), \quad \forall t \geq 0.$$

Proof. Denote by P_n and Q_n the solutions of problems (2.7) and (5.1) with A replaced by the Yosida approximants A_n, $n \in \mathbf{N}$. By Theorem 2.1 we have:

$$\lim_{n\to\infty} P_n = P, \quad \lim_{n\to\infty} Q_n = Q$$

in $C_s([0,T];\Sigma^+(H))$ for all $T > 0$. By the previous proposition it follows that

$$P_n^{-1}(t) = Q_n(t), \quad \forall t \geq 0.$$

Letting n tend to infinity on the equality

$$P_n(t)Q_n(t)x = Q_n(t)P_n(t)x = x, \quad x \in H,$$

the conclusion follows. ∎

6 The linear quadratic control problem with finite horizon

6.1 The main result

In this section we consider the control problem (1.1)-(1.2). We assume that $(\mathcal{H})$ is verified and denote by $P \in C_s([0,\infty[;\Sigma^+(H))$ the mild solution of the Riccati equation (2.7). We first consider the closed loop equation

$$\begin{cases} x'(t) = Ax(t) - BB^\star P(T-t)x(t), & t \in [0,T] \\ x(0) = x_0 \in H \end{cases} \tag{6.1}$$

Proposition 6.1 *Assume that* $(\mathcal{H})$ *is verified and let* $x_0 \in H$. *Then equation* (6.1) *has a unique mild solution* $x \in C([0,T];H)$.

Proof. The conclusion follows from the Proposition 3.6 (Volume I, Chapter 1, §3). ∎

We now prove a basic identity.

Proposition 6.2 *Assume that* $(\mathcal{H})$ *is verified and let* $u \in L^2(0,T;U)$, $x_0 \in H$. *Let* x *be the mild solution of the state equation* (1.1) *and let* P *be the mild solution of Riccati equation* (2.7). *Then the following identity holds*

$$J(u) = \int_0^T |u(s) + B^\star P(T-s)x(s)|^2\, ds + (P(T)x_0, x_0). \qquad (6.2)$$

Proof. Let P_n be the solution to (2.10) and let x_n be the solution to the problem

$$\begin{cases} x_n'(t) = A_n x_n(t) + Bu(t), & t \in [0,T] \\ x_n(0) = x_0 \in H, \end{cases}$$

where the A_n's are the Yosida approximations of A. We follow here a classical argument, see for instance R.V. Brockett [1]. By computing the derivative

$$\frac{d}{ds}(P_n(T-s)x_n(s), x_n(s))$$

and completing the squares, we obtain the identity

$$\frac{d}{ds}(P_n(T-s)x_n(s), x_n(s))$$
$$= |u(s) + B^\star P_n(T-s)x_n(s)|^2 - |Cx_n(s)|^2 - |u(s)|^2.$$

Integrating from 0 to T and letting n tend to infinity we obtain (6.2). ∎

We are now ready to prove the following result.

Theorem 6.1 *Assume that* $(\mathcal{H})$ *is verified and let* $x_0 \in H$. *Then there exists a unique optimal pair* $(u^\star, x^\star)$. *Moreover the following statements hold:*

(i) $x^\star \in C([0,T];H)$ *is the mild solution to the closed loop equation* (6.1),

(ii) $u^\star \in C([0,T];U)$ *is given by the feedback formula*

$$u^\star(t) = -B^\star P(T-t)x^\star(t), \quad t \in [0,T], \tag{6.3}$$

(iii) The optimal cost $J(u^\star)$ *is given by*

$$J(u^\star) = (P(T)x_0, x_0). \tag{6.4}$$

Proof. We first remark that by identity (6.2) it follows that

$$J(u) \geq (P(T)x_0, x_0), \tag{6.5}$$

for any control $u \in L^2(0,T;U)$. Let now $x^\star$ be the mild solution to (6.1) and let $u^\star$ be given by (6.3). Setting in (6.2) $u = u^\star$ and taking into account (6.5) it follows that $(u^\star, x^\star)$ is an optimal pair and that (6.4) holds. It remains to prove uniqueness. Let $(\overline{u}, \overline{x})$ be another optimal pair. Setting in (6.2), $u = \overline{u}$ and $x = \overline{x}$ we obtain

$$\int_0^T |\overline{u}(s) + B^\star P(T-s)\overline{x}(s)|^2 \, ds = 0,$$

so that $\overline{u}(s) = -B^\star P(T-s)\overline{x}(s)$ for almost every s in $[0,T]$. But this implies that $\overline{x}$ is a mild solution of (6.1) so that $\overline{x} = x^\star$ and, consequently, $\overline{u} = u^\star$. ∎

6.2 The case of unbounded observation

We now consider the case of unbounded observation, assuming that $(\mathcal{H})$-(i)-(ii)-(iii)-(v) are verified. It is convenient to introduce the linear operator:

$$\xi \mapsto L_T(\xi) : D(A) \to L^2(0,T;H)$$

where $(L_T\xi)(t) = Ce^{tA}\xi$. Since, by $(\mathcal{H})$-(v)

$$\|L_T\xi\|^2_{L^2(0,T;H)} \leq K(T)|\xi|^2,$$

L_T has an extension to the whole space H, which will still be denoted by L_T.

Let $x_0 \in H$, $u \in L^2(0,T;U)$ and let x be the corresponding solution of (1.1). We want to define the cost functional $J(u)$; this is not a priori defined because $x(t)$ does not necessarily belong to $D(C)$ and the term $\int_0^T |Cx(t)|^2 dt$ is not well-defined. We set

$$Cx(t) = (L_T x_0)(t) + \int_0^t L_T(Bu(s))(t-s)\,ds.$$

This definition is meaningful in virtue of the following lemma.

Lemma 6.1 *Let $z \in L^2(0,T;H)$, and set*

$$w(t) = \int_0^t L_T(z(s))(t-s)\,ds.$$

Then $w \in L^2(0,T;H)$ and the following estimate holds

$$\|w\|^2_{L^2(0,T;H)} \leq K(T)\,\|z\|^2_{L^2(0,T;H)}.$$

Proof. It is sufficient to verify the estimate for $z \in L^2(0,T;D(A))$; in this case we have

$$w(t) = \int_0^t Ce^{(t-s)A}z(s)\,ds.$$

Let $\varphi \in L^2(0,T;H)$, then we have

$$\int_0^T (w(t),\varphi(t))\,dt = \int_0^T ds \int_s^T (Ce^{(t-s)A}z(s),\varphi(t)\,)\,dt.$$

It follows that

$$\begin{aligned}
&\left|\int_0^T (w(t),\varphi(t))\,dt\right| \\
&\leq \int_0^T ds \left[\int_s^T |Ce^{(t-s)A}z(s)|^2 dt\right]^{\frac{1}{2}} \left[\int_s^T |\varphi(t)|^2 dt\right]^{\frac{1}{2}} \\
&\leq K(T)\,\|\varphi\|^2_{L^2(0,T;H)}\,\|z\|^2_{L^2(0,T;H)}.
\end{aligned}$$

Since this is true for any φ, we obtain the estimate. ■

Now the following result is proved as Theorem 6.1

Theorem 6.2 *Assume that* $(\mathcal{H})$*-(i)-(ii)-(iii)-(v) are verified and let* $x_0 \in H$*. Then there exists a unique optimal pair* $(u^\star, x^\star)$*. Moreover the following statements hold:*

(i) $x^\star \in C([0,T];H)$ *is the mild solution to the closed loop equation* (6.1),

(ii) $u^\star \in C([0,T];U)$ *is given by the feedback formula* (6.3),

(iii) the optimal cost $J(u^\star)$ *is given by* (6.4).

6.3 Regularity properties of the optimal control

We give here some regularity results for the optimal pair $(u^\star, x^\star)$.

Proposition 6.3 *Assume that* $(\mathcal{H})$ *is verified. Let* $P_0 \in D(\mathcal{A})$ *and* $x \in D(A)$*, where* $\mathcal{A}$ *is the linear operator defined by* (3.3)*. Then*

$$x^\star \in C^1([0,T];H) \cap C([0,T];D(A)), \quad u^\star \in C^1([0,T];H).$$

Proof. By Proposition 3.2 we know that the mild solution P to the Riccati equation (2.7) is a strict solution, so that

$$P \in C_s^1([0,T];\Sigma(H)) \cap C_s([0,T];D(\mathcal{A})).$$

Now the assertion concerning $x^\star$ follows from Proposition 3.3 (Volume I, Chapter 1). ∎

Proposition 6.4 *Assume that* $(\mathcal{H})$ *is verified and that* e^{tA} *is an analytic semigroup. Then the following statements hold:*

(i) for any $\varepsilon \in]0, \frac{T}{2}[$ *,* $x^\star$ *(resp.* $u^\star$*) belongs to* $C^\infty(]\varepsilon, T-\varepsilon[;H)$ *(resp.* $C^\infty(]\varepsilon, T-\varepsilon[;U)$*).*

(ii) $u^\star$ *and* $x^\star$ *are analytic in* $]0,T[$*.*

Proof. (i) We first remark that, by (6.3) $u^* \in C([0,T];U)$. Moreover, since $(x^\star)' = Ax^\star + Bu^\star$ and $x^\star(0) = x_0$, we have, by Proposition 3.10 (Volume I, Chapter 1), $x^\star \in C^\alpha([\varepsilon,T];H)$ (for any $\alpha \in]0,1[$) and $\varepsilon \in [0,T]$. By (6.3) and Theorem 3.1 we have $u^\star \in C^\alpha([\varepsilon,T-\varepsilon];U)$. Then, by Proposition 3.9 (Volume I, Chapter 1), $x^\star$ belongs to $C^1([\varepsilon,T-\varepsilon];H)$. By iterating this argument the assertion (i) follows. Finally (ii) is a consequence of Theorem 3.2. ∎

6.4 Hamiltonian systems

Assume that $(\mathcal{H})$ is verified, let P be the mild solution of (2.7) and let $(u^\star, x^\star)$ be an optimal pair. In (6.1) it is useful to introduce the following function

$$p^\star(t) = P(T-t)x^\star(t)$$

called the *adjoint variable*. If we differentiate $p^\star(t)$, assuming for a while that the conditions of Proposition 6.3 hold, we easily deduce from (2.7) and (6.1) that $p^\star(t)$ satisfies the equations

$$\begin{cases} -p'(t) = A^\star p(t) + C^\star Cx(t) \\ \\ P(T) = P_0 x(T) \end{cases} \tag{6.6}$$

and thus considering (6.6) togheter with

$$\begin{cases} x'(t) = Ax(t) - BB^\star p(t) \\ \\ x(0) = x_0 \end{cases} \tag{6.7}$$

we obtain a system (the two point boundary value problem) whose solution is $(x^\star, p^\star)$. The system (6.6)–(6.7) is called the *Hamiltonian system*. Once we have established (6.6)–(6.7) for regular x_0, P_0, it is easy to extend it to the general case.

Remark 6.1 One can notice the analogy between (6.6)–(6.7) and (2.24), which was the source of the explicit formula (2.22) for $P(t)$.

∎

7 Some generalizations and complements

7.1 Non homogeneous state equation

Consider the following optimal control problem: to minimize

$$J(u) = \int_0^T \left\{ |Cx(t)|^2 + |u(t)|^2 \right\} dt + (P_0 x(T), x(T)) \tag{7.1}$$

over all controls $u \in L^2(0,T;U)$ subject to the differential equation constraint

$$\begin{cases} x'(t) = Ax(t) + f(t) + Bu(t), \quad t \in [0,T] \\ \\ x(0) = x_0 \in H. \end{cases} \tag{7.2}$$

We assume that hypothesis $(\mathcal{H})$ holds and that $f \in L^2(0,T;H)$. Moreover we denote by P the mild solution of the Riccati equation (2.7). We want to show that it is possible to generalize the Dynamic Programming approach to this more general situation. The main difference is the introduction of a dual variable r, defined by the following backward Cauchy problem: for all t in $[0,T]$

$$\begin{cases} r'(t) + (A^\star - P(T-t)BB^\star)r(t) + P(T-t)f(t) = 0, \\ \\ r(T) = 0. \end{cases} \tag{7.3}$$

Notice that, by the change of variable $t \to T - t$, problem (7.3) reduces to an initial value problem which has a unique mild solution.

The following identity generalizes identity (6.2).

Lemma 7.1 *Let $x_0 \in H$, $f \in L^2(0,T;H)$, $u \in L^2(0,T;U)$. Then we have*

$$\begin{aligned} J(u) &= (P(T)x_0, x_0) + 2(r(0), x_0) \\ &\quad + \int_0^T \left\{2(r(s), f(s)) - |B^\star r(s)|^2\right\} ds \\ &\quad + \int_0^T |u(s) + B^\star r(s) + B^\star P(T-s)x(s)|^2\, ds, \end{aligned} \tag{7.4}$$

where x and r are respectively the mild solutions of (7.2) *and* (7.3).

Proof. Let P_n be the solution of (2.10) and let x_n and r_n be the solutions of the problems

$$\begin{cases} x_n'(t) = A_n x_n(t) + f(t) + Bu(t), \\ x_n(0) = x_0 \end{cases}$$

$$\begin{cases} r_n'(t) = -(A_n^\star - P_n(T-t)BB^\star)r_n(t) - P_n(T-t)f(t), \\ r_n(T) = 0. \end{cases}$$

Then, by integrating the identity

$$\frac{d}{ds}\{(P_n(T-s)x_n(s), x_n(s)) + 2(r_n(s), x_n(s))\}$$

$$= |B^\star P_n(T-s)x_n(s) + B^\star r_n(s) + u(s)|^2$$

$$+2(r_n(s), f(s)) - |B^\star r_n(s)|^2 - |Cx_n(s)|^2 - |u(s)|^2$$

between 0 and T and by letting n tend to infinity we obtain (7.4). ∎

By using identity (7.4) we can easily generalize Theorem 6.1. The proof is similar and will be omitted.

Theorem 7.1 *Assume that* $(\mathcal{H})$ *is verified, let* $x_0 \in H$ *and* $f \in L^2(0,T;H)$. *Then there exists a unique optimal pair* $(u^\star, x^\star)$ *for problem* (7.1)–(7.2). *Moreover the following statements hold:*
(i) $x^\star$ *is the mild solution to the closed loop equation*

$$\begin{aligned} &x'(t) = (A - BB^\star P(T-t))x(t) - BB^\star r(t) + f(t), \quad t \in [0,T], \\ &x(0) = x_0, \end{aligned}$$

(ii) $u^\star$ *is given by the feedback formula*

$$u^\star(t) = -B^\star[P(T-t)x^\star(t) + r(t)],$$

where r *is the mild solution to* (7.3),
(iii) the optimal cost is given by

$$J(u^\star) = (P(T)x_0, x_0) + 2(r(0), x_0) + \int_0^T [(2r(s), f(s)) - |B^\star r(s)|^2] ds.$$

Remark 7.1 *Tracking Problem*
Consider the following optimal control problem: to minimize

$$\begin{aligned} K(u) = & \int_0^T \left\{|Cx(t) - \xi(t)|^2 + |u(t)|^2\right\} dt \\ & +(P_0(x(T) - \xi(T)), (x(T) - \xi(T))) \end{aligned} \tag{7.5}$$

over all controls $u \in L^2(0,T;U)$ subject to the differential equation constraint

$$\begin{cases} z'(t) = Az(t) + Bu(t), \quad t \in [0,T] \\ \\ z(0) = z_0 \in H. \end{cases} \tag{7.6}$$

Here $\xi \in W^{1,2}([0,T];H) \cap L^2(0,T;D(A))$ is a given function. Now, by setting

$$x = z - \xi,\ f = A\xi - \xi',\ x_0 = z_0 - x(0),$$

equation (7.6) reduces to equation (7.2) and problem (7.5)-(7.6) to problem (7.1)-(7.2). ∎

7.2 Time dependent state equation and cost function

Consider the system

$$\begin{cases} x'(t) = A(t)x(t) + B(t)u(t) \\ \\ x(0) = x_0 \end{cases} \tag{7.7}$$

where $A(t) : D(A(t)) \subset H \to H$ and $B(t) \in \mathcal{L}(U;H)$ are linear operators.

We make the following assumptions on the families $\{A(t)\}_{t\in[0,T]}$ and $\{B(t)\}_{t\in[0,T]}$:

(i) $A(t) : D(A(t)) \subset H \to H$ *generates a* C_0 *semigroup in* H.
(ii) there exists a strongly continuous mapping

$$U_A(\cdot,\cdot) : \{(t,s) \in \mathbf{R}^2 : t \geq s\} \to \mathcal{L}(H)$$

such that $U_A^*(\cdot,\cdot)$ *is also strongly continuous and*

$$\frac{\partial}{\partial t} U_A(t,s)x = A(t)U_A(t,s)x,\ U_A(s,s)x = x,$$

$\forall\, x \in D(A(t)),\ 0 \leq s \leq t \leq T.$
(iii) we have $\lim_{n\to\infty} U_{A_n}(t,s)x = U_A(t,s)x, \forall\, x \in H$ *uniformly on the bounded sets of* $\{(t,s) \in \mathbf{R}^2 : t \geq s\}$, *where* $U_{A_n}(t,s)$ *is the evolution operator generated by the Yosida approximations of* $A(t)$.
(iv) $B(\cdot)u$ *is continuous for all* $u \in U$.

(7.8)

Under these assumptions problem (7.7) has a unique mild solution given by:

$$x(t) = U(t,0)x_0 + \int_0^t U(t,s)Bu(s)\,ds.$$

Assumptions (7.8) are verified in many problems both parabolic and hyperbolic (see for instance P. ACQUISTAPACE and B. TERRENI [1], A. LUNARDI [1], A. PAZY [1] and H. TANABE [1]).

We want to minimize the *cost function*

$$J(u) = \int_0^T \left\{|C(s)x(s)|^2 + |u(s)|^2\right\} ds + (P_0 x(T), x(T))$$

over all controls $u \in L^2(0,T;U)$ subject to the differential equation constraint (7.7). Concerning the operators B, C and P_0 we shall assume that

$$\begin{array}{l} P_0 \in \Sigma^+(H), \quad C \in C_s([0,T];\mathcal{L}(H;Y)), \\ B \in C_s([0,T];\mathcal{L}(U;H)). \end{array} \tag{7.9}$$

It is convenient to consider the backward Riccati equation

$$\left\{\begin{array}{l} Q' + A^\star Q + QA - QBB^\star Q + C^\star C = 0 \\ \\ Q(T) = P_0. \end{array}\right. \tag{7.10}$$

Its corresponding mild form is

$$\begin{aligned} Q(t)x \quad &= \quad U^\star(T,t)P_0U(T,t)x + \int_t^T U^\star(s,t)C^\star(s)C(s)U(s,t)x\,ds \\ &\quad - \int_t^T U^\star(s,t)Q(s)B(s)B^\star(s)Q(s)U(s,t)x\,ds, \quad x \in H, \end{aligned}$$

where $U^\star(s,t)$ is the adjoint of $U(s,t)$.

The following theorem is proved as Theorem 2.1 .

Theorem 7.2 *Assume* (7.8) *and* (7.9). *Then problem* (7.10) *has a unique mild solution* $Q \in C_s([0,\infty[;\Sigma^+(H))$ *and*

$$\lim_{n\to\infty} Q_n = Q \text{ in } C_s([0,\infty[;\Sigma(H)),$$

for any $T > 0$, *where* Q_n *is the solution to problem* (7.10) *with* $A(t)$ *replaced by* $A_n(t)$.

The following monotonicity property of the solutions of Riccati equations is proved as Proposition 2.2

Proposition 7.1 *Consider the Riccati equations*

$$\begin{cases} Q_i' + A_i^\star Q_i + Q_i A_i - Q_i B_i B_i^\star Q_i + C_i^\star C_i = 0 \\ \\ Q_i(T) = Q_{i,0}, \quad i = 1,2. \end{cases} \tag{7.11}$$

Assume that $(A_i, B_i, C_i, P_{i,0})$ *verify* (7.8) *and* (7.9) *for* $i = 1, 2$, *and that, in addition,*

$$P_{1,0} \le P_{2,0},$$
$$C_1^\star(t)C_1(t) \le C_2^\star(t)C_2(t),\ B_2(t)B_2^\star(t) \le B_1(t)B_1^\star(t), \quad t \in [0,T].$$

Then we have

$$P_1(t) \le P_2(t), \quad t \ge 0. \tag{7.12}$$

By using Theorem 7.2 , we can easily generalize Theorem 6.1.

Theorem 7.3 *Assume* (7.8) *and* (7.9) *and let* $x_0 \in H$. *Then there exists a unique optimal pair* $(u^\star, x^\star)$ *and* $u^\star \in C([0,T]; U)$ *is related to* $x^\star$ *by the feedback formula*

$$u^\star(t) = -B^\star(t)Q(t)x^\star(t), \quad t \in [0,T]$$

Finally,the optimal cost $J(u^\star)$ *is given by*

$$J(u^\star) = (Q(0)x_0, x_0).$$

Remark 7.2 For the time-varying variational case the reader is referred to the book of J.-L. LIONS [1]. ■

7.3 Dual Riccati equation

We assume here that $(\mathcal{H})$ is verified and consider the Riccati equation

$$\begin{cases} P' = A^*P + PA - PBB^*P + C^*C, \quad t \ge 0 \\ \\ P(0) = P_0 \end{cases} \tag{7.13}$$

and the dual equation

$$\begin{cases} Q' = AQ + QA^* - QC^*CQ + BB^*, & t \geq 0 \\ Q(0) = Q_0, \end{cases} \tag{7.14}$$

where $Q_0 \in \Sigma^+(H)$. Clearly the set (A^*, C^*, B^*, Q_0) also verifies assumption $(\mathcal{H})$, thus the equations (7.13) and (7.14) have unique solutions $P, Q \in C_s([0, +\infty[\,; \Sigma^+(H))$ respectively.

We want to present here a simple formula which gives P in terms of Q. This formula will turn out to be useful in Chapter 3 below and it was proved in V. Barbu and G. Da Prato [1], by variational methods.

Fix $T > 0$ and set

$$\tilde{Q}_T(s) = Q(T-s), \quad s \in [0, T]. \tag{7.15}$$

and

$$G_T(s) = A^* - C^*CQ(T-s), \quad s \in [0, T]; \tag{7.16}$$

we shall denote by U_{G_T} the evolution operator associated with G_T.

We first prove a lemma.

Lemma 7.2 *Fix $T > 0$ and set*

$$Z_T(t) = [I + P(t)\tilde{Q}_T(t)]^{-1}P(t) = P(t)[I + \tilde{Q}_T(t)P(t)]^{-1}.$$

Then for any $x \in H$ we have

$$\begin{aligned} Z_T(t)x = \; & U_{G_T}(t,0)Z_T(0)U^*_{\tilde{G}_T}(t,0)x \\ & + \int_0^t U_{G_T}(t,s)C^*CU^*_{\tilde{G}_T}(t,s)x\, ds. \end{aligned} \tag{7.17}$$

Proof. We assume that A is bounded, otherwise we replace A by its Yosida approximations A_n and then we let n tend to infinity. We first remark that the definition of Z_T is meaningful by Proposition 1.1 in Appendix A. Setting for simplicity $Z = Z_T$ and $\tilde{Q} = \tilde{Q}_T$ we have.

$$(I + P\tilde{Q})Z = P$$

and so

$$(P'\tilde{Q} + P\tilde{Q}')Z + (I + P\tilde{Q})Z' = P'$$

which implies

$$\begin{aligned}(I+P\tilde{Q})Z' &= P' - (P'\tilde{Q} + P\tilde{Q}')[I+P\tilde{Q}]^{-1}P \\ &= P' - (P'\tilde{Q} + P\tilde{Q}')P[I+\tilde{Q}P]^{-1} \\ &= [P'(I+\tilde{Q}P) - (P'\tilde{Q} + P\tilde{Q}')P][I+\tilde{Q}P]^{-1} \\ &= [P' - P\tilde{Q}'P][I+\tilde{Q}P]^{-1}.\end{aligned}$$

It follows that

$$(I+P\tilde{Q})Z'[I+\tilde{Q}P] = P' - P\tilde{Q}'P.$$

Then, by substituting P' and Q' with the expressions given by (7.13) and (7.14), we have

$$\begin{aligned}&[I+P\tilde{Q}]Z'[I+\tilde{Q}P] \\ &= A^*P + PA - PBB^*P + C^*C \\ &\quad + P[A\tilde{Q} + \tilde{Q}A^* - \tilde{Q}C^*C\tilde{Q} + BB^*]P \\ &= [I+P\tilde{Q}](A^* - C^*C\tilde{Q})P + P(A - \tilde{Q}C^*C)[I+\tilde{Q}P] \\ &\quad + [I+P\tilde{Q}]C^*C[I+\tilde{Q}P].\end{aligned}$$

Therefore

$$Z' = (A^* - C^*C\tilde{Q})Z + Z(A - \tilde{Q}C^*C) + C^*C$$

and the conclusion follows. ∎

Theorem 7.4 *Assume that* $(\mathcal{H})$ *is verified, let* P *and* Q *be the respective solutions in* $C_s([0,+\infty[\,;\Sigma^+(H))$ *of equations* (7.13) *and* (7.14) *with* $Q_0 = 0$. *Let* $T > 0$ *and set*

$$G_T(t) = A^* - C^*C\tilde{Q}_T(t) = A^* - C^*C\,Q(T-t), \quad t \in [0,T].$$

Then we have

$$\begin{aligned}P(T) &= U_{G_T}(T,0)[I + P_0Q(T)]^{-1}P_0U^*_{G_T}(T,0) \\ &\quad + \int_0^T U_{G_T}(T,s)C^*CU^*_{G_T}(T,s)\,ds.\end{aligned} \tag{7.18}$$

Proof. Since
$$Z_T(0) = [I + P_0 Q(T)]^{-1} P_0$$
and $Z_T(T) = P(T)$, the conclusion follows from (7.17). ∎

8 Examples of controlled systems

8.1 Parabolic equations

Let Ω be an open bounded set of $\mathbf{R}^n$ with regular boundary $\partial\Omega$. Consider the state equation

$$\begin{cases} \dfrac{\partial x}{\partial t}(t,\xi) = (\Delta_\xi + c)x(t,\xi) + (Bu(t,\cdot))(\xi), \text{ in }]0,T] \times \Omega \\ x(t,\xi) = 0 \text{ on }]0,T] \times \partial\Omega \\ x(0,\xi) = x_0(\xi) \text{ in } \Omega, \end{cases} \tag{8.1}$$

where $c \in \mathbf{R}$ and $B \in \mathcal{L}(L^2(\Omega))$. We choose $H = U = Y = L^2(\Omega)$ as space of states, controls and observations. The control u is said to be a *distributed control.* We denote by A the linear self-adjoint operator in H:

$$\begin{cases} Ax = \Delta_\xi x + cx, \quad \forall\, x \in D(A) \\ D(A) = H^2(\Omega) \cap H_0^1(\Omega). \end{cases}$$

Setting $x(t) = x(t,\cdot), u(t) = u(t,\cdot)$ we can write problem (8.1) in the abstract form (1.1). By Proposition 2.11 (Volume I, Chapter 1), A is the infinitesimal generator of an analytic semigroup in H. Let C and P_0 be non negative linear bounded operators in $L^2(\Omega)$. Consider the following problem: to minimize

$$\begin{aligned} J(u) &= \int_0^T \int_\Omega \left\{ |(Cx(t,\cdot))(\xi)|^2 + |u(t,\xi)|^2 \right\} dt d\xi \\ &+ \int_\Omega (P_0 x(T,\cdot)(\xi) x(T,\xi) d\xi \end{aligned}$$

subject to state equation (8.1). We remark that assumption $(\mathcal{H})$ is fulfilled. Due to Theorem 2.1, the Riccati equation (2.7) has a unique solution $P \in C_s([0,T];\Sigma^+(H))$. Then, by Theorem 6.1, there exists a unique optimal pair $(u^\star, x^\star)$, where $x^\star \in C([0,T];L^2(\Omega))$ is the classical solution in $]0,T[$ to the closed loop equation

$$\begin{cases} \dfrac{\partial x}{\partial t}(t,\xi) = (\Delta_\xi + c)x(t,\xi) - (BB^\star(P(T-t)x(t,\cdot))(\xi), \\ \qquad\qquad\qquad\qquad\qquad\qquad \text{in }]0,T] \times \Omega \\ x(t,\xi) = 0 \;\text{ on }\;]0,T] \times \partial\Omega \\ x(0,\xi) = x_0(\xi) \;\text{ in }\; \Omega \end{cases}$$

that is,

$$\forall \varepsilon \in]0,T[, \quad \frac{\partial x}{\partial t} \text{ and } \frac{\partial^2 x}{\partial \xi^2} \in L^2([\varepsilon, T-\varepsilon] \times \Omega).$$

Moreover $u^\star$ is given by the formula

$$u^\star(t,\xi) = -(B^\star(P(T-t)x^\star(t,\cdot))(\xi).$$

Consider now a problem with unbounded observation: to minimize

$$\begin{aligned} J(u) &= \int_0^T \int_\Omega \left\{ |(\nabla_\xi x(t,\xi)|^2 + |u(t,\xi)|^2 \right\} dt d\xi \\ &\quad + \int_\Omega |x(T,\xi)|^2 d\xi \end{aligned}$$

subject to state equation (8.1). In this case the linear operator C is given by

$$Cx = -\sqrt{\Delta_\xi x}$$

and Proposition 4.2 can be applied. ∎

8.2 Wave equation

Let Ω be as in the previous example and consider the problem:

$$\begin{cases} \dfrac{\partial^2 x}{\partial t^2}(t,\xi) = \Delta_\xi x(t,\xi) + (Bu(t,\cdot))(\xi), \quad \text{in }]0,T] \times \Omega \\ x(t,\xi) = 0 \ \text{ on }]0,T] \times \partial\Omega \\ x(0,\xi) = x_0(\xi), \ \text{ in } \Omega \\ \dfrac{\partial x}{\partial t}(0,\xi) = x_1(\xi) \ \text{ in } \Omega. \end{cases} \tag{8.2}$$

Set

$$H = Y = H_0^1(\Omega) \oplus L^2(\Omega), \quad U = L^2(\Omega) \tag{8.3}$$

and denote by $X = \begin{bmatrix} x^0 \\ x^1 \end{bmatrix}$, the generic element of H. The inner product in H is defined by

$$\left(\begin{bmatrix} x^0 \\ x^1 \end{bmatrix}, \begin{bmatrix} z^0 \\ z^1 \end{bmatrix} \right) = \int_\Omega (\nabla_\xi x^0 \cdot \nabla_\xi z^0 + x^1 z^1)\, d\xi. \tag{8.4}$$

Let Λ be the self-adjoint positive operator on $L^2(\Omega)$ defined by

$$D(\Lambda) = H^2(\Omega) \cap H_0^1(\Omega), \quad \Lambda x = -\Delta_\xi x. \tag{8.5}$$

Then we have

$$(X, Z) = (\sqrt{\Lambda} x^0, \sqrt{\Lambda} z^0) + (x^1, z^1).$$

Define the linear operator A on H:

$$\begin{cases} AX = \begin{bmatrix} 0 & 1 \\ -\Lambda & 0 \end{bmatrix} \begin{bmatrix} x^0 \\ x^1 \end{bmatrix}, \quad \forall\, X \in D(A) \\ D(A) = H^2(\Omega) \cap H_0^1(\Omega) \oplus H_0^1(\Omega) \end{cases} \tag{8.6}$$

We have $A^\star = -A$ and, by Proposition 2.12 (Volume I, Chapter 1) we know that A is the infinitesimal generator of a contraction group in H. Let finally $B \in \mathcal{L}(U; H)$ be defined as

$$Bu = \begin{bmatrix} 0 \\ u \end{bmatrix}, \quad u \in U,$$

then $B^\star \in \mathcal{L}(H,U)$ is given by

$$B^\star \begin{bmatrix} x^0 \\ x^1 \end{bmatrix} = x^1, \quad \begin{bmatrix} x^0 \\ x^1 \end{bmatrix} \in H,$$

and we have

$$BB^\star \begin{bmatrix} x^0 \\ x^1 \end{bmatrix} = \begin{bmatrix} 0 \\ x^1 \end{bmatrix} = \begin{bmatrix} 0 & 0 \\ 0 & 1 \end{bmatrix} \begin{bmatrix} x^0 \\ x^1 \end{bmatrix}.$$

Setting

$$x^0(t) = x(t,\cdot), \quad x^1(t) = \frac{\partial x}{\partial t}(t,\cdot), \quad u(t) = u(t,\cdot)$$

we can write (8.2) in the abstract form

$$Y' = AY + Bu, \quad Y(0) = Y_0, \tag{8.7}$$

where

$$Y(t) = \begin{bmatrix} x^0(t) \\ x^1(t) \end{bmatrix}, \quad Y_0 = \begin{bmatrix} x_0 \\ x_1 \end{bmatrix}.$$

We are interested in the following optimal control problem: to minimize

$$\begin{aligned} J(u) = \int_0^T \int_\Omega \left(|\nabla_\xi x(t,\xi)|^2 + \left|\frac{\partial x}{\partial t}(t,\xi)\right|^2 + |u(t,\xi)|^2 \right) dt d\xi \\ + \int_\Omega \left(|\nabla_\xi x(T,\xi)|^2 + \left|\frac{\partial x}{\partial t}(T,\xi)\right|^2 \right) d\xi, \end{aligned} \tag{8.8}$$

over all $u \in L^2([0,\infty] \times \Omega)$ subject to equation (8.2). The cost function $J(u)$ can be written as

$$J(u) = \int_0^T \left\{ |Y(t)|_H^2 + |u(t)|_U^2 \right\} dt + |Y(T)|_H^2.$$

The Riccati equation is

$$P' = \begin{bmatrix} 0 & -1 \\ \Lambda & 0 \end{bmatrix} P + P \begin{bmatrix} 0 & 1 \\ -\Lambda & 0 \end{bmatrix} - P \begin{bmatrix} 0 & 0 \\ 0 & 1 \end{bmatrix} P + \begin{bmatrix} 1 & 0 \\ 0 & 1 \end{bmatrix},$$

$$P(0) = \begin{bmatrix} 1 & 0 \\ 0 & 1 \end{bmatrix}.$$

We can represent P as

$$P = \begin{bmatrix} P_{11} & P_{12} \\ P_{21} & P_{22} \end{bmatrix},$$

where

$$P_{11} \in \mathcal{L}(H_0^1(\Omega)), \quad P_{12} \in \mathcal{L}(L^2(\Omega); H_0^1(\Omega))$$

$$P_{21} \in \mathcal{L}(H_0^1(\Omega); L^2(\Omega)), \quad P_{22} \in \mathcal{L}(L^2(\Omega)),$$

and the following identities hold:

$$P_{11}^\star = \Lambda P_{11}\Lambda^{-1}, \quad P_{12}^\star = P_{21}\Lambda^{-1}, \quad P_{21}^\star = \Lambda P_{12}, \quad P_{22}^\star = P_{22}.$$

Setting $C = I$, we can apply Theorem 2.1 and conclude that the Riccati equation has a unique mild solution $P \in C_s([0,\infty[\,;\Sigma^+(H))$. Moreover by Theorem 6.1, there exists a unique optimal pair $(u^\star, x^\star)$ and

$$u^\star(t,\xi) = -(P_{21}(T-t)x^\star(t,\cdot))(\xi) - (P_{22}(T-t)\frac{\partial x^\star}{\partial t}(t,\cdot))(\xi).$$

The closed loop equation is

$$\begin{cases} \dfrac{\partial^2 x}{\partial t^2}(t,\xi) = \Delta_\xi x(t,\xi) - (P_{21}(T-t)x(t,\cdot))(\xi) \\ \qquad\qquad -(P_{22}(T-t)\dfrac{\partial x}{\partial t}(t,\cdot))(\xi) \ \text{ in }]0,T[\,\times\Omega, \\ x(t,\xi) = 0 \text{ on }]0,T[\times\partial\Omega, \\ x(0,\xi) = x_0(\xi), \\ \dfrac{\partial x}{\partial t}(t,\xi) = x_1(\xi) \ \ \textit{in } \Omega. \end{cases}$$

Remark 8.1 Similar results can be obtained if one replaces in problem (8.2) the Dirichlet by the Neumann boundary condition.

$$\begin{cases} \dfrac{\partial^2 x}{\partial t^2}(t,\xi) = \Delta_\xi x(t,\xi) + (Bu(t,\cdot))(\xi), \\ \qquad\qquad \text{in }]0,T] \times \Omega \\ \dfrac{\partial x}{\partial \nu}(t,\xi) = 0 \;\text{ on }\;]0,T] \times \partial\Omega \\ x(0,\xi) = x_0(\xi), \;\text{ in }\; \Omega \\ \dfrac{\partial x}{\partial t}(0,\xi) = x_1(\xi) \;\text{ in }\; \Omega, \end{cases} \tag{8.9}$$

where ν is the outward normal to $\partial\Omega$. In this case we set

$$H = Y = H^1(\Omega) \oplus L^2(\Omega), \quad U = L^2(\Omega) \tag{8.10}$$

and define the scalar product in H by

$$\begin{aligned} \left(\begin{bmatrix} x^0 \\ x^1 \end{bmatrix}, \begin{bmatrix} z^0 \\ z^1 \end{bmatrix} \right) &= \int_\Omega (\nabla_\xi x^0 \cdot \nabla_\xi z^0 + x^1 z^1)\, d\xi \\ &\quad + \int_\Omega x^0 z^0 \, d\xi. \end{aligned} \tag{8.11}$$

Moreover we define a self-adjoint positive operator Λ_1 on $L^2(\Omega)$ by

$$\begin{cases} D(\Lambda_1) = \{x \in H^2(\Omega) : \dfrac{\partial x}{\partial \nu} = 0 \;\text{ on }\; \partial\Omega\} \\ \Lambda_1 x = -\Delta_\xi x, \end{cases} \tag{8.12}$$

and the linear operator A_1 on H:

$$\begin{cases} A_1 X = \begin{bmatrix} 0 & 1 \\ -\Lambda_1 & 0 \end{bmatrix} \begin{bmatrix} x^0 \\ x^1 \end{bmatrix}, \quad \forall X \in D(A_1) \\ D(A_1) = \{x \in H^2(\Omega) : \dfrac{\partial x}{\partial \nu} = 0 \;\text{ on }\; \partial\Omega\} \oplus H_0^1(\Omega). \end{cases} \tag{8.13}$$

Now all previous considerations can be easily generalized. ■

8.3 Delay equations

We start with a simple example given in §2.10 (Volume I, Chapter 1). A more elaborate treatment of differential delay systems with delays in the control and observation variables will be given after using the general state space frameworks developed in Chapter 4 of Volume I.

Let D, $E \in \mathcal{L}(\mathbf{C}^n)$, $r > 0$ and consider the problem:

$$\begin{cases} z'(t) = Dz(t) + Ez(t-r) + u(t), \quad t \geq 0 \\ z(0) = h_0 \in \mathbf{C}^n \\ z(\theta) = h_1(\theta), \quad a.e.\ \theta \in [-r, 0] \end{cases} \tag{8.14}$$

where $h_1 \in L^2(-r, 0; \mathbf{C}^n)$ and $u \in L^2(0, T; \mathbf{C}^n)$. We shall use here the notation of §2.10 (Chapter 1 in Volume I). We set

$$H = L^2(-r, 0; \mathbf{C}^n), \quad U = \mathbf{C}^n, \quad \tilde{x}(t) = (z(t), z_t).$$

Then problem (8.14) is equivalent to

$$\tilde{x}' = A\tilde{x} + Bu, \quad \tilde{x}(0) = (h_0, h_1),$$

where A is defined by (2.104) in Chapter 1 of Volume I and B is the linear operator

$$u \mapsto Bu = (u, 0) \,:\, \mathbf{C}^n \to H.$$

We remark that the adjoint operator $B^\star$ is given by

$$(x_0, x_1) \mapsto B^\star(x_0, x_1) = x_0 \,:\, H \to \mathbf{C}^n.$$

Consider now the problem: to minimize

$$J(u) = \int_0^T \left\{ |z(t)|^2 + |u(t)|^2 \right\} dt + |z(T)|^2$$

over all $u \in L^2(0, T; \mathbf{C}^n)$ subject to equation (8.14). $J(u)$ can be written as

$$J(u) = \int_0^T \left\{ |Cx(t)|^2 + |u(t)|^2 \right\} dt + (P_0 z(T), z(T))_{\mathbf{C}^n \times L^2(-r,0;\mathbf{C}^n)},$$

where

$$C(x_0, x_1) = (x_0, 0) \text{ and } P_0(x_0, x_1) = (x_0, 0).$$

The Riccati equation is

$$P' = A^\star P + PA - P^2 + C^\star C, \quad P(0) = P_0,$$

since, as easily seen, $BB^\star = \begin{bmatrix} I & 0 \\ 0 & 0 \end{bmatrix}$.

We can write P as

$$P = \begin{bmatrix} P_{01} & P_{01} \\ P_{10} & P_{11} \end{bmatrix},$$

where

$$\begin{aligned} &P_{00} \in \mathcal{L}(\mathbf{C}^n), \quad P_{01} \in \mathcal{L}(L^2(0,T;\mathbf{C}^n);\mathbf{C}^n)), \\ &P_{10} \in \mathcal{L}(\mathbf{C}^n;(L^2(0,T;\mathbf{C}^n)), \quad P_{11} \in \mathcal{L}(L^2(0,T:);\mathbf{C}^n)). \end{aligned}$$

Then the optimal control is given by

$$u^\star(t) = -P_{00}(T-t)z^\star(t) - P_{01}(T-t)z_t^\star.$$

We now consider the more general situation. The abstract theory developped in Chapter 4 of Volume I (Theorem 4.1) for unbounded observation operators applies to a class of delay systems with delayed observations where the control operator of the state equation is bounded. Following the notation and definitions of §5 (Chapter 4, Volume I) a delay differential system with controls and observations is given by

$$\begin{cases} \dot{x}(t) = Lx_t + Bu_t, \quad t > 0 \\ \\ (x(0), x_0, u_0) = (\varphi^0, \varphi^1, w) \in M^2 \times L^2(-h, 0; \mathbf{R}^m) \end{cases} \tag{8.15}$$

$$y(t) = Cx_t, \quad t > 0. \tag{8.16}$$

The spaces of controls and observation are $U = \mathbf{R}^m$ and $Y = \mathbf{R}^k$, respectively. By introducing the *extended structural state* $\widehat{x}(t)$ in $H = M^2 \times L^2(-h, 0; Y)$ (cf. Volume I, Chapter 4, §6.2, Theorem

6.1, eqs (6.11) to (6.14)), system (8.15)–(8.16) can be transformed in the form

$$\begin{cases} \frac{d}{dt}\tilde{\jmath}^* \widehat{x}(t) = (\widetilde{A}^\top)^* \widehat{x}(t) + (\widetilde{B}^\top)^* u(t), \quad t > 0 \\ \\ \widehat{x}(0) = \xi \end{cases} \tag{8.17}$$

where $\widetilde{B}^\top : D(\widetilde{A}^\top) \to U$ and $\widehat{C} : D((\widetilde{A}^\top)^*) \to Y$ are continuous linear maps when the domains are endowed with their respective graph norm topologies. The operator $\widetilde{B}^\top$ is bounded and continuous on H when it is of the form

$$\widetilde{B}^\top(\psi^0, \psi^1, v) = B_0^\top \psi^0 + \int_{-h}^0 B_1^\top(\theta)\psi^1(\theta)\, d\theta \tag{8.18}$$

for some matrices B_0 and $B_1(\theta)$, $\theta \in I(-h,0)$, such that the elements of $B_1(\cdot)$ belong to $L^2(-h,0)$. Then

$$(\widetilde{B}^\top)^* u = (B_0 u, B_1(\cdot)u, 0). \tag{8.19}$$

So we specialize to systems (8.15) of the form

$$\begin{cases} \dot{x}(t) = Lx_t + B_0 u(t) + \int_{-h}^0 B_1(\theta)\psi^1(\theta) u(t+\theta)\, d\theta, \ t > 0 \\ \\ (x(0), x_0, u_0) = (\varphi^0, \varphi^1, w). \end{cases} \tag{8.20}$$

The operator $\widehat{C}$ is of the general form (cf. Volume I, Chapter 4, §6.2.2)

$$\begin{aligned} &\xi = (\varphi_0, \overline{L}\varphi_1 + \zeta, \overline{C}\varphi_1 + \lambda) \mapsto \widehat{C}\xi = C\varphi_1 + \lambda(0) \\ &: D((\widetilde{A}^\top)^*) \to Y \end{aligned} \tag{8.21}$$

where $(\varphi, \zeta, \lambda)$ is a representation of the elements of $D((\widetilde{A}^\top)^*)$.

So we only need to check part (ii) of condition $\mathcal{H}$–(v)

$$\int_0^t |\widehat{C} e^{s(\widetilde{A}^\top)^*}\xi|_Y^2\, ds \le K(t)|\xi|_H^2, \quad \forall \xi \in D((\widetilde{A}^\top)^*).$$

In the notation of Volume I, Chapter 4, $e^{t(\widetilde{A}^\top)^*}$ is $(\widetilde{S}^\top)^*(t)$ and

$$\begin{aligned} \widehat{C}(\widetilde{S}^\top)^*(t)\xi &= C\widehat{x}(t) = Cx_t + [\tau(t)\lambda](0) \\ &= (Ce_+^0 x)(t) + (e_+^{-h}\xi^2)(-t) \end{aligned}$$

which is continuous as an element of $L^2(0,T;Y)$ with respect to ξ and $(e_+^{-h}\xi^2)(-t)$ is the shift of an L^2–function. This holds for each finite $T > 0$.

As a result we have the following proposition.

Proposition 8.1 *When the mapping B is of the form (8.18), the Riccati equation*

$$\begin{cases} P' = \widetilde{A}^\top P + P(\widetilde{A}^\top)^* - P(\widetilde{B}^\top)^*\widetilde{B}^\top P + \widehat{C}^*\widehat{C}, \\ \\ P(0) = P_0 \end{cases} \tag{8.22}$$

has a unique mild solution P in $C_s([0,\infty[;\Sigma^+(H))$, where $H = M^2 \times L^2(-h,0;Y)$.

The Riccati equation (8.22) on $[0,T]$, $T > 0$, is associated with the minimization of the cost function

$$J(u) = \int_0^T [|Cx_t|^2 + |u(t)|^2]dt + Fx(T)\cdot x(T) \tag{8.23}$$

over all u in $L^2(0,T;Y)$, where

$$P_0(\xi^0,\xi^1,\xi^2) = (F\xi^0,0,0), \quad \forall\, \xi = (\xi^0,\xi^1,\xi^2).$$

The optimal control u^* can be expressed in feedback form

$$u^*(t) = -\widetilde{B}^\top P(T-t)\widehat{x}(t), \quad t > 0,$$

where for the initial conditions in (8.15)

$$\widehat{x}(t) = (x(t), \overline{L}x_t + \overline{B}u_t^*, \overline{C}x_t).$$

Since $P(t)$ can be decomposed into a 3×3 matrix of operators on the state space $H = \mathbf{R}^n \times L^2(-h,0,\mathbf{R}^n) \times L^2(-h,0,Y)$

$$P(t) = \begin{bmatrix} P_{00}(t) & P_{01}(t) & P_{02}(t) \\ P_{10}(t) & P_{11}(t) & P_{12}(t) \\ P_{20}(t) & P_{21}(t) & P_{22}(t) \end{bmatrix}.$$

and $\widetilde{B}^\top$ has the special form (8.18)

$$\begin{aligned}
&\widetilde{B}^\top P(T-t)\widetilde{x}(t)\\
&= B_0^\top\{P_{00}(T-t)x(t)+P_{01}(T-t)x(t)[\overline{L}x_t+\overline{B}u_t^*]\\
&\qquad + P_{02}(T-t)\overline{C}x_t\}\\
&\quad + \int_{-h}^0 B_1^\top(\theta)\quad \{P_{10}(T-t)x(t)+P_{11}(T-t)x(t)[\overline{L}x_t+\overline{B}u_t^*]\\
&\qquad\qquad + P_{12}(T-t)\overline{C}x_t\}(\theta)\,d\theta.
\end{aligned}$$

It is important to notice that the feedback is through the structural operators $\overline{L}$, $\overline{B}$ and $\overline{C}$ on x_t and u_t^*. This state captures the minimal information necessary for the feedback synthesis. Without this notion of state, determining directly expressions as above is almost hopeless in the general case. We leave it to the reader to specify $\overline{L}$, $\overline{B}$ and $\overline{C}$ in special cases!

The same problem can also be formulated by using the *extended state* $\widetilde{x}(t)$ in $H = M^2 \times L^2(-h,0;U)$ (cf. Volume I, Chapter 4, §§ 5.2 and 6.1, eqs (6.7) to (6.10)) which verifies the state equation

$$\left\{\begin{aligned}
&\frac{d}{dt}\widetilde{\imath}^*\widetilde{x}(t) = (\widetilde{A}^*)^*\widetilde{x}(t) + (\widehat{B}^\top)^*u(t)\\
&\widetilde{x}(0) = (\varphi^0\varphi^1, w)
\end{aligned}\right. \tag{8.24}$$

with the observation equation

$$y(t) = \widetilde{C}\widetilde{x}(t), \tag{8.25}$$

(cf. Volume I, Chapter 4, §6, equations (6.7) and (6.8)) where $\widehat{B}^\top : D(\widetilde{A}^*) \to U$ and $\widetilde{C} : D(\widetilde{A}) \to Y$ are continuous linear maps when the domains are endowed with their respective graph norm topologies. When the control operator is given as in (8.20), the operator $\widehat{B}^\top$ is equal to

$$\widehat{B}^\top(\xi) = B_0^\top\psi(0) + \int_{-h}^0 B_1^\top(\theta)\psi(\theta)\,d\theta + \lambda(0) \tag{8.26}$$

where ψ, ζ, λ is a representation of an arbitrary element ξ of $D(\widetilde{A}^*)$

$$\xi = (\psi(0), \overline{L}^\top\psi + \zeta, \overline{B}^\top\psi + \lambda).$$

The operator $\widehat{B}^\top$ is unfortunately not bounded on H. As for $\widetilde{C}$ it takes the simple form

$$\widetilde{C}(\varphi(0), \varphi, w) = C\varphi \tag{8.27}$$

and part (ii) of condition $\mathcal{H}$–(v)

$$\int_0^t |\widetilde{C}e^{s\widetilde{A}}\xi|_Y^2\, ds \le K(t)|\xi|_H^2, \quad \forall\, \xi \in D(\widetilde{A})$$

is verified. In the notation of Volume I, Chapter 4, $e^{t\widetilde{A}}$ is $\widetilde{S}(t)$ and

$$\widetilde{C}\widetilde{S}(t)\xi = \widetilde{C}(x(t), x_t, (e_-^0 w)_t) = (Cx)(t)$$

which is continuous as an element of $L^2(0,T;Y)$ with respect to ξ in $M^2 \times L^2(0,T;U)$. However $\widehat{B}^\top$ is not bounded on H since it contains a delta function. So we cannot use the general theory of §4.

However we have seen in §6.3 of Chapter 4 in Volume I that $\hat{x}$ and $\tilde{x}$ are intertwined via the structural operator

$$F = \begin{bmatrix} I & 0 & 0 \\ 0 & \overline{L} & \overline{B}_0 \\ 0 & \overline{C} & \overline{B}_1 \end{bmatrix}. \tag{8.28}$$

(cf. (6.74) and (6.77) to (6.80) in Theorem 6.2). Then for the optimal control u^*

$$\hat{x}(t; F(\phi^0, \phi^1, w), u^*) = F\tilde{x}(t; (\phi^0, \phi^1, w), u^*).$$

So it is not too difficult to guess that for the formulation (8.23)-(8.24)-(8.25), there will be a Riccati equation and a solution $\Pi(t)$ related to $P(t)$ through the identity

$$\Pi(t) = F^\top P(t) F.$$

To see that multiply the Riccati equation (8.22) by $F^\top$ on the left and F on the right and use the intertwining identities

$$\widehat{C}F = \widetilde{C},\ \widehat{B}^\top F^\top = \widetilde{B}^\top,\ F^\top \widehat{A}^\top = \widetilde{A}^* F^\top,\ \widehat{A}^{\top *} F = F\widetilde{A}$$

(cf. Volume II, Chapter 4, (6.79), (6.80) and from (6.78)) to obtain

$$\begin{cases} \Pi' = \widetilde{A}^*\Pi + \Pi\widetilde{A} - \Pi(\widehat{B}^\top)^*\widehat{B}^\top\Pi + \widetilde{C}^*\widetilde{C}, \\ \\ \Pi(0) = F^\top P_0 F. \end{cases} \tag{8.29}$$

It corresponds to the control of an infinite dimensional system with both unbounded control and observation operators. We know what the Riccati equation associated with the state $\tilde{x}(t)$ will look like but it remains to give a precise meaning to such an equation and establish the properties of its solution. The necessary techniques which would allow us to do it are different than the ones which will be developped in the next two chapters.

The last equation is associated with the same problem (8.23) with

$$\Pi_0(\xi^0, \xi^1, \xi^2) = (F\xi^0, 0, 0), \quad \forall\, \xi = (\xi^0, \xi^1, \xi^2). \tag{8.30}$$

The optimal control u^* can be expressed in feedback form

$$u^*(t) = -\widehat{B}^\top\Pi(T-t)\widetilde{x}(t), \quad t > 0,$$

where for the initial conditions in (8.15)

$$\widetilde{x}(t) = (x(t), x_t, u_t).$$

However in view of (8.26) the interpretation of $\widehat{B}^\top\Pi(T-t)$ is not clear unless some additional properties of $\Pi(t)$ are obtained. In that respect the *extended state* is simpler to define but since it does not incorporate the delay structure of the problem, the feedback law is difficult to explicit. For a simple example of this property the reader is referred to M.C. DELFOUR, E.B. LEE and A. MANITIUS [1] and M.C. DELFOUR [1]. For a more detailed bibliography on the control of delay systems the reader is referred to the references in Chapter 4 of Volume I. Though the general case with delays in control and observation has a unique solution, it was not possible here to detail all special cases and the history of the problems

8.4 Evolution equations in noncylindrical domains

Consider the state equation

$$\begin{cases} \dfrac{\partial x}{\partial t}(t,\xi) = \Delta_\xi x(t,\xi) + u(t,\xi), \quad t \in]0,T],\ \xi \in \Omega_t, \\ \\ x(t,\xi) = 0,\ t \in]0,T],\ \xi \in \partial\Omega_t \\ \\ x(0,\xi) = x_0(\xi),\quad \xi \in \Omega_0 \end{cases} \tag{8.31}$$

where for any $t \in [0,T]$, Ω_t is a bounded set in $\mathbf{R}^n$ with a regular boundary $\partial\Omega_t$.

We want to minimize the cost

$$\begin{aligned} J(u) \;=\; & \int_0^T dt \int_{\Omega_t} \left\{ |x(t,\xi)|^2 + |u(t,\xi)|^2 \right\} d\xi \\ & + \int_{\Omega_T} |x(T,\xi)|^2 d\xi \end{aligned} \tag{8.32}$$

over all controls $u \in L^2([0,T] \times \Omega_t)$ subject to state equation (8.31). By $L^2([0,T] \times \Omega_t)$ we mean the set of all functions u

$$(t,\xi) \mapsto u(t,\xi) \;:\; \{(t,\xi) :\ t \in [0,T],\ \xi \in \Omega_t\} \to \mathbf{R},$$

such that $t \mapsto \int_{\Omega_t} |u(t,\xi)|^2 d\xi$ is measurable and

$$\int_0^T dt \int_{\Omega_t} |u(t,\xi)|^2\, d\xi < +\infty.$$

We first reduce problem (8.31) to an evolution equation in a cylindrical domain, following G. DA PRATO and J.-P. ZOLÉSIO [1]. To this end we introduce, as in J.-P. ZOLÉSIO [1], a suitable change of variables:

$$x(t,\xi) = z(t, T_t(\xi)), \quad u(t,\xi) = v(t, T_t(\xi))$$

where T_t is a regular mapping such that

$$T_t(\Omega_0) = \Omega_t, \quad t \in [0,T].$$

Then problem (8.31) reduces to

$$\begin{cases} z_t = J_t^{-1} \ \mathrm{div}\ [J_t((DT_t)^{-1})^\star (DT_t)^{-1}\nabla z] \\ \qquad + ((DT_t)^{-1})^\star \nabla z \cdot \dfrac{d}{dt}T_t + v, \quad t\in[0,T],\ \xi\in\Omega_0 \\ z(t,\xi) = 0, \quad t\in[0,T],\ \xi\in\partial\Omega_0 \\ z(0,\xi) = x_0(\xi), \quad \xi\in\Omega_0, \end{cases} \tag{8.33}$$

where J_t is the determinant of the Jacobian matrix DT_t of T_t.

We set $H = U = L^2(\Omega_0)$, $z(t,\cdot) = z(t)$, $v(t,\cdot) = v(t)$ and write (8.33) as an abstract evolution equation in H

$$z'(t) = A(t)z(t) + v(t), \quad z(0) = x_0$$

where $A(t)$ is defined by

$$\begin{cases} A(t)w = J_t^{-1} \ \mathrm{div}\ [J_t((DT_t)^{-1})^\star (DT_t)^{-1}\nabla w] \\ \qquad + ((DT_t)^{-1})^\star \nabla w \cdot \dfrac{d}{dt}T_t \\ D(A(t)) = H^2(\Omega_0) \cap H_0^1(\Omega_0). \end{cases}$$

$A(t)$ is the realization of an elliptic operator under Dirichlet boundary conditions, and consequently it is the infinitesimal generator of an analytic semigroup in H. Moreover we have $A \in C^1([0,T];\mathcal{L}(D;H))$ so that by H. TANABE [1], assumptions (7.8) hold. The cost function (8.32) becomes

$$J(v) = \int_0^T dt \int_{\Omega_0} J_t(T_t^{-1}(x))(|z(t,\xi)|^2 + |v(t,\xi)|^2)\, d\xi + \int_{\Omega_0} |z(T,\xi)|^2\, d\xi.$$

and we can study the problem as in §7.2.

Remark 8.2 It is also possible to study problem (8.31) without changing variables. By proceeding as in P. CANNARSA, G. DA PRATO and J.-P. ZOLÉSIO [1], we assume that there exists a bounded open set D that contains all sets Ω_t, $t \in [0,T]$, and introduce unbounded operators $\{A(t)\}$, $t \in [0,T]$ with domain depending on time, but in the fixed space $H = L^2(D)$.

$$\begin{cases} D(A(t)) = \{y \in L^2(D) \,:\, y|_{\Omega_t} \in H^2(\Omega_t),\, y|_{D\backslash\Omega_t} \in H^2(D\backslash\Omega_t), \\ \qquad\qquad\qquad\qquad\qquad y = 0 \text{ on } \partial D,\, y = 0 \text{ on } \partial\Omega_t\} \\ A(t)u = \Delta u, \quad \forall\, u \in D(A(t)). \end{cases}$$

Then we write problem (8.31) in the abstract form (7.7) and prove that (7.8) hold. ∎

Remark 8.3 An optimal control problem for the wave equation in moving domain has been studied with similar techniques in G. Da Prato and J.-P. Zolésio [2]. ∎

Chapter 2

Systems with Unbounded Control Operators: Parabolic Equations with Control on the Boundary

1 Introduction

As in the previous chapter, we shall denote by H, U and Y the Hilbert spaces of states, controls and observations, respectively. We consider a dynamical system, whose state $x(t)$ is subject to the following equation

$$x'(t) = Ax(t) + Bu(t), \quad x(0) = x_0 \in H,$$

where $u \in L^2(0, T; U)$ and $A : D(A) \subset H \to H$ generates an analytic semigroup in H. However, in the present case, the linear operator B is not supposed to be bounded from U into H. This situation has been discussed at lenght in Chapter 1 (Part II). However some of the key constructions will be repeated here as needed. In that case many possibilities could be considered. However in practice, it will be natural to consider situations where B maps U into the dual space $(D(A^\star))'$ of $D(A^\star)$. This will be apparent in the following Examples 1.1 and 1.2. Equivalently, B is supposed to be of the form $B = (\lambda_0 - A)D$ where $D \in \mathcal{L}(U; H)$ and λ_0 is an element in $\rho(A)$.

Under these assumptions we write the state equation as

$$x'(t) = Ax(t) + (\lambda_0 - A)Du(t), \quad x(0) = x_0,$$

or in the mild form

$$x(t) = e^{tA}x_0 + (\lambda_0 - A)\int_0^t e^{(t-s)A}Du(s)\,ds. \tag{1.1}$$

Remark that formula (1.1) is meaningful and $x \in L^2(0,T;H)$, see Volume I. This formula will represent the state of our system.

The above representation of the operator B is also discussed in Chapter 2 of Volume I. The differential equation for the state still makes sense for a control operator $B \in \mathcal{L}(U; D(A^*)')$ and a representation formula can be obtained for $x(t)$ (cf. Vol. I, Chapt. 4, Thm 1.1). These results can be sharpened when A is the generator of an analytic semigroup (cf. Thms 2.2 and 2.3) and the natural operator $B \in \mathcal{L}(U, D(A^*)')$ is $D = [\lambda_0 - A]^{-1}B$ through formula (1.1).

We shall assume that

$$(\mathcal{HP})_1 \begin{cases} (i)\ A \text{ generates an analytic semigroup } e^{tA} \text{ of type } \omega_0 \\ \qquad \text{and } \lambda_0 \text{ is a real number in } \rho(A) \text{ such that } \omega_0 < \lambda_0. \\ (ii)\ \exists\, \alpha \in]0,1[\text{ such that } D \in \mathcal{L}(U, D(A^\alpha)), \end{cases}$$

where $D(A^\alpha) = D([\lambda_0 - A]^\alpha)$ is the domain of the fractional power $[\lambda_0 - A]^\alpha$ of the operator $\lambda_0 - A$, as defined in Volume I, Chapter 1, §5. For more details on the above techniques the reader is referred to Chapter 1 (Part II).

Assumption $(\mathcal{HP})_1$ is equivalent to say that

$$B = [\lambda_0 - A]\,D = [\lambda_0 - A]^{1-\alpha}\,[\lambda_0 - A]^\alpha D \tag{1.2}$$

belongs to $\mathcal{L}(U, D((A^*)^{1-\alpha})')$. In the variational case

$$D((A^*)^{1-\alpha}) = [D(A^*), H]_\alpha \quad \Rightarrow \quad B \in \mathcal{L}(U, [D(A^*), H]'_\alpha)$$

and the regularity results of Theorem 2.2 (Volume I, Chapter 4) would apply:

$$\begin{aligned} &\forall u \in L^2(0,T;U), \quad \forall x(0) \in [[D(A), H]_\alpha, [D(A^*), H]'_{1-\alpha}]_{\frac{1}{2}} \\ &\Rightarrow \quad x \in W(0,T; [D(A), H]_\alpha, [D(A^*), H]'_{1-\alpha}). \end{aligned} \tag{1.3}$$

Remark 1.1 Assume that $(\mathcal{HP})_1$ is verified and let x be defined by (1.1). By the Closed Graph Theorem, $[\lambda_0 - A]^\alpha D$ is a bounded operator. Moreover, since

$$(\lambda_0 - A)e^{(t-s)A}Du(s) = [\lambda_0 - A]^{1-\alpha}e^{(t-s)A}[\lambda_0 - A]^\alpha Du(s),$$

there exists a constant $k_\alpha > 0$ such that

$$|(\lambda_0 - A)e^{(t-s)A}Du(s)| \leq k_\alpha(t-s)^{\alpha-1}|u(s)|, \quad s \in [0,t].$$

Thus the state $x(\cdot)$ can also be written as

$$x(t) = e^{tA}x_0 + \int_0^t (\lambda_0 - A)e^{(t-s)A}Du(s)\,ds,$$

and the following estimate holds

$$|x(t)| \leq |e^{tA}x_0| + k_\alpha \int_0^t (t-s)^{\alpha-1}|u(s)|\,ds.$$

If $\alpha > \frac{1}{2}$, from the Hölder estimate, it follows that $x \in L^\infty(0,T;H)$; this implies that $x \in C([0,T];H)$ by a standard density argument. In fact setting

$$x_k(t) = e^{tA}x_0 + \int_0^t (\lambda_0 - A)e^{(t-s)A}kR(k,A)Du(s)\,ds,$$

we clearly have $x_k \in C([0,T];H)$ and it is easy to check that $x_k(t) \to x(t)$ uniformly in t.

If $\alpha \leq \frac{1}{2}$ then $x \notin C([0,T];H)$ in general; however, arguing as before, we can easily check that $[\lambda_0 - A]^{-\beta}x \in C([0,T];H)$ for all $\beta > \frac{1}{2} - \alpha$. In the sequel, we shall discuss separately the cases $\alpha > \frac{1}{2}$ and $\alpha \leq \frac{1}{2}$; as we shall see, the second case is much more difficult to deal with. ■

Consider the following *optimal control problem*: to minimize the cost function

$$J(u) = \int_0^T \{|Cx(t)|^2 + |u(t)|^2\}\,dt + (P_0x(T), x(T)), \tag{1.4}$$

over all $u \in L^2(0,T;U)$, subject to the differential equation constraint (1.1). We assume, besides $(\mathcal{HP})_1$,

$$(\mathcal{HP})_2 \begin{cases} (i) \text{ that } C \in \mathcal{L}(H;Y),\ P_0 \in \Sigma^+(H), \\ (ii) \text{ and that if } \alpha \le \frac{1}{2} \text{ there exists } \beta \in]\frac{1}{2} - \alpha, \frac{1-\alpha}{2}[\\ \quad \text{such that the following holds} \\ x \in D([\lambda_0 - A]^\beta) \Rightarrow P_0\,[\lambda_0 - A]^\beta x \in D([\lambda_0 - A^\star]^\beta), \\ \quad \text{and } [\lambda_0 - A^\star]^\beta P_0\,[\lambda_0 - A]^\beta \text{ is bounded.} \end{cases}$$

If assumption $(\mathcal{HP})_2$ holds, we shall use the notation

$$P_\beta = \text{closure}\,[\lambda_0 - A^\star]^\beta P_0[\lambda_0 - A]^\beta,$$

so that $P_\beta \in \Sigma^+(H)$ and

$$P_0 = [\lambda_0 - A^\star]^{-\beta} P_\beta [\lambda_0 - A]^{-\beta}.$$

We remark that if $\alpha > \frac{1}{2}$, then we do not require additional conditions on P_0, except that $P_0 \in \Sigma^+(H)$, whereas if $\alpha \le \frac{1}{2}$ we need the assumption $(\mathcal{HP})_2$-(ii) to make sense of the term

$$(P_0 x(T), x(T)) = (P_\beta[\lambda_0 - A]^{-\beta} x(T), [\lambda_0 - A]^{-\beta} x(T)),$$

in the cost functional. In fact, if $\beta > \frac{1}{2} - \alpha$, then $[\lambda_0 - A]^{-\beta} x \in C([0,T];H)$ (see Remark 2.1), and the trace value $[\lambda_0 - A]^{-\beta} x(T)$ is meaningful.

By $(\mathcal{HP})$ we mean the set of all the assumptions $(\mathcal{HP})_1$ and $(\mathcal{HP})_2$. We shall also say that (A, D, C, P_0) fulfill $(\mathcal{HP})$. The definitions of *optimal control, optimal state* and *optimal pair* are given as in Chapter 1.

We shall study the optimization problem (1.1)–(1.4), by using Dynaming Programming and by proceeding into the usual two steps.

First step. We consider the Riccati equation, that formally reads as follows

$$\begin{cases} P' = A^\star P + PA - P(\lambda_0 - A)D((\lambda_0 - A)D)^\star P + C^\star C \\ \\ P(0) = P_0. \end{cases} \tag{1.5}$$

We set

$$E = [\lambda_0 - A]^\alpha D, \quad V = [\lambda_0 - A^\star]^{1-\alpha} P.$$

Then $E \in \mathcal{L}(U;H)$ by virtue of hypothesis $(\mathcal{HP})_1$-(ii). Thus the meaningless term $P(\lambda_0 - A)D((\lambda_0 - A)D)^\star P$, can be written as $V^\star EE^\star V$ and the Riccati equation rewritten as follows

$$\begin{cases} P' = A^\star P + PA - V^\star EE^\star V + C^\star C, \\ \\ P(0) = P_0, \end{cases} \tag{1.6}$$

or in the integral form

$$\begin{aligned} P(t)x &= e^{tA^\star} P_0 e^{tA} x + \int_0^t e^{(t-s)A^\star} C^\star C e^{(t-s)A} x\, ds \\ &\quad - \int_0^t e^{(t-s)A^\star} V^\star(s) EE^\star V(s) e^{(t-s)A} x\, ds, \ x \in H. \end{aligned} \tag{1.7}$$

We look for a solution $P(\cdot)$ of (1.7) such that $V(t) \in \mathcal{L}(H)$ for all $t > 0$.

Second step. We shall show that the optimal control $u^\star$ is related to the optimal state $y^\star$ by the feedback formula

$$\begin{aligned} u^\star(t) &= -D^\star(\lambda_0 - A^\star)P(T-t)x^\star(t) \\ &= -E^\star V(T-t)x^\star(t). \end{aligned} \tag{1.8}$$

Moreover $x^\star(t)$ is the solution of the *closed loop equation*

$$\begin{aligned} x(t) &= e^{tA} x_0 \\ &\quad + \int_0^t [\lambda_0 - A]^{1-\alpha} e^{(t-s)A} EE^\star V(T-s)x(s)\, ds, \end{aligned} \tag{1.9}$$

which can be solved by a fixed point argument.

We now give some bibliographical comments. Early papers treating boundary control problems are due to A.V. BALAKRISHNAN [1] [2] and D.C. WASHBURN [1], who developed further an old idea in H.O. FATTORINI [1]. Simplifications and refinements through domains of fractional powers were introduced in R. TRIGGIANI [1]. Another approach is due to A. CHOJNOWSKA-MICHALIK [1] and J. ZABCZYK [1] (see also Chapter 3 in Volume I for an alternate interpretation of their constructions). The Riccati equation (1.7) was first studied by A.V. BALAKRISHNAN [2] when $P_0 = 0$ and by I. LASIECKA and R. TRIGGIANI [3] when P_0 is a positive multiple of the identity. They were able to build an explicit solution of the Riccati equation by generalizing the representation formula (2.22) of Chapter 1; we remark, however, that this method does not give uniqueness. A direct approach to solve (1.7), was first used by F. FLANDOLI [1] under the assumption that $V_0 = [\lambda_0 - A^\star]^{1-\alpha} P_0$ is bounded; he showed that, in this case, there exists a unique solution $P(t)$ such that $[\lambda_0 - A^\star]^{1-\alpha} P(t)$ is bounded for any t. Moreover the closed loop equation (1.9) can be solved directly and the feedback formula (1.8) holds true. The same direct approach was used by G. DA PRATO and A. ICHIKAWA [1] which proved that equation (1.7) has a unique solution for any $P_0 \in \Sigma^+(H)$ if $\alpha > \frac{1}{2}$, whereas if $\alpha \leq \frac{1}{2}$ they need in addition that the linear operator $[\lambda_0 - A^\star]^\gamma P_0$ is bounded for some $\gamma > 1 - 2\alpha$. For some other results in this direction see M.C. DELFOUR and M. SORINE [1] and M. SORINE [1] using J.-L. Lions' direct method, and A.J. PRITCHARD and D. SALOMON [1].

In this Chapter we assume $(\mathcal{HP})$ following the recent paper of F. FLANDOLI [6], and show the existence and uniqueness of the Riccati equation (see §2.2 below). Then, we solve the control problem in §2.3.

We shall not study the Riccati equation (1.7) when $\alpha \leq \frac{1}{2}$ and P_0 only belongs to $\Sigma^+(H)$. In this case the cost functional (1.4) is not defined on the whole $L^2(0,T;U)$, (see Remark 1.1), and the Dynamic Programming approach, based on the direct solution of equation (1.7), does not work. Concerning the existence (but not the uniqueness) of a solution to (1.7) we mention the following results.

(i) If P_0 is a positive multiple of the identity, there exists a solution of (1.7) and one can also solve the corresponding control problem (1.1)–(1.4) (see I. LASIECKA and R. TRIGGIANI [3], F. FLANDOLI [4]).

(ii) There exists $P_0 \in \Sigma^+(H)$ such that equation (1.7) does not a have a solution (see F. FLANDOLI [4]).

(iii) A sharp sufficient condition, which is "almost" necessary, for the existence of a solution of (1.7) is given in I. LASIECKA and R. TRIGGIANI [8].

It is also possible to study generalizations as in §1.7 of Chapter 1 to

(i) non homogeneneous state equation,
(ii) tracking problem,
(iii) time dependent state equation and cost function.

The generalizations (i) and (ii) are straightforward, and they will be left to the reader. For the point (iii), which is much more technical, see the papers by P. ACQUISTAPACE, F. FLANDOLI and B. TERRENI [1], P. ACQUISTAPACE and B. TERRENI [2] and P. ACQUISTAPACE [1].

We end this section by giving two examples concerning the *heat equation*. For a result concerning the *strongly damped wave equation* see F. BUCCI [1]. Several other examples can be found in the lecture notes by I. LASIECKA and R. TRIGGIANI [7].

Another approach to boundary control problems using a quadratic cost function not necessarily coercive can be found in L. PANDOLFI [1].

Example 1.1 *Dirichlet boundary condition*
Let Ω be a bounded open set in $\mathbf{R}^N$ with a smooth boundary $\partial\Omega$. Consider the optimal control problem: to minimize

$$
\begin{aligned}
J(u) &= \int_0^T \int_\Omega |x(t,\xi)|^2 \, d\xi + \int_0^T \int_{\partial\Omega} |u(t,\xi)|^2 \, d\sigma \\
&\quad + \int_\Omega |\Gamma x(T,\cdot)(\xi)|^2 \, d\xi,
\end{aligned}
$$

over all controls $u \in L^2([0,T] \times \partial\Omega)$ subject to the state constraints

$$
\left\{
\begin{array}{l}
\dfrac{\partial x}{\partial t}(t,\xi) = \Delta_\xi x(t,\xi), \quad (t,\xi) \in]0,T[\times \Omega, \\[2ex]
x(0,\xi) = x_0(\xi), \quad \xi \in \Omega, \\[2ex]
x(t,\xi) = u(t,\xi), \quad (t,\xi) \in]0,T[\times \partial\Omega,
\end{array}
\right.
\tag{1.10}
$$

where $x_0 \in L^2(\Omega)$,

$$
\Delta_\xi = \sum_{i=1}^N \frac{\partial^2 x}{\partial \xi_i^2}
$$

is the Laplace operator and $\Gamma \in \mathcal{L}(L^2(\partial\Omega))$.

We set $H = Y = L^2(\Omega)$, $U = L^2(\partial\Omega)$ and introduce the Dirichlet realization A of the Laplace operator

$$D(A) = H^2(\Omega) \cap H_0^1(\Omega), \quad Ax = \Delta_\xi x, \quad \forall x \in D(A).$$

As is well known (see for instance S. AGMON [1]), A is a strictly negative self adjoint operator in $L^2(\Omega)$, so that hypothesis $(\mathcal{HP})_1$-(i) holds with $\lambda_0 = 0$. Moreover we have

$$D((-A)^\alpha) = \begin{cases} H^{2\alpha}(\Omega), & \text{if } \alpha \in]0, \frac{1}{4}[\\ \{u \in H^{2\alpha}(\Omega) : \; u = 0 \text{ on } \partial\Omega\}, & \text{if } \alpha \in]\frac{1}{4}, 1[. \end{cases}$$

Let us introduce now the *Dirichlet mapping*

$$v \mapsto Dv = w \; : \; L^2(\partial\Omega) \to L^2(\Omega),$$

where

$$\Delta_\xi w = 0 \text{ in } \Omega, \quad w(\xi) = v(\xi) \text{ in } \partial\Omega.$$

As proved in J.-L. LIONS and E. MAGENES [1], we have $D \in \mathcal{L}(L^2(\partial\Omega); H^{\frac{1}{2}}(\Omega))$; therefore $D \in \mathcal{L}(L^2(\partial\Omega); D((-A)^\alpha)))$ for any $\alpha \in]0, \frac{1}{4}[$. Thus assumptionss $(\mathcal{HP})$ are verified provided that $\alpha \in]0, \frac{1}{4}[$, $\beta \in]\frac{1}{2} - \alpha, \frac{1}{2}[$, and $\Gamma \in \mathcal{L}(L^2(\partial\Omega); D((-A)^\beta)))$, as noted in R. TRIGGIANI [1]. We now show, following A.V. BALAKRISHNAN [1], that problem (1.10) can be set in the form (1.1).

First we assume that $u \in W^{1,2}(0, T; L^2(\partial\Omega))$ and introduce a new variable by setting

$$y(t, \xi) = x(t, \xi) - (Du(t, \cdot))(\xi).$$

We obtain

$$\begin{cases} \dfrac{\partial y}{\partial t}(t, \xi) = \Delta_\xi y(t, \xi) - \dfrac{\partial Du}{\partial t}(t, \xi), & (t, \xi) \in]0, T[\times \Omega, \\ y(0, \xi) = x_0(\xi) - (Du(0, \cdot))(\xi), & \xi \in \Omega, \\ y(t, \xi) = 0, & (t, \xi) \in]0, T[\times \partial\Omega. \end{cases}$$

Then, we set $y(t) = y(t, \cdot)$, $u(t) = u(t, \cdot)$, and write the above problem as

$$y(t) = e^{tA}(x_0 - Du(0)) - \int_0^t e^{(t-s)A} Du'(s)\, ds.$$

By performing an integration by parts, it is not difficult to check that x fulfills (1.1). Finally, the hypothesis $u \in W^{1,2}([0,T]; L^2(\partial\Omega))$ can be easily removed by regularization. ∎

Example 1.2 *Neumann boundary condition*
Let Ω be a bounded open set in $\mathbf{R}^N$ with a smooth boundary $\partial\Omega$. Consider the optimal control problem: to minimize

$$\begin{aligned} J(u) &= \int_0^T \int_\Omega |x(t,\xi)|^2\, d\xi + \int_0^T \int_{\partial\Omega} |u(t,\xi)|^2\, d\sigma \\ &\quad + \int_\Omega |\Gamma x(T,\cdot)(\xi)|^2\, d\xi, \end{aligned}$$

over all controls $u \in L^2([0,T] \times \partial\Omega)$, subject to the state constraints

$$\begin{cases} \dfrac{\partial x}{\partial t}(t,\xi) = \Delta_\xi x(t,\xi), \quad (t,\xi) \in]0,T[\times\Omega, \\[2ex] x(0,\xi) = x_0(\xi), \quad \xi \in \Omega, \\[2ex] \dfrac{\partial x}{\partial \nu}(t,\xi) = u(t,\xi), \quad (t,\xi) \in]0,T[\times\partial\Omega, \end{cases} \tag{1.11}$$

where $x_0 \in L^2(\Omega)$, $\Gamma \in \mathcal{L}(L^2(\Omega))$ and ν represents the outward normal to $\partial\Omega$.

We set $H = Y = L^2(\Omega)$, $U = L^2(\partial\Omega)$ and introduce the Neumann realization A of the Laplace operator

$$\begin{cases} D(A) = \{x \in H^2(\Omega) \;:\; \dfrac{\partial x}{\partial \nu} = 0\}, \\ Ax = \Delta_\xi x, \quad \forall\, x \in D(A). \end{cases}$$

As is well known (see for instance S. Agmon [1]), A is a non positive self adjoint operator in $L^2(\Omega)$, so that hypothesis $(\mathcal{HP})_1$-(i) holds with any $\lambda_0 > 0$; we choose $\lambda_0 = 1$. We have

$$D([\lambda_0 - A]^\alpha) = \begin{cases} H^{2\alpha}(\Omega), & \text{if } \alpha \in]0, \tfrac{3}{4}[\\ \{x \in H^{2\alpha}(\Omega) : \dfrac{\partial x}{\partial \nu} = 0\}, & \text{if } \alpha \in]\tfrac{3}{4}, 1[. \end{cases}$$

Introduce the *Neumann mapping*

$$v \mapsto Nv = w \;:\; L^2(\partial\Omega) \to L^2(\Omega),$$

where

$$\Delta_\xi w - w = 0 \text{ in } \Omega, \quad \frac{\partial w}{\partial \nu}(\xi) = v(\xi) \text{ in } \partial\Omega.$$

As proved in J.-L. LIONS and E. MAGENES [1], we have $N \in \mathcal{L}(L^2(\partial\Omega); H^{\frac{3}{2}}(\Omega))$; thus $N \in \mathcal{L}(L^2(\partial\Omega); D([\lambda_0 - A]^\alpha))$ for any $\alpha \in]0, \frac{3}{4}[$; and assumptions $(\mathcal{HP})$ are verified. Moreover, by proceeding as in Example 1.1 we see that problem (1.11) is equivalent to problem (1.1) with $D = N$ and $\lambda_0 = 1$. ∎

Remark 1.2 One obtains similar results for an elliptic operator with general boundary conditions, (see I. LASIECKA [1].) ∎

2 Riccati equation

2.1 Notation

We use here the notation introduced in §2.1 of Chapter 1 and we assume that $(\mathcal{HP})$ (that is both $(\mathcal{HP})_1$ and $(\mathcal{HP})_2$) hold. If $P \in C_s([a,b]; \Sigma^+(H))$, we set

$$V_P(t) = [\lambda_0 - A^\star]^{1-\alpha} P(t), \quad t \in [a,b].$$

When no confusion may arise, we will drop index P, writing $V_P = V$.

We are going to study the Riccati equation

$$\begin{cases} P' = A^\star P + PA - V^\star E E^\star V + C^\star C, \\ \\ P(0) = P_0, \end{cases} \tag{2.1}$$

where

$$E = [\lambda_0 - A]^\alpha D.$$

We also consider the integral form

$$\begin{aligned} P(t)x &= e^{tA^\star} P_0 e^{tA} x + \int_0^t e^{(t-s)A^\star} C^\star C e^{(t-s)A} x \, ds \\ &\quad - \int_0^t e^{(t-s)A^\star} V^\star(s) E E^\star V(s) e^{(t-s)A} x \, ds, \end{aligned} \tag{2.2}$$

where $x \in H$.

It is useful to introduce the approximating problem for $k > \lambda_0$:

$$\begin{cases} P_k' = A^\star P_k + P_k A - V_k^\star E_k E_k^\star V_k + C^\star C, \\ \\ P_k(0) = P_0, \end{cases} \tag{2.3}$$

where

$$E_k = kR(k,A)E, \quad V_k = V_{P_k} = [\lambda_0 - A^\star]^{1-\alpha} P_k$$

and $R(k,A) = (k-A)^{-1}$. Problem (2.3) can also be written in mild form as follows

$$\begin{aligned} P_k(t)x &= e^{tA^\star} P_0 e^{tA} x + \int_0^t e^{(t-s)A^\star} C^\star C e^{(t-s)A} x\, ds \\ &\quad - \int_0^t e^{(t-s)A^\star} V_k^\star(s) E_k E_k^\star V_k(s) e^{(t-s)A} x\, ds, \end{aligned} \tag{2.4}$$

where $x \in H$.

Remark 2.1 The operators (A, B_k, C, P_0),

$$B_k = [\lambda_0 - A^\star]^{1-\alpha} kR(k,A)[\lambda_0 - A^\star]^\alpha D,$$

fulfill assumptions $(\mathcal{H})$ of Chapter 1. So, by Theorem 2.6 of Chapter 1, equation (2.4) has a unique mild solution $P_k \in C_s([0,\infty[\,;\Sigma^+(H))$. ∎

Finally, in order to make simpler some estimates, we introduce a constant L such that

$$\begin{cases} (i)\ \ L \ge 1, \quad \|(\lambda_0 - A)^{-1}\| \le L, \\ \quad\ \ \|(\lambda_0 - A)^{-\gamma}\| \le L, \quad \forall\, \gamma \in]0,1[\,, \\ \quad\ \ \|E_k E_k^\star\| \le L, \quad \forall\, k \in \mathbf{N}. \\ \\ (ii)\ \ \|e^{t(A-\lambda_0)}\| \le L, \\ \quad\ \ t\|(A-\lambda_0)e^{t(A-\lambda_0)}\| \le L, \\ \quad\ \ t^\gamma \|(\lambda_0 - A)^{-\gamma} e^{t(A-\lambda_0)}\| \le L, \quad \forall\, \gamma \in]0,1[\,, \forall\, t > 0. \end{cases}$$

2.2 Riccati equation for $\alpha > \frac{1}{2}$

Assume that the assumptions $(\mathcal{HP})$ are verified with $\alpha > \frac{1}{2}$, and set, for the sake of simplicity, $\lambda_0 = 0$. We shall look for a solution of the Riccati equation (2.1) in the following functional space. For any interval $[a,b]$ we denote by $C_{s,\alpha}([a,b];\Sigma(H))$ the set of all $P \in C_s([a,b];\Sigma(H))$ such that

$$\begin{cases} (i) \quad P(t)x \in D((-A^\star)^{1-\alpha}), \quad \forall x \in H,\ \forall t \in]a,b], \\ (ii) \ (-A^\star)^{1-\alpha}P \in C(]a,b];\mathcal{L}(H)), \\ (iii) \ \lim_{t\to a}(t-a)^{1-\alpha}(-A^\star)^{1-\alpha}P(t)x = 0, \quad \forall x \in H. \end{cases} \tag{2.5}$$

Define

$$\|P\|_1 = \sup_{t\in]a,b]} \|(t-a)^{1-\alpha}(-A^\star)^{1-\alpha}P(t)\|.$$

$C_{s,\alpha}([a,b];\Sigma(H))$ endowed with the norm

$$\|P\|_\alpha = \|P\| + \|P\|_1$$

is a Banach space. We set

$$C_{s,\alpha}([a,b];\Sigma^+(H)) = \{P \in C_{s,\alpha}([a,b];\Sigma(H)) : \ P(t) \geq 0,\ \forall\, t \in]a,b]\}$$

and denote by $C_{s,\alpha}([0,\infty];\Sigma(H))$ the set of all $P \in C_s([0,\infty];\Sigma(H))$ such that $P \in C_{s,\alpha}([0,T];\Sigma(H))$ for all $T > 0$.

Definition 2.1 *A* mild solution *of problem* (2.1) *in the interval* $[0,T]$ *is an operator valued function* $P \in C_{s,\alpha}([0,T];\Sigma(H))$ *that verifies the integral equation*

$$\begin{aligned} P(t)x &= e^{tA^\star}P_0e^{tA}x + \int_0^t e^{(t-s)A^\star}C^\star Ce^{(t-s)A}x\,ds \\ &\quad - \int_0^t e^{(t-s)A^\star}V^\star(s)EE^\star V(s)e^{(t-s)A}x\,ds, \end{aligned} \tag{2.6}$$

where $x \in H$. *A* weak solution *of problem* (2.1) *in* $[0,T]$ *is an operator valued function* $P \in C_{s,\alpha}([0,T]\ ;\Sigma(H))$ *such that, for any* $x,y \in$

$D(A)$, $(P(\cdot)x, y)$ *is differentiable in* $[0, T]$ *and verifies*

$$\begin{cases} \dfrac{d}{dt}(P(t)x, y) = (P(t)x, Ay) + (P(t)Ax, y) + (Cx, Cy) \\ \qquad\qquad - (E^\star V(t)x, E^\star V(t)y), \\ \\ P(0) = P_0. \end{cases} \tag{2.7}$$

Proposition 2.1 *Let* $P \in C_{s,\alpha}([0, T]; \Sigma(H))$*. Then* P *is a mild solution of problem* (2.1) *if and only if* P *is a weak solution.*

Proof. It is completely similar to the proof of Proposition 2.3 of Chapter 1, and so, it will be omitted. ∎

We need, as in Chapter 1, a generalization of the Contraction Mapping Principle. Let $T > 0$, and let $\{\gamma_k\}$ be a sequence of mappings from $C_{s,\alpha}([0, T]; \Sigma(H))$ into itself, such that

$$\|\gamma_k(P) - \gamma_k(Q)\|_\alpha \le a\|P - Q\|_\alpha$$

for some $a \in [0, 1[$ and all $P, Q \in C_{s,\alpha}([0, T]; \Sigma(H))$. Moreover assume that there exists a mapping γ from $C_{s,\alpha}([0, T]; \Sigma(H))$ into itself such that $\forall\, P \in C_{s,\alpha}([0, T]; \Sigma(H))$, $\forall\, m \in \mathbf{N}$, $\forall\, x \in H$

$$\begin{cases} \lim\limits_{k\to\infty} \gamma_k^m(P)x = \gamma^m(P)x, \quad \text{in} \quad C([0, T]; H) \\ \\ \lim\limits_{k\to\infty} t^{1-\alpha}(-A^\star)^{1-\alpha}\gamma_k^m(P)x = t^{1-\alpha}(-A^\star)^{1-\alpha}\gamma^m(P)x, \\ \text{in} \quad C([0, T]; H), \end{cases} \tag{2.8}$$

Then, by the Contraction Mapping Principle, there exist unique P_k and P in $C_{s,\alpha}([0, T]; \Sigma(H))$ such that

$$\gamma_k(P_k) = P_k, \quad \gamma(P) = P.$$

The following result can be proved as Lemma 2.1 of Chapter 1.

Lemma 2.1 *Under the previous assumptions on the sequence of mappings* $\{\gamma_k\}$

$$\begin{cases} \lim\limits_{k\to\infty} P_k(\cdot)x = P(\cdot)x \quad \textit{in} \quad C([0, T]; H), \\ \\ \lim\limits_{k\to\infty} t^{1-\alpha}(-A^\star)^{1-\alpha}P_k(\cdot)x = t^{1-\alpha}(-A^\star)^{1-\alpha}P(\cdot)x, \\ \textit{in} \quad C([0, T]; H), \quad \forall\, x \in H,\ T > 0. \end{cases} \tag{2.9}$$

The main result of this section is the following

Theorem 2.1 *Assume that assumptions $(\mathcal{HP})$ are verified with $\alpha > \frac{1}{2}$. Then, problem (2.1) has a unique mild solution $P \in C_{s,\alpha}([0,T];\Sigma(H))$. Moreover, the solution P_k of problem (2.4) also belongs to $C_{s,\alpha}([0,T];\Sigma(H))$ for all $k \in \mathbf{N}$ and (2.9) holds for all $x \in H$ and $T > 0$.*

Proof. We write problem (2.1) as

$$P = \gamma(P) = F + H - \varphi(P),$$

where

$$F(t) = e^{tA^\star} P_0 e^{tA},$$

$$H(t) = \int_0^t e^{(t-s)A^\star} C^\star C e^{(t-s)A} ds,$$

$$(\varphi(P))(t) = \int_0^t e^{(t-s)A^\star} V^\star(s) E E^\star V(s) e^{(t-s)A} ds.$$

Analogously, we write problem (2.4) as

$$P_k = \gamma_k(P_k) = F + H - \varphi_k(P_k)$$

where

$$(\varphi_k(P))(t)x = \int_0^t e^{(t-s)A^\star} V_k^\star(s) E_k E_k^\star V_k(s) e^{(t-s)A} x\, ds, \ x \in H$$

and $V_k(s) = V_{P_k}(s)$.

In the sequel of the proof $T > 0$, $\alpha \in]0,1[$ and $L \geq 1$ are fixed. We proceed in several steps.

Step 1. $F, H \in C_{s,\alpha}([0,T];\Sigma(H))$ and

$$\|F\|_\alpha \leq 2L^2\|P_0\|, \quad \|H\|_\alpha \leq (1 + \frac{1}{\alpha})L^2\|C\|^2 T. \tag{2.10}$$

In fact, we clearly have $\|F\| \leq L^2\|P_0\|$. Moreover

$$\|t^{1-\alpha}(-A^\star)^{1-\alpha} F(t)\| = \|t^{1-\alpha}(-A^\star)^{1-\alpha} e^{tA^\star} P_0 e^{tA}\| \leq L^2\|P_0\|$$

so that $\|F\|_1 \leq L^2\|P_0\|$ and the first inequality in (2.10) follows. Concerning the second, it suffices to remark that $\|H(t)\| \leq L^2\|C\|^2T$ and, for all $x \in H$,

$$|t^{1-\alpha}(-A^\star)^{1-\alpha}H(t)| = \left|t^{1-\alpha}(-A^\star)^{1-\alpha}\int_0^t e^{(t-s)A^\star}C^\star Ce^{(t-s)A}\,ds\right|$$

$$\leq L^2\|C\|^2t^{1-\alpha}\int_0^t (t-s)^{\alpha-1}\,ds\,|x| \leq \frac{1}{\alpha}L^2\|C\|^2T.$$

Step 2. φ and φ_k map $C_{s,\alpha}([0,T];\Sigma(H))$ into itself. Moreover there exists a constant $C_{1,T} > 0$ such that, for all $k \in \mathbf{N}$ and $P \in C_{s,\alpha}([0,T];\Sigma(H))$

$$\|\varphi(P)\|_\alpha + \|\varphi_k(P)\|_\alpha \leq C_{1,T}\|P\|_\alpha^2, \tag{2.11}$$

where

$$C_{1,T} = 2\left[\frac{1}{2\alpha-1} + \beta(2\alpha-1,\alpha)\right]T^{2\alpha-1}L^3,$$

and β represents the Euler function,

$$\beta(x,y) = \int_0^1 t^{x-1}(1-t)^{y-1}dt.$$

We only estimate $\|\varphi(P)\|_\alpha$. The estimate for $\|\varphi_k(P)\|_\alpha$ is similar. Let $P \in C_{s,\alpha}([0,T];\Sigma(H))$ and $x \in H$; then $\varphi(P) \in C_s([0,T];\Sigma(H))$ and we have

$$|(\varphi(P))(t)| \leq L^3\int_0^t s^{2\alpha-2}\,ds\,\|P\|_1^2 \leq \frac{1}{2\alpha-1}L^3\|P\|_1^2T^{2\alpha-1}$$

which implies

$$\|\varphi(P)\| \leq \frac{1}{2\alpha-1}L^3\|P\|_1^2T^{2\alpha-1}. \tag{2.12}$$

By arguing as in Lemma 3.1 of Chapter 1, it is not difficult to show that $\varphi(P) \in C_{s,\alpha}([0,T];\Sigma(H))$. Moreover we have

$$|t^{1-\alpha}(-A^\star)^{1-\alpha}(\varphi(P))(t)| \leq L^3t^{1-\alpha}\int_0^t (t-s)^{\alpha-1}s^{2\alpha-2}\,ds\|P\|_1^2$$

$$\leq L^3\|P\|_1^2T^{2\alpha-1}\int_0^1 (1-\sigma)^{\alpha-1}\sigma^{2\alpha-2}\,d\sigma|x|$$

$$= L^3\|P\|_1^2T^{2\alpha-1}\beta(2\alpha-1,\alpha),$$

which yields

$$\|\varphi(P)\|_1 \le T^{2\alpha-1}L^3\beta(2\alpha-1,\alpha)\|P\|_1^2. \tag{2.13}$$

Now, (2.11) follows from (2.12) and (2.13).

Step 3. There exists a constant $C_{2,T} > 0$ such that, for all $P, Q \in C_{s,\alpha}([0,T];\Sigma(H))$ we have

$$\begin{aligned} &\|\varphi(P)-\varphi(Q)\|_\alpha + \|\varphi_k(P)-\varphi_k(Q)\|_\alpha \\ &\le C_{2,T}(\|P\|_\alpha + \|Q\|_\alpha)\|P-Q\|_\alpha, \end{aligned} \tag{2.14}$$

where

$$C_{2,T} = 2\left[\frac{1}{2\alpha-1} + \beta(2\alpha-1,\alpha)\right] T^{2\alpha-1}L^3.$$

In fact, let $P, Q \in C_{s,\alpha}([0,T];\Sigma(H))$, $V = V_P$, $Z = V_Q$ and $x \in H$; we have

$$\begin{aligned} &|t^{1-\alpha}(\varphi(P))(t)x - t^{1-\alpha}(\varphi(Q))(t)| \\ &\le \left|t^{1-\alpha}\int_0^t e^{(t-s)A^\star}[(V(s)-Z(s))^\star EE^\star V(s)]\, e^{(t-s)A}\, ds\right| \\ &\quad + \left|t^{1-\alpha}\int_0^t e^{(t-s)A^\star}[Z^\star(s)EE^\star(V(s)-Z(s))]\, e^{(t-s)A}\, ds\right|, \end{aligned}$$

and the conclusion follows by arguing as in the second step.

We denote by $B(r,t)$ the ball

$$B(r,t) = \{P \in C_{s,\alpha}([0,t];\Sigma(H)) \,:\, \|P\|_\alpha \le r\}.$$

Step 4. For all $p > 0$, there exist $\tau = \tau(p) > 0$ and $r = r(p) > 0$, such that

$$\|P_0\| \le p \quad \Rightarrow \quad \gamma(B(r,\tau)) \cup \gamma_k(B(r,\tau)) \subset B(r,\tau), \tag{2.15}$$

and

$$\begin{aligned} &\|\gamma(P)-\gamma(Q)\|_\alpha \le \tfrac{1}{2}\|P-Q\|_\alpha, \\ &\|\gamma_k(P)-\gamma_k(Q)\|_\alpha \le \tfrac{1}{2}\|P-Q\|_\alpha, \end{aligned} \tag{2.16}$$

for all P, $Q \in \gamma(B(r,\tau))$. This follows easily from the inequalities (2.10), (2.11) and (2.14).

Step 5. Existence and positivity.

Choose $p > \|P_0\|$ and let $\tau = \tau(p)$. By (2.15), (2.16) and the Contraction Mapping Principle, it follows that there exist unique P and P_k in $C_{s,\alpha}([0,T];\Sigma(H))$ such that

$$\gamma(P) = P, \quad \gamma_k(P_k) = P_k.$$

So problems (2.1) and (2.3) have unique solutions P and P_k respectively in $[0,\tau]$. Moreover, by Lemma 2.1 it follows that (2.9) holds for all $x \in H$ and $T > 0$. Since, by Theorem 2.6 of Chapter 1, $P_k(t) \geq 0$ for all $t \in [0,\tau]$, we finally find

$$P(t) \geq 0, \quad \text{for all } t \in [0,\tau].$$

Step 6. Global existence and uniqueness.

We first remark that we have shown existence in the interval $[0,\tau]$ which only depends on the norm of P_0. Thus, in order to prove global existence we only have to estimate the norm of $P(t)$, $t \in [0,\tau[$. Let $P \in C_{s,\alpha}([0,\tau];\Sigma(H))$ be the solution of (2.1) in $[0,\tau]$. Since $P(t) \geq 0$ we have

$$\begin{aligned} (P(t)x,x) &\leq (P_0e^{tA}x, e^{tA}x) + \int_0^t |Ce^{(t-s)A}x|^2 ds \\ &\leq L^2\{\|P_0\| + T\|C\|^2\}|x|^2, \end{aligned}$$

which implies that

$$\|P(t)\| \leq \|L\|^2\{\|P_0\| + T\|C\|^2\}, \text{ for all } t \in [0,\tau].$$

Set now $p_1 = L^2\{\|P_0\| + T\|C\|^2\}$ and $\tau_1 = \tau(p_1)$. Let P be the solution to the Riccati equation (2.1) in $[0,\tau_1]$. Proceeding as above, we can solve the Riccati equation

$$\begin{aligned} Q(t) &= e^{(t-\tau_1+\varepsilon)A^\star}P(\tau_1-\varepsilon)e^{(t-\tau_1+\varepsilon)A} \\ &\quad + \int_{\tau_1-\varepsilon}^t e^{(t-s)A^\star}C^\star Ce^{(t-s)A}\,ds \\ &\quad - \int_{\tau_1-\varepsilon}^t e^{(t-s)A^\star}V^\star_{Q(s)}E^\star EV_{Q(s)}e^{(t-s)A}\,ds, \end{aligned}$$

in the interval $[\tau_1-\varepsilon, 2\tau_1-\varepsilon]$ (here ε is a small positive number to be chosen later) and we find a solution $Q \in C_{s,\alpha}([\tau_1-\varepsilon, 2\tau_1-\varepsilon]; \Sigma(H))$ such that

$$\|Q(t)\| \le L^2\left\{\|P_0\| + T\|C\|^2\right\}, \quad \forall t \in [\tau_1-\varepsilon, 2\tau_1-\varepsilon].$$

Now, it ts easy to check that setting

$$\overline{P}(t) = \begin{cases} P(t), & \text{if } t \in [0,\tau_1], \\ Q(t), & \text{if } t \in [\tau_1, 2\tau_1-\varepsilon], \end{cases}$$

then $\overline{P}$ is the unique solution of (2.1) in $[0, 2\tau_1-\varepsilon]$; moreover $\overline{P}(t) \ge 0$ in $[0, 2\tau_1-\varepsilon]$ and

$$\|\overline{P}(t)\| \le L^2\left\{\|P_0\| + T\|C\|^2\right\}, \quad t \in [0, 2\tau_1-\varepsilon].$$

Thus we can repeat this argument successively in the intervals $[2\tau_1 - 2\varepsilon, 3\tau_1-2\varepsilon]$, $[3\tau_1-3\varepsilon, 4\tau_1-3\varepsilon]$ and so on. So we get the conclusions in N steps, provided N and ε are chosen such that $N\tau_1 > T$ and

$$\varepsilon \le \frac{N\tau_1 - T}{N-1}.$$

The proof is complete. ■

We now prove the *continuous dependence* with respect to P_0 and C of the solutions of (2.1). Consider a sequence of Riccati equations:

$$\begin{cases} \dfrac{dP^h}{dt} = A^\star P^h + P^h A - (V^h)^\star E E^\star V^h + (C^h)^\star C^h, \\ \\ P^h(0) = P_0^h, \end{cases} \tag{2.17}$$

where $V^h = V^{P^h}$, $h \in \mathbf{N}$. Assume that

$$\begin{cases} (i) & \textit{For any } h \in \mathbf{N},\ (A, D, C^h, P_0^h) \textit{ fulfill} \\ & (\mathcal{HP}) \textit{ with } \alpha > \frac{1}{2}, \\ (ii) & \lim\limits_{h\to\infty} (C^h)^\star C^h x = C^\star C x, \textit{ for all } x \in H, \\ (iii) & \lim\limits_{h\to\infty} P_0^h x = P_0 x \textit{ for all } x \in H, \end{cases} \tag{2.18}$$

and prove the following result.

Proposition 2.2 *Assume that* $(\mathcal{HP})$ *and* (2.18) *are verified with* $\alpha > \frac{1}{2}$. *Let* P *and* P^h *be the mild solutions to* (2.1) *and* (2.17) *respectively. Then, for any* $x \in H$ *and any* $T > 0$ *we have*

$$\begin{cases} \lim_{h\to\infty} P^h(\cdot)x = P(\cdot)x, \ \ in \ \ C([0,T];H), \\ \\ \lim_{h\to\infty} t^{1-\alpha}V^h(\cdot)x = t^{1-\alpha}V(\cdot)x \ \ in \ \ C([0,T];H). \end{cases} \tag{2.19}$$

Proof. Fix $T > 0$. By the Uniform Boundnedess Theorem there exist positive numbers q and c, such that

$$\|P_0^h\| \le q, \quad \|(C^h)^\star C^h\| \le c, \quad \forall\, h \in \mathbf{N}.$$

Set $p = L^2(q + Tc)$, $\tau = \tau(p)$; then, arguing as in the proof of Theorem 2.7 of Chapter 1, we first show that (2.19) holds in $[0,\tau]$ and then we prove that this argument can be iterated in the interval $[\tau, 2\tau]$ and so on. ∎

We finally show a monotonicity property of the solutions of the Riccati equation

Proposition 2.3 *Assume that* $(A, D, C_i, P_{i,0})$ *fulfill* $(\mathcal{HP})$ *for* $i = 1, 2$, *with* $\alpha > \frac{1}{2}$, *and, in addition, that*

$$P_{1,0} \le P_{2,0}, \quad C_1^\star C_1 \le C_2^\star C_2.$$

Let $P_i, i = 1,2$ *be the mild solution of the Riccati equations*

$$\begin{cases} \frac{d}{dt}P_i = A^\star P_i + P_i A - V_i^\star E E^\star V_i + (C_i)^\star C_i, \\ \\ P_i(0) = P_{i,0} \end{cases} \tag{2.20}$$

where $V_i = V_{P_i}$. *Then we have*

$$P_1(t) \le P_2(t), \quad t \ge 0.$$

Proof. For any $k \in \mathbf{N}$, $i = 1, 2$, let $P_{i,k}$ be the mild solution to the Riccati equation

$$\begin{cases} \frac{d}{dt}P_{i,k} = A^\star P_{i,k} + P_{i,k} A - V_{i,k}^\star E E^\star V_{i,k} + (C_i)^\star C_i, \\ \\ P_{i,k}(0) = P_{i,0}, \end{cases}$$

where $V_{i,k} = (-A_k^\star)^{1-\alpha} P_{i,k}$. Then, by Proposition 2.2 of Chapter 1,

$$P_{1,k}(t) \le P_{2,k}(t), \quad t \ge 0, k \in \mathbf{N}.$$

The conclusion now follows from Theorem 2.1. ∎

2.3 Solution of the Riccati equation for $\alpha \le \frac{1}{2}$

We assume here that assumptions $(\mathcal{HP})$ are verified with $\alpha \le \frac{1}{2}$ and some $\beta \in]\frac{1}{2} - \alpha, \frac{1-\alpha}{2}[$. We set again $\lambda_0 = 0$ and denote by $P_\beta \in \Sigma^+(H)$ the closure of $(-A^\star)^\beta P_0 (-A)^\beta$, so that

$$P_0 = (-A^\star)^{-\beta} P_\beta (-A)^{-\beta}.$$

One can see very easily that the proof of Theorem 2.1 cannot be repeated, since $2 - 2\alpha \ge 1$ and some integrals would be divergent. Thus we introduce, following F. FLANDOLI [6], a more complicated functional space and also some additional notations. If $[a, b]$ is any interval, we set

$$\mu_{[a,b]}(t) = (t-a)^{1-\alpha-\beta}, \quad t \in [a, b]$$

and

$$\nu_{[a,b]}(t, s) = \frac{(s-a)^{1-\alpha-\beta}(t-s)^\beta}{(t-a)^\beta}, \quad a \le s \le t \le b.$$

If $P \in C_s([a, b]; \Sigma^+(H))$, we set

$$V_P(t) = (-A^\star)^{1-\alpha} P(t), \quad t \in [a, b]$$

$$W_P(t, s) = (-A^\star)^{1-\alpha} P(s)(-A)^\beta e^{(t-s)A}, \quad a \le s \le t \le b.$$

When no confusion may arise, we will drop indexes $[a, b]$ and P, in the formulas above.

For any interval $[a, b]$ we denote by $C_{s,\alpha,\beta}([a, b]; \Sigma(H))$ the set of all $P \in C_s([a, b]; \Sigma(H))$ such that (2.5)-(i)-(ii) hold and moreover

$$\begin{cases} (i) \ \ P(t)x \in D((-A^\star)^{1-\alpha}), \quad \forall x \in H, \ \forall t \in]a, b], \\ (ii) \ \ V_P \in C(]a, b]; \mathcal{L}(H)), \\ (iii) \ \ \lim_{t \to a} \mu_{[a,b]}(t) V_P(t)x = 0, \quad \forall x \in H. \\ (iv) \ \ \lim_{(t,s) \to (a,a)} \nu_{[a,b]}(t, s) W_P(t, s)x = 0, \quad \forall x \in H. \end{cases} \tag{2.21}$$

Define

$$\|P\|_2 = \sup_{t\in\,]a,b]} \|\mu_{[a,b]}(t)V_P(t)\|,$$

and

$$\|P\|_3 = \sup_{a\le s<t\le b} \|\nu_{[a,b]}(t,s)W_P(t,s)\|.$$

$C_{s,\alpha,\beta}([a,b];\Sigma(H))$, endowed with the norm

$$\|P\|_{\alpha,\beta} = \|P\| + \|P\|_2 + \|P\|_3,$$

is a Banach space. We set

$$C_{s,\alpha,\beta}([a,b];\Sigma^+(H))$$

$$= \{P \in C_{s,\alpha,\beta}([a,b];\Sigma(H)) \,:\, P(t)\ge 0, \quad \forall\, t\in\,]a,b]\},$$

and let $C_{s,\alpha,\beta}([0,\infty[\,;\Sigma(H))$ be the set of all $P \in C_s([0,\infty[\,;\Sigma(H))$ such that $P \in C_{s,\alpha,\beta}([0,T];\Sigma(H))$ for all $T > 0$.

Definition 2.2 *A* mild solution *of problem* (2.1) *in the interval* $[0,T]$ *is an operator valued function* $P \in C_{s,\alpha,\beta}([0,T];\Sigma(H))$ *which verifies the integral equation* (2.6). *A* weak solution *of problem* (2.1) *in* $[0,T]$ *is an operator valued function* $P \in C_{s,\alpha,\beta}([0,T];\Sigma(H))$ *such that, for any* $x,y \in D(A)$, $(P(\cdot)x,y)$ *is differentiable in* $[0,T]$ *and verifies* (2.7).

Proposition 2.4 *Let* $P \in C_{s,\alpha,\beta}([0,T];\Sigma(H))$. *Then* P *is a mild solution of problem* (2.1) *if and only if* P *is a weak solution.*

Proof. It is completely similar to the proof of Proposition 2.3 of Chapter 1. ∎

We shall need, as in §2, a generalization of the Contraction Mapping Principle. Let $T > 0$, and let $\{\gamma_k\}$ be a sequence of mappings from $C_{s,\alpha,\beta}([0,T];\Sigma(H))$ into itself, such that

$$\|\gamma_k(P) - \gamma_k(Q)\|_\alpha \le a\|P-Q\|_\alpha,$$

for some $a \in [0,1[$ and all $P,Q \in C_{s,\alpha,\beta}([0,T];\Sigma(H))$. Assume moreover that there exists a mapping γ from $C_{s,\alpha,\beta}([0,T];\Sigma(H))$ into itself such that:

$$\begin{cases} \lim\limits_{k\to\infty} \gamma_k^m(P)x = \gamma^m(P)x \text{ in } C([0,T];H), \\ \\ \lim\limits_{k\to\infty} \mu(\cdot)V_{\gamma_k^m(P)}x = \mu(\cdot)V_{\gamma^m(P)}x \text{ in } C([0,T];H), \\ \\ \lim\limits_{k\to\infty} \nu(\cdot,\cdot)W_{\gamma_k^m(P)}x = \nu(\cdot,\cdot)W_{\gamma^m(P)}x \text{ in } C(\Delta_T;H), \end{cases}$$

where $\Delta_T = \{(t,s) : 0 \le s \le t \le T\}$, $\forall P \in C_{s,\alpha,\beta}([0,T];\Sigma(H))$, $\forall m \in \mathbf{N}$ and $x \in H$. Then, by the Contraction Mapping Principle, there exist unique P_k and P in $C_{s,\alpha,\beta}([0,T];\Sigma(H))$ such that

$$\gamma_k(P_k) = P_k, \quad \gamma(P) = P$$

The following result can be proved as Lemma 2.1 of Chapter 1.

Lemma 2.2 *Under the previous assumptions on the sequence of mappings $\{\gamma_k\}$, for all $x \in H$*

$$\begin{cases} \lim\limits_{k\to\infty} P_k(\cdot)x = P(\cdot)x \;\; in \;\; C([0,T];H), \\ \\ \lim\limits_{k\to\infty} \mu(\cdot)V_{P_k}(\cdot)x = \mu(\cdot)V_P(\cdot)x \; in \;\; C([0,T];H), \\ \\ \lim\limits_{k\to\infty} \nu(\cdot,\cdot)W_{P_k}(\cdot,\cdot)x = \nu(\cdot,\cdot)W_P(\cdot,\cdot)x \;\; in \;\; C(\Delta_T;H). \end{cases} \tag{2.22}$$

Prior to state the main result of this section, we need another lemma

Lemma 2.3 *If $P \in C_{s,\alpha,\beta}([0,T];\Sigma(H))$, we have*

$$\|(-A^\star)^{1-\alpha}P(s)(-A)^{1-\alpha}e^{(t-s)A}\| \le 2^{1-\alpha}L\frac{t^\beta}{s^{1-\alpha-\beta}(t-s)^{(1-\alpha)}}\|P\|_3.$$

Proof. Let $t \in]0,T]$, $s \in [0,t[$. We have

$$(-A^\star)^{1-\alpha}P(s)(-A)^{1-\alpha}e^{(t-s)A}$$

$$= (-A^\star)^{1-\alpha}P(s)(-A)^\beta e^{(\frac{t+s}{2}-s)A}(-A)^{1-\alpha-\beta}e^{(\frac{t+s}{2}-s)A}$$

It follows that

$$\|(-A^\star)^{1-\alpha}P(s)(-A)^{1-\alpha}e^{(t-s)A}\|$$

$$\leq L(t+s)^{\beta}s^{\alpha+\beta-1}(t-s)^{-\beta}(\tfrac{t-s}{2})^{\alpha+\beta-1}\|P\|_3$$

which implies the conclusion. ∎

The main result of this section is the following.

Theorem 2.2 *Assume that assumptions ($\mathcal{HP}$) are verified for $\alpha \leq \frac{1}{2}$ and some $\beta \in]\frac{1}{2}-\alpha, \frac{1-\alpha}{2}[$. Then the Riccati equation (2.1) has a unique mild solution $P \in C_{s,\alpha,\beta}([0,\infty[\,;\Sigma^+(H))$. Moreover, the solution P_k of problem (2.4) also belongs to $C_{s,\alpha,\beta}([0,\infty[\,;\Sigma(H))$ for all $k \in \mathbf{N}$ and (2.22) holds $\forall x \in H$.*

Proof. We write problems (2.1) and (2.4) as

$$P = \gamma(P) = F + H - \varphi(P),\ P_k = \gamma_k(P_k) = F + H - \varphi_k(P_k),\ k \in \mathbf{N},$$

where F, H, φ and φ_k are defined as in the proof of Theorem 2.1. We fix $T > 0$ and proceed in several steps.

Step 1. $F \in C_{s.\alpha,\beta}([0,T];\Sigma(H))$ and

$$\|F\|_{\alpha,\beta} \leq L^2\|P_0\| + 2L^3\|P_\beta\|. \tag{2.23}$$

We have indeed

$$\begin{aligned}&\|t^{1-\alpha-\beta}(-A^\star)^{1-\alpha}F(t)\|\\ &= \|t^{1-\alpha-\beta}(-A^\star)^{1-\alpha-\beta}e^{tA^\star}P_\beta(-A)^{-\beta}e^{tA}\|\\ &\leq L^3\|P_\beta\|\end{aligned}$$

which implies

$$\|F\|_2 \leq L^3\|P_\beta\|. \tag{2.24}$$

Moreover, if $t \in]0,T]$ and $s \in]0,t]$, we have

$$\|\nu(t,s)(-A^\star)^{1-\alpha}F(s)(-A)^{\beta}e^{(t-s)A}\|$$

$$= \|\nu(t,s)(-A^\star)^{1-\alpha-\beta}e^{sA^\star}P_\beta e^{tA}\|$$

$$\leq L^2[(t-s)/t]^{\beta}\|P_\beta\| \leq L^3\|P_\beta\|$$

since $(t-s)/t \le 1$. It follows that

$$\|F\|_3 \le L^3 \|P_\beta\|. \tag{2.25}$$

Now (2.23) follows from (2.24) and (2.25) and the estimate $\|F\| \le L^2\|P_0\|$.

Step 2. $H \in C_{s,\alpha,\beta}([0,T];\Sigma(H))$ and

$$\begin{aligned}\|H\|_{\alpha,\beta} &\le L^2\|C\|^2 T + \tfrac{1}{\alpha} L^2 \|C\|^2 T^{1-\beta} \\ &\quad + L^2\|C\|^2 T^{1-2\beta+2\varepsilon} \int_0^{+\infty} \rho^{\varepsilon-1}(1+\rho)^\beta \, d\rho,\end{aligned} \tag{2.26}$$

where $\varepsilon = \frac{1}{2} \min\ \{\alpha, \beta\}$.

Let $x \in H$, then

$$\begin{aligned}&|t^{1-\alpha-\beta}(-A^\star)^{1-\alpha} H(t)x| \\ &= \left| t^{1-\alpha-\beta}(-A^\star)^{1-\alpha} \int_0^t e^{(t-s)A^\star} C^\star C e^{(t-s)A} x \, ds \right| \\ &\le L^2 \|C\|^2 t^{1-\alpha-\beta} \int_0^t (t-s)^{\alpha-1} \, ds \, |x| \le \frac{1}{\alpha} L^2 \|C\|^2 T^{1-\beta} |x|.\end{aligned}$$

It follows that

$$\|H\|_2 \le \frac{1}{\alpha} L^2 \|C\|^2 T^{1-\beta}. \tag{2.27}$$

Moreover

$$\begin{aligned}&|\nu(t,s)(-A^\star)^{1-\alpha} H(s)(-A)^\beta e^{(t-s)A} x| \\ &= \left| \nu(t,s)(-A^\star)^{1-\alpha} \int_0^s e^{(s-\sigma)A^\star} C^\star C(-A)^\beta e^{(t-\sigma)A} \, d\sigma \, x \right| \\ &\le L^2\|C\|^2 \nu(t,s) \int_0^s (s-\sigma)^{\alpha-1}(t-\sigma)^{-\beta} \, d\sigma \, |x| \\ &= L^2\|C\|^2 \nu(t,s) \int_0^{s/(t-s)} (\rho)^{\alpha-1}(1+\rho)^{-\beta} \, d\rho \, |x|.\end{aligned}$$

Since $\rho^{\alpha-\varepsilon} \le [s/(t-s)]^{\alpha-\varepsilon}$, we have

$$|\nu(t,s)(-A^\star)^{1-\alpha}H(s)(-A)^\beta e^{(t-s)A}x|$$

$$\le L^2\|C\|^2\frac{(t-s)^\varepsilon s^{1-\beta-\varepsilon}}{t^\beta}\int_0^{s/(t-s)}\frac{d\rho}{\rho^{1+\varepsilon}(1+\rho)^\beta}|x|$$

$$\le L^2\|C\|^2T^{1-2\beta+2\varepsilon}\int_0^{+\infty}\frac{d\rho}{\rho^{1+\varepsilon}(1+\rho)^\beta}|x|$$

which implies

$$\|H\|_3 \le L^2\|C\|^2T^{1-2\beta}\int_0^{+\infty}\frac{d\rho}{\rho^{1+\varepsilon}(1+\rho)^\beta}. \tag{2.28}$$

Since $\|H\| \le L^2\|C\|^2T$, the conclusion follows from (2.27) and (2.28).

Step 3. φ and φ_k map $C_{s,\alpha,\beta}([0,T];\Sigma(H))$ into itself. Moreover there exists two constants $C_1 > 0$ and $C_2 > 0$ such that for all $P \in C_{s,\alpha,\beta}([0,T];\Sigma(H))$

$$\begin{aligned}&\|\varphi(P)\|_{\alpha,\beta} + \|\varphi(P_k)\|_{\alpha,\beta}\\ &\le T^{2\alpha+2\beta-1}C_1\|P\|^2_{\alpha,\beta} + C_2(\|P\|_2\|P\|_3 + \|P\|_3^2).\end{aligned} \tag{2.29}$$

Let $P \in C_{s,\alpha,\beta}([0,T];\Sigma(H))$; we have

$$\begin{aligned}\|\varphi(P)\| &\le L^3\|P\|_2^2\int_0^t (t-s)^{2\alpha+2\beta-2}ds\\ &= \tfrac{1}{2\alpha+2\beta-1}L^3\|P\|_2^2T^{2\alpha+2\beta-1},\end{aligned} \tag{2.30}$$

since $2\alpha+2\beta-1>0$. Moreover for all $x \in H$ we have

$$|t^{1-\alpha-\beta}(\varphi(P))(t)x|$$

$$= \left|t^{1-\alpha-\beta}\int_0^t e^{(t-s)A^\star}V_P^\star(s)EE^\star V_P(s)e^{(t-s)A}x\,ds\right|$$

$$\begin{aligned}= \Big|t^{1-\alpha-\beta}\int_0^t &[(-A^\star)^{1-\alpha}P(s)(-A)^{1-\alpha}e^{(t-s)A}]^\star\\ &EE^\star[(-A^\star)^{1-\alpha}P(s)]e^{(t-s)A}x\,ds\Big|.\end{aligned}$$

By using Lemma 2.3, we obtain

$$|t^{1-\alpha-\beta}(\varphi(P))(t)x|$$

$$\leq 2^{1-\alpha}L^3t^{1-\alpha}\int_0^t s^{2\alpha+2\beta-2}(t-s)^{\alpha-1}ds\|P\|_2\|P\|_3|x|$$

$$= 2^{1-\alpha}L^3t^{2\alpha+2\beta-1}\int_0^1 \sigma^{2\alpha+2\beta-2}(1-\sigma)^{\alpha-1}d\sigma\|P\|_2\|P\|_3|x|.$$

Hence

$$\|\varphi(P)\|_2$$
$$\leq 2^{1-\alpha}L^3T^{2\alpha+2\beta-1}\int_0^1 \sigma^{2\alpha+2\beta-2}(1-\sigma)^{\alpha-1}d\sigma\;\|P\|_2\|P\|_3. \tag{2.31}$$

Finally, using once again Lemma 2.3, we find

$$|\nu(t,s)(-A^\star)^{1-\alpha}(\varphi(P))(s)(-A)^\beta e^{(t-s)A}x|$$

$$= \Big|\nu(t,s)\int_0^s [(-A^\star)^{1-\alpha}P(\sigma)(-A)^{1-\alpha}e^{(s-\sigma)A}]^\star$$

$$EE^\star(-A^\star)^{1-\alpha}P(\sigma)(-A)^\beta e^{(t-\sigma)A}]x\,d\sigma\Big|$$

$$\leq 2^{1-\alpha}L^3s^{1-\alpha}(t-s)^\beta$$

$$\int_0^s \sigma^{2\alpha+2\beta-2}(t-\sigma)^{-\beta}(s-\sigma)^{1-\alpha}\,d\sigma\;\|P\|_3^2|x|.$$

Since $(t-s)^\beta \leq (t-\sigma)^\beta$ we obtain

$$|\nu(t,s)(-A^\star)^{1-\alpha}(\gamma(P))(s)(-A)^\beta e^{(t-s)A}x|$$

$$\leq 2^{1-\alpha}L^3s^{1-\alpha}\int_0^s \sigma^{2\alpha+2\beta-2}(s-\sigma)^{1-\alpha}d\sigma\;\|P\|_3^2|x|$$

$$\leq 2^{1-\alpha}L^3s^{2\alpha+2\beta-1}\int_0^1 \rho^{2\alpha+2\beta-2}(1-\rho)^{\alpha-1}d\rho\;\|P\|_3^2|x|$$

so that

$$\begin{aligned}&\|\gamma(P)\|_3\\&\leq 2^{1-\alpha}L^3 s^{2\alpha+2\beta-1}\int_0^1 \rho^{2\alpha+2\beta-2}(1-\rho)^{1-\alpha}d\rho\,\|P\|_3^2\end{aligned} \tag{2.32}$$

and (2.29) follows from (2.30), (2.31) and (2.32).

Step 4. There exists a constant $C_3 > 0$ such that for all $P, Q \in C_{s,\alpha,\beta}([0,T];\Sigma(H))$ we have

$$\begin{aligned}&\|\gamma(P)-\gamma(Q)\|_{\alpha,\beta}+\|\gamma(P_k)-\gamma(Q_k)\|_{\alpha,\beta}\\&\leq C_3 T^{2\alpha+2\beta-1}(\|P\|_{\alpha,\beta}+\|Q\|_{\alpha,\beta})\,\|P-Q\|_{\alpha,\beta}.\end{aligned} \tag{2.33}$$

The proof is similar to the previous one in step 3.

We denote now by $B(r,t)$ the ball

$$B(r,t) = \{P \in C_{s,\alpha,\beta}([0,t];\Sigma(H))\ :\ \|P\|_{\alpha,\beta} \leq r\}.$$

Step 5. For all $p > 0$, there exist $\tau = \tau(p) > 0$ and $r = r(p) > 0$, such that

$$\|P_\beta\| \leq p \quad \Rightarrow \quad \gamma(B(r,\tau)) \cup \gamma_k(B(r,\tau)) \subset B(r,\tau), \tag{2.34}$$

and for all $P, Q \in B(r,\tau)$

$$\begin{cases} \|\gamma(P)-\gamma(Q)\|_{\alpha,\beta} \leq \frac{1}{2}\|P-Q\|_{\alpha,\beta}, \\ \\ \|\gamma_k(P)-\gamma_k(Q)\|_{\alpha,\beta} \leq \frac{1}{2}\|P-Q\|_{\alpha,\beta}. \end{cases} \tag{2.35}$$

This follows easily from the inequalities (2.23), (2.26), (2.33) and (2.35).

Step 6. Local existence, convergence and positivity.
Choose $p > \|P_\beta\|$ and $\tau = \tau(p)$. Then, by (2.34), (2.35) and the Contractions Mapping Principle, there exist unique P and P_k in $C_{s,\alpha,\beta}([0,\tau];\Sigma(H))$ such that

$$\gamma(P) = P, \quad \gamma_k(P_k) = P_k \tag{2.36}$$

and problems (2.1) and (2.4) have unique solutions P and P_k respectively in $[0,\tau]$.

Moreover, by Lemma 2.2, (2.22) follows. Finally, by Theorem 2.6 of Chapter 1, we have $P_k(t) \geq 0$ for all $t \in [0, \tau]$; this yields

$$P(t) \geq 0, \quad \forall t \in [0, \tau].$$

Step 7. Conclusion.
Since the interval $[0, \tau]$ depends only on the norm of P_β, in order to prove global existence, we have to find an estimate for the norm of $(P(t))_\beta$, $t \in [0, \tau[$. Let $P \in C_{s,\alpha,\beta}([0, \tau]; \Sigma(H))$ be the solution of (2.1) in $[0, \tau]$. Let $x \in D((-A)^\beta)$, since $P(t) \geq 0$, we have

$$\begin{aligned}
&(P(t)(-A)^\beta x, (-A)^\beta x) \\
&\leq (P_0(-A)^\beta e^{tA} x, (-A)^\beta e^{tA} x) + \int_0^t \left|C(-A)^\beta e^{sA} x\right|^2 ds \\
&\leq L^2\{\|P_\beta\| + \tfrac{1}{\beta} T^\beta \|C\|^2\}|x|^2
\end{aligned}$$

which implies

$$\|(-A^\star)^\beta P(t)(-A)^\beta\| \leq L^2\{\|P_\beta\| + \frac{1}{\beta} T^\beta \|C\|^2\}|x|^2, \tag{2.37}$$

for all $t \in [0, \tau]$. Set now

$$p = L^2\{\|P_\beta\| + \frac{1}{\beta} T^\beta \|C\|^2\}|x|^2, \quad \tau_1 = \tau(p);$$

and let P be the solution to the Riccati equation (2.1) in $t \in [0, \tau_1]$. Proceeding as in Step 6 of the proof of Theorem 2.1, we prove that there exists a unique solution of (2.1) in $[0, 2\tau_1 - \varepsilon]$ and so on. The proof is complete. ∎

We now prove the *continuous dependence* with respect to P_β and C of the solutions of (2.1). Consider the sequence (2.17) of the Riccati equations. We assume that

$$\begin{cases}
(i) & \textit{for any } h \in \mathbf{N},\ (A, D, C^h, P_0^h) \textit{ fulfil } (\mathcal{HP}) \\
& \textit{with } \alpha \leq \frac{1}{2}, \\
(ii) & \lim_{n\to\infty} (C^h)^\star C^h x = C^\star C x, \\
& \textit{for all } x \in H, \\
(iii) & \lim_{n\to\infty} P_\beta^h x = P_\beta x \textit{ for all } x \in H.
\end{cases} \tag{2.38}$$

and we prove the result

Proposition 2.5 *Assume that* $(\mathcal{HP})$ *and* (2.38) *are verified for* $\alpha \leq \frac{1}{2}$. *Let* P *and* P^h *be the mild solutions to* (2.1) *and* (2.17), *respectively. Then, for any* $x \in H$ *and any* $T > 0$ *we have the following properties:*

$$\begin{cases} \lim_{h\to\infty} P^h(\cdot)x = P(\cdot)x \ \ in \ \ C([0,T];H) \\ \\ \lim_{h\to\infty} \mu(\cdot)V_{P^h}(\cdot)x = \mu(\cdot)V_P(\cdot)x \ \ in \ \ C([0,T];H) \\ \\ \lim_{h\to\infty} \nu(\cdot,\cdot)W_{P^k}(\cdot,\cdot)x = \nu(\cdot,\cdot)W_P(\cdot,\cdot)x \ \ in \ \ C(\Delta_T;H) \end{cases} \tag{2.39}$$

$\forall x \in H$.

Proof. Fix $T > 0$. By the Uniform Boundnedess Theorem there exist positive numbers q_β and c, such that

$$\forall\, h \in \mathbf{N}, \quad \|P^h_\beta\| \leq q_\beta, \quad \|(C^h)^\star C^h\| \leq c.$$

Set

$$p = L^2\left(q_\beta + \frac{1}{\beta}T^\beta c\right), \quad \tau = \tau(p).$$

Then, arguing as in the proof of Theorem 2.2 we first show that $P_k \to P$ in $C_s([0,\tau];\Sigma(H))$ and then we show that this argument can be iterated in the interval $[\tau, 2\tau]$ and so on. ∎

Finally the following result can be proved as Proposition 2.3.

Proposition 2.6 *Assume that* $(A, D, C_i, P_{i,0})$ *fulfill* $(\mathcal{HP})$ *for* $i = 1, 2$, *with* $\alpha \leq \frac{1}{2}$, *and, in addition, that*

$$P_{1,0} \leq P_{2,0}, \quad C_1^\star C_1 \leq C_2^\star C_2,$$

Let $P_i, i = 1, 2$ *,be the mild solution of the Riccati equations* (2.20). *Then we have*

$$P_1(t) \leq P_2(t), \quad t \geq 0.$$

3 Dynamic programming

In this section, we are going to solve problem (1.1)-(1.4). We assume that $(\mathcal{HP})$ is verified with $\lambda_0 = 0$, and denote by P and P_k the solutions of the Riccati equations (2.1) and (2.4), respectively given by Theorems 2.1 and 2.2. We set also $V = V_P$, $V_k = V_{P_k}$. We shall extend the arguments of §1.5 of Chapter 1 based on the fundamental identity (3.1) and the solution of the closed loop equation (3.4) below.

Proposition 3.1 *Let $u \in L^2(0,T;U)$ and let x be the solution to state equation* (1.1). *Then the following identity holds*

$$\begin{aligned}(P(t)x_0, x_0) &+ \int_0^t |u(s) + E^\star V(t-s)x(s)|^2\, ds \\ &= \int_0^t \{|Cx(s)|^2 + |u(s)|^2\} ds + (P_0 x(t), x(t)).\end{aligned} \tag{3.1}$$

Proof. Let $u \in L^2(0,T;U)$ and let x be the solution to (1.1). Let $\{u_k\}$ be a sequence in $W^{1,2}(0,T;U)$ such that $u_k \to u$ in $L^2(0,T;U)$, and let x_k be the solution to (1.1) corresponding to u_k. We have, as can be easily checked,

$$x_k(t) = e^{tA}x_0 + Du_k(t) - \int_0^t e^{(t-s)A} Du_k'(s)\, ds.$$

By Proposition 3.8 in Volume I, Chapter 1, it follows that $x_k(t)$ is differentiable for any $t \in]0,T]$ and

$$x_k'(t) = Ae^{tA}x_0 - A\int_0^t e^{(t-s)A} Du_k'(s)\, ds.$$

Now we compute the derivative

$$\frac{d}{ds}(P_k(T-s)x_k(s), x_k(s))$$

and, integrating from 0 and T and letting k tend to infinity, we obtain (3.1).

■

Let us consider now the *closed loop equation*

$$x(t) = e^{tA}x_0 - \int_0^t (-A)^{1-\alpha} e^{(t-s)A} EE^\star V(T-s)x(s)\, ds. \tag{3.2}$$

Remark that the right-hand side of (3.2) is meaningful for any $t \in [0, T[$ since

$$\|(-A)^{1-\alpha} e^{(t-s)A} EE^{\star} V(T-s)\|$$

$$\leq \begin{cases} L^2 \|P\|_{\alpha} [(T-s)(t-s)]^{\alpha-1} & \text{if } \alpha > \frac{1}{2} \\ L^2 \|P\|_{\alpha,\beta} (T-s)^{\alpha+\beta-1} (t-s)^{\alpha-1} & \text{if } \alpha \leq \frac{1}{2}. \end{cases} \tag{3.3}$$

Proposition 3.2 *Assume that $(\mathcal{HP})$ holds. Then the following statements hold true:*
(i) If $\alpha > \frac{1}{2}$, there exists a unique solution of (3.2), $x \in C([0,T]; H)$.
(ii) If $\alpha \leq \frac{1}{2}$, there exists a unique solution of (3.2), $x \in C([0,T[; H)$ and a constant $C > 0$ such that

$$|x(t)| \leq C(T-t)^{-\beta}, \quad t \in [0, T[. \tag{3.4}$$

Proof. For any $y \in C([0,T]; H)$ we set

$$\lambda(y)(t) = \int_0^t (-A)^{1-\alpha} e^{(t-s)A} EE^{\star} V(T-s) x(s)\, ds, \quad t \in [0,T].$$

We first prove (i). Assume $\alpha > \frac{1}{2}$, then by (3.3) using the Hölder estimate, we have

$$|(\lambda(y))(t)| \leq L^2 \|P\|_{\alpha} \int_0^t (t-s)^{\alpha-1} (T-s)^{\alpha-1} ds\ \|y\|_{C([0,t];H)}$$

$$\leq L^2 \|P\|_{\alpha} \left\{ \int_0^t (t-s)^{2\alpha-2} ds \int_0^t (T-s)^{2\alpha-2} ds \right\}^{\frac{1}{2}} \|y\|_{C([0,t];H)}$$

$$\leq \tfrac{1}{2\alpha-1} L^2 \|P\|_{\alpha} t^{2\alpha-1} \|y\|_{C([0,t];H)},$$

so that $\lambda \in \mathcal{L}(C([0,t]; H))$ for any $t \in]0,T]$ and

$$\|\lambda\|_{\mathcal{L}(C([0,t];H))} \leq \frac{1}{2\alpha - 1} L^2 \|P\|_{\alpha} t^{2\alpha-1}.$$

Thus, if t is small, λ is a contraction and equation (3.2) has a unique solution in $C([0,t]; H)$. Now this argument can be repeated in $[t, 2t]$ and so on, giving the conclusion.

We now prove (ii). Fix $T_1 \in]0,T[$, and let $\alpha \le \frac{1}{2}$ and $\beta \in]\frac{1}{2}-\alpha, \frac{1-\alpha}{2}[$. By (3.3) we have

$$|(\lambda(y))(t)| \le L^2\|P\|_{\alpha,\beta}\int_0^t (t-s)^{\alpha-1}(T-s)^{\alpha+\beta-1}ds\;\|y\|_{C([0,t];H)}$$

$$\le L^2\|P\|_{\alpha,\beta}(T-t)^{\alpha+\beta-1}\int_0^t s^{\alpha-1}ds\;\|y\|_{C([0,t];H)}$$

$$\le \tfrac{1}{\alpha}L^2\|P\|_{\alpha,\beta}(T-T_1)^{\alpha+\beta-1}t^{\alpha}\|y\|_{C([0,t];H)}, \quad t\in[0,T],$$

so that $\lambda \in \mathcal{L}(C([0,t];H))$ for any $t \in]0,T_1]$ and

$$\|\lambda\|_{\mathcal{L}(C([0,t];H))} \le \frac{1}{\alpha}L^2\|P\|_{\alpha,\beta}(T-T_1)^{\alpha+\beta-1}t^{\alpha}.$$

Thus, if t is small, λ is a contraction and equation (3.2) has a unique solution in $C([0,t];H)$, and proceeding by iteration, we prove the first part of (ii).

It remains to show (3.4). Following F. FLANDOLI [6], we set

$$z(t) = V(T-t)x(t), \quad t\in[0,T[. \tag{3.5}$$

Then z verifies the equation

$$z(t) = V(T-t)e^{tA}x_0 - \int_0^t V(T-t)(-A)^{1-\alpha}e^{(t-s)A}EE^{\star}z(s)\,ds.$$

Taking into account Lemma 2.3 it follows that

$$(T-t)^{1-\alpha-\beta}|z(t)|$$

$$\le L|x_0|\|P\|_{\alpha,\beta} + 2^{1-\alpha}L^2\|P\|_{\alpha,\beta}\int_0^t (T-s)^{\beta}(t-s)^{\alpha-1}|z(s)|\,ds.$$

Finally set $w(t) = (T-t)^{1-\alpha-\beta} z(t)$; then we have

$$
\begin{aligned}
|w(t)| &\le L|x_0||P\|_{\alpha,\beta} \\
&\quad + 2^{1-\alpha} L^2 \|P\|_{\alpha,\beta} \int_0^t (T-s)^{\alpha+2\beta-1} (t-s)^{\alpha-1} |w(s)|\, ds \\
&\le L|x_0||P\|_{\alpha,\beta} \\
&\quad + 2^{1-\alpha} L^2 |P\|_{\alpha,\beta} \int_0^t (t-s)^{2\alpha+2\beta-2} |w(s)|\, ds.
\end{aligned}
\tag{3.6}
$$

Since $2 - 2\alpha - 2\beta < 1$, by the Gronwall's lemma we have $|w(t)| \le K$, for some $K > 0$ and all $t \in [0, T[$. It follows that

$$|z(t)| \le K(T-t)^{\alpha+\beta-1}, \quad t \in [0, T[. \tag{3.7}$$

Now, from (3.2), we have

$$
\begin{aligned}
|x(t)| &\le L|x_0| + L^2 K \int_0^t (t-s)^{\alpha-1} (T-s)^{\alpha+\beta-1}\, ds \\
&= L|x_0| + L^2 K (T-t)^{2\alpha+\beta-1} \int_0^{\frac{t}{T-t}} \sigma^{\alpha-1} (1+\sigma)^{\alpha+\beta-1}\, d\sigma.
\end{aligned}
$$

Since $\sigma^{2\alpha+2\beta-1} \le (\frac{t}{T-t})^{2\alpha+2\beta-1}$ it follows that

$$|x(t)| \le L|x_0| + L^2 K T^{2\alpha+2\beta-1} (T-t)^{-\beta} \int_0^{\infty} \sigma^{\alpha+2\beta} (1+\sigma)^{\alpha+\beta-1} d\sigma$$

and the proof is complete. ∎

We now prove the main result of this section.

Theorem 3.1 *Assume that $(\mathcal{HP})$ is verified and let $x_0 \in H$. Then there exists a unique optimal pair $(u^\star, x^\star)$. Moreover the following statements hold:*

(i) $x^\star$ is the mild solution to the closed loop equation (3.2) and $x^\star \in C([0,T]; H)$ if $\alpha \in]\frac{1}{2}, 1[$ whereas $x^\star \in C(]0,T]; H) \cap L^2(0,T;H)$ if $\alpha \in]0, \frac{1}{2}[$.

(ii) $u^\star \in C([0,T]; U)$ is given by the feedback formula:

$$u^\star(t) = -E^\star V(T-t) x^\star(t). \tag{3.8}$$

(iii) the optimal cost $J(u^\star)$ is given by

$$J(u^\star) = (P(T)x_0, x_0). \tag{3.9}$$

Proof. Let $u \in L^2(0,T;U)$ and let y be the solution to state equation (1.1). Then, by (3.1), we have

$$\begin{aligned}(P(T)x_0, x_0) + \int_0^T |u(s) + E^\star EV(T-s)x(s)|^2 ds \\ = J(u).\end{aligned} \tag{3.10}$$

It follows that

$$(P(T)x_0, x_0) \le J(u)$$

for all $u \in L^2(0,T;U)$. Now, let $x^\star$ be the solution of the closed loop equation (3.2). By Proposition 3.2, $x^\star$ belongs to $C([0,T];H)$ if $\alpha < \frac{1}{2}$ and to $L^2(0,T;H)$ if $\alpha > \frac{1}{2}$ (due to the estimate (3.4)). Now, let $u^\star$ be defined by (3.8) and set $u = u^\star$, $x = x^\star$ in (3.10). We find $(P(T)x_0, x_0) = J(u^\star)$, so that $u^\star$ is optimal. Finally, the uniqueness of the optimal control can be proved as in the proof of Theorem 5.3 in Chapter 1. ∎

Chapter 3

Systems with Unbounded Control Operators: Hyperbolic Equations with Control on the Boundary

1 Introduction

As before, we shall denote by H, U and Y the Hilbert spaces of states, controls and observations respectively, and consider a dynamical system, whose state $x(\cdot)$ is the solution of the following equation

$$\begin{cases} x'(t) = Ax(t) + Bu(t), \quad t \in [0,T], \\ x(0) = x_0 \in H, \end{cases}$$

where $u \in L^2(0,T;U)$, and $A : D(A) \subset H \to H$ generates a strongly continuous group on H. We identify the elements of H' with those of H so that the linear operator $(A^*)^* : H \to D(A^*)'$ is a linear extension of the linear operator $A : D(A) \to H$. As in the previous chapter, the linear operator B is not supposed to be bounded from U into H, but it belongs to $\mathcal{L}(U; D(A^*)')$, or equivalently, B is of the form $B = (\lambda_0 - A)E$ where $E \in \mathcal{L}(U;H)$ and λ_0 is an element in $\rho(A)$. More precisely, following I. LASIECKA and R. TRIGGIANI [1], [2], [7] we shall assume that

$$(\mathcal{HH})_1 \begin{cases} (i) & A \text{ generates a strongly continuous group } e^{tA} \text{ in } H \\ & \text{of type } \omega_0 \text{ and } \lambda_0 \text{ is a real number in } \rho(A) \text{ such that} \\ & \omega_0 < \lambda_0, \\ (ii) & E \in \mathcal{L}(U;H), \\ (iii) & \exists\ K > 0 \text{ such that} \\ & \displaystyle\int_0^T |E^\star A^\star e^{sA^\star} x|^2 ds \le K^2 |x|^2, \quad \forall x \in D(A^\star). \end{cases}$$

If assumptions $(\mathcal{HH})_1$ hold, then we can give a precise meaning to the state equation. We have in fact the following result due to I. LASIECKA and R. TRIGGIANI [1], [2], [7].

Proposition 1.1 *Assume that $(\mathcal{HH})_1$ is verified and for each control function $u \in L^2(0,T;U)$ define*

$$z(t) = \int_0^t e^{(t-s)A} Eu(s)\, ds, \quad t \in [0,T]. \tag{1.1}$$

Then $z(t) \in D(A)$ for all $t \in [0,T]$ and $Az \in C([0,T];H)$. Moreover the following inequality holds

$$|Az(t)| \le K \|u\|_{L^2(0,T;U)}, \quad t \in [0,T]. \tag{1.2}$$

Proof. We divide the proof into two steps.

Step 1. $z(t) \in D(A)$ for all $t \in [0,T]$.
If $x \in D(A^\star)$ we have

$$(z(t), A^\star x) = \int_0^t (u(s), E^\star A^\star e^{(t-s)A^\star} x)\, ds.$$

It follows that

$$\begin{aligned} |(z(t), A^\star x)|^2 &\le \|u\|^2_{L^2(0,T;U)} \int_0^t |E^\star A^\star e^{(t-s)A^\star} x|^2 ds \\ &\le K^2 \|u\|^2_{L^2(0,T;U)} |x|^2. \end{aligned}$$

If we identify elements of H' with those of H, then this inequality implies that the linear form $x \mapsto (z(t), A^\star x)$ on H, is continuous, and hence that $z(t) \in D(A)$ and (1.2) holds.

Step 2. $Az \in C([0,T];H)$.
Let $\{u_n\}$ be a sequence in $W^{1,2}(0,T;U)$ such that u_n converges to u in $L^2(0,T;U)$. Set

$$z_n = \int_0^t e^{(t-s)A} E u_n(s)\, ds.$$

It follows that $z_n \in C([0,T];D(A))$ (cf. Volume I, Chapter 1, Proposition 3.3). Now setting $w_n(t) = Az_n(t)$, $w_n \in C([0,T];H)$ and we have

$$|w(t) - w_n(t)|^2 = |Az(t) - Az_n(t)|^2 \le K\|u - u_n\|^2_{L^2(0,T;U)} \to 0$$

as $n \to \infty$ uniformly in $[0,T]$. The proof is complete. ∎

In the sequel we shall write the solution of the state equation as follows:

$$\begin{aligned} x(t) &= e^{tA}x_0 + G(u)(t) \\ &= e^{tA}x_0 + (\lambda_0 - A)\int_0^t e^{(t-s)A} E u(s)\, ds. \end{aligned} \tag{1.3}$$

By Proposition 1.1, $G \in \mathcal{L}(L^2(0,T;U); C([0,T];H))$.

We now consider the following optimal control problem: to minimize

$$J(u) = \int_0^T \{|Cx(t)|^2 + |u(t)|^2\}\, dt + (P_0 x(T), x(T)), \tag{1.4}$$

over all $u \in L^2(0,T;U)$, subject to the state equation constraint (1.3). In addition to the assumptions $(\mathcal{HH})_1$, we assume that

$(\mathcal{HH})_2$ $\quad C \in \mathcal{L}(H;Y)$, $P_0 \in \Sigma^+(H)$.

In the sequel $(\mathcal{HH})$ will mean that both assumptions $(\mathcal{HH})_1$ and $(\mathcal{HH})_2$ are verified. We shall also say that (A, E, C, P_0) fulfill $(\mathcal{HH})$. From now on we assume for simplicity that $\lambda_0 = 0$.

The definitions of *optimal control*, *optimal state* and *optimal pair* are as in the previous chapters.

We again study the optimization problem (1.3)-(1.4) by using Dynamic Programming. However in the present case, new technical difficulties arise. Consider in fact the Riccati equation that we formally write as follows:

$$\begin{cases} P' = A^\star P + PA - P(E^\star A^\star)^\star E^\star A^\star P + C^\star C, \\ P(0) = P_0. \end{cases} \tag{1.5}$$

If the data C and P_0 are regular or more precisely if the linear operators $\sqrt{P_0}A$ and CA admit bounded extensions, $\overline{\sqrt{P_0}A}$ and $\overline{CA}$ then problem (1.5) can be easily solved. In fact by setting $R = A^\star PA$ it reduces to

$$\begin{cases} R' = A^\star R + RA - REE^\star R + (\overline{CA})^\star \overline{CA}, \\ R(0) = \left(\overline{\sqrt{P_0}A}\right)^\star \overline{\sqrt{P_0}A}, \end{cases}$$

which has a unique solution by Theorem 2.1 of Chapter 1.

However these conditions on C and P_0 are not natural, so we prefer to proceed in the following different way.

(i) We consider the approximating problem

$$\begin{cases} P_n' = A^\star P_n + P_n A - P_n(E^\star A_n^\star)^\star E^\star A_n^\star P_n + C^\star C, \\ P(0) = P_0, \end{cases} \tag{1.6}$$

where $A_n = nAR(n,A)$ are the Yosida approximations of A, and prove the convergence in $C_s([0,T];\Sigma(H))$ of the sequence $\{P_n\}$ to an element P.

(ii) For any $n \in \mathbf{N}$ we consider the control problem: to minimize

$$J_n(u) = \int_0^T \{|Cx_n(t)|^2 + |u(t)|^2\}dt + (P_0 x_n(T), x_n(T)), \tag{1.7}$$

over all $u \in L^2(0,T;U)$, subject to the constraint

$$\begin{cases} x_n'(t) = Ax_n(t) - A_n Eu(t), \quad t \in [0,T], \\ x_n(0) = x_0. \end{cases} \tag{1.8}$$

Denote by $(u_n^\star, x_n^\star)$ the corresponding optimal pair given by Theorem 6.1 of Chapter 1. Then we prove that $(u_n^\star, x_n^\star)$ converges to the optimal pair of problem (1.3)-(1.4).

Notice that we avoid the difficulty of giving a general definition of a solution of the Riccati equation (1.5); for us the solution will namely be the limit of the sequence $\{P_n\}$. The proof of the existence of this limit is given in §2, whereas in §3 we shall see how to use this result to study the initial optimal control problem.

For an existence and uniqueness result, under suitable regularity assumptions on C and P_0 (weaker than the ones discussed before), see G. DA PRATO, I. LASIECKA and R. TRIGGIANI [1]. Finally we notice that in the previously quoted papers by I. LASIECKA and R. TRIGGIANI explicit formulas for the solutions to the Riccati equation and the optimal pair (even under more general assumptions) are constructed by generalizing the representation formula (2.22) of Chapter 1.

In §5 we shall consider, following V. BARBU and G. DA PRATO [1], the case when e^{tA} is a general semigroup, and a condition similar to $(\mathcal{HH})$ holds. In this case we are able to define a solution of the Riccati equation by using the dual Riccati equation introduced in Chapter 1, §7.3.

2 Riccati equation

We assume here that $(\mathcal{HH})$ is verified and consider the Riccati equation (1.6), whose mild solution is denoted by P_n. For the sake of simplicity, we assume that $P_0 \in \mathcal{L}_r(H)$, that is, that P_0 is invertible with a bounded inverse (cf. Chapter 1, §5); at the end of this section we shall explain how to handle the general case.

Following F. FLANDOLI [2], we consider the Riccati equation

$$\begin{cases} Q' = -AQ - QA^\star - QC^\star CQ + (E^\star A^\star)^\star E^\star A^\star \\ Q(0) = P_0^{-1}. \end{cases} \tag{2.1}$$

and

$$\begin{cases} Q_n' = -AQ_n - Q_nA^\star - Q_nC^\star CQ_n + (E^\star A_n^\star)^\star E^\star A_n^\star \\ \\ Q_n(0) = P_0^{-1}, \end{cases} \tag{2.2}$$

where $A_n = nAR(n, A)$ are the Yosida approximations of A. Since the linear operator $E^\star A^\star$ fulfills the assumptions of Theorem 4.1 of Chapter 1, problem (2.1) has a unique mild solution Q which belongs to $C_s([0,T];\Sigma(H))$ and we have

$$\lim_{n\to\infty} Q_n = Q, \text{ in } \quad C_s([0,T];\Sigma(H)) \tag{2.3}$$

and

$$Q_n(t) = P_n^{-1}(t), \quad \forall\, t \geq 0. \tag{2.4}$$

Theorem 2.1 *Assume that* $(\mathcal{HH})$ *is verified and that* $P_0 \in \mathcal{L}_r(H)$. *Let* P_n *be the solution to the Riccati equation* (1.6). *Then we have*

$$\lim_{n\to\infty} P_n = P \quad \text{in } \; C_s([0,T];\Sigma(H)), \tag{2.5}$$

where $P(t) = Q^{-1}(t)$, $t \in [0,T]$ *and* Q *is the mild solution of* (2.1).

We shall call P the *mild solution* of equation (1.5).
Proof. It is sufficient to prove that there exists a constant $C_1 > 0$ such that

$$\|P_n(t)\| \leq C_1, \quad \forall\, t \in [0,T],\, \forall n \in \mathbf{N}. \tag{2.6}$$

In fact, if (2.6) holds, then we have

$$\|Q_n^{-1}(t)\| \leq C_1, \quad \forall\, t \in [0,T],\, \forall n \in \mathbf{N}.$$

Thus $Q(t) \in \mathcal{L}_r(H)$, $\forall t \in [0,T]$. Moreover from the identity

$$Q^{-1}(t) - P_n(t) = P_n(t)[Q_n(t) - Q(t)]Q^{-1}(t),$$

one can prove the existence of the limit (2.5). Finally to prove (2.6), it is sufficient to observe that

$$P_n(t) \leq e^{tA^\star} P_0 e^{tA} + \int_0^t e^{(t-s)A^\star} C^\star C e^{(t-s)A} ds.$$

The proof is complete. ∎

Remark 2.1 Assume that $(\mathcal{HH})$ is verified and that $P_0, \overline{P}_0 \in \mathcal{L}_r(H)$ with $P_0 \leq \overline{P}_0$. Let P and $\overline{P}$ be the corresponding mild solutions of (1.5). Then we have $P(t) \leq \overline{P}(t)$, $\forall t \geq 0$. ■

Remark 2.2 Assume that P_0 does not have a bounded inverse and denote by P_ε the mild solution of (1.5) with P_0 replaced by $P_0 + \varepsilon I$. Then by the previous remark P_ε is decreasing and so the limit of P_ε exists in $C_s([0,T];\Sigma(H))$ as ε goes to zero. ■

3 Dynamic programming

Here we assume that $(\mathcal{HH})$ is verified. We denote by P and P_n the mild solutions of equations (1.5) and (1.6), and by $(u^\star, x^\star)$ (resp. $(u_n^\star, x_n^\star)$) the optimal pair for the control problem (1.3)-(1.4) (resp. (1.7)-(1.8)).

We first consider the case when $P_0 \in \mathcal{L}_r(H)$.

Theorem 3.1 *Assume that $(\mathcal{HH})$ is verified and that $P_0 \in \mathcal{L}_r(H) \cap \Sigma^+(H)$. Then we have*

(i) $\lim_{n\to\infty} u_n^\star = u^\star$ *in* $L^2(0,T;U)$,

(ii) $\lim_{n\to\infty} x_n^\star = x^\star$ *in* $C([0,T];H)$.

Proof. Let $u \in L^2(0,T;U)$ and let x be given by (1.3) and x_n by (1.8). We remark that, since $x_n(t) = e^{tA}x_0 + n(nI - A)^{-1}G(u)$, we have $x_n \to x$ in $C([0,T];H)$. Now by formula (6.2) in Chapter 1, it follows that

$$
\begin{aligned}
&(P_n(T)x_0, x_0) + \int_0^T |u(s) + E^\star A_n^\star P_n(T-s)x_n(s)|^2 ds \\
&= \int_0^T [|Cx_n(s)|^2 + |u(s)|^2] ds + (P_0 x_n(T), x_n(T)).
\end{aligned}
\tag{3.1}
$$

Setting $u = u_n^\star$ in (3.1), we find

$$
\begin{aligned}
&(P_n(T)x_0, x_0) \\
&= \int_0^T [|Cx_n^\star(s)|^2 + |u_n^\star(s)|^2] ds + (P_0 x_n^\star(T), x_n^\star(T)).
\end{aligned}
\tag{3.2}
$$

Thus we have

$$(P(T)x_0, x_0) \le J(u), \quad \forall\, u \in L^2(0,T;U). \tag{3.3}$$

Moreover from identities (3.1) and (3.2) the sequence $\{u_n^*\}$ is bounded in $L^2(0,T;U)$ and so there exists a subsequence $\{u_{n_k}^*\}$ which weakly converges to an element $\widetilde{u}$ of $L^2(0,T;U)$. Since

$$u_{n_k}^*(\cdot) = S(\cdot)x_0 - n_k R(n_k, A) G(u_{n_k}^*)$$

and

$$G(u_{n_k}^*) \rightharpoonup G(\widetilde{u}) \;\text{ in }\; L^2(0,T;H),$$

we have

$$x_{n_k}^* \rightharpoonup \widetilde{x} \;\text{ in }\; L^2(0,T;H),$$

where

$$\widetilde{x} = S(\cdot)x_0 - G(\widetilde{u}),$$

and the symbol $\rightharpoonup$ denotes the weak convergence.

By letting n go to infinity in (3.2) we have

$$(P(T)x_0, x_0) \ge J_T(\widetilde{u}),$$

which, along with (3.3), yields $\widetilde{u} = u^*$, $\widetilde{x} = x^*$ and

$$(P(T)x_0, x_0) = J_T(u^*).$$

Thus the sequence $\{u_n^*\}$ weakly converges to u^* and it only remains to prove that $u_n^* \to u^*$. By (3.2) we have

$$\lim_{n\to\infty} \left[\|u_n^*\|^2_{L^2(0,T;U)} + \|x_n^*\|^2_{L^2(0,T;H)} + |\sqrt{P_0}x_n^*(T)|^2\right]$$

$$\le J(u^*) = \|u^*\|^2_{L^2(0,T;U)} + \|x^*\|^2_{L^2(0,T;H)} + |\sqrt{P_0}\, x^*(T)|^2.$$

Since

$$u_n^* \rightharpoonup u^* \;\text{ in }\; L^2(0,T;U),$$
$$x_n^* \rightharpoonup x^* \;\text{ in }\; L^2(0,T;H),$$
$$\sqrt{P_0}\, x_n^*(T) \rightharpoonup \sqrt{P_0}\, x^*(T) \;\text{ in }\; H,$$

as $n \to \infty$, it follows that all the above convergences are strong. The proof is complete. ■

We now consider the case where $P_0 \in \Sigma^+(H)$ is not invertible. Define for any $\varepsilon > 0$ the cost

$$J_\varepsilon(u) = \int_0^T \{|Cx(t)|^2 + |u(t)|^2\}dt + (P_0x(T), x(T)) + \varepsilon|x(T)|^2,$$

and let $(u_\varepsilon^\star, x_\varepsilon^\star)$ be an optimal pair for the corresponding control problem. Let moreover P_ε be the mild solution of the Riccati equation (1.5) with P_0 replaced by $P_0 + \varepsilon I$. As we know (see Remark 3.1), P_ε is convergent in $C_s([0,T];\Sigma(H))$ to an element $P \in C_s([0,T];\Sigma(H))$, which has been defined as the mild solution of (1.5)

Theorem 3.2 *Assume that $(\mathcal{HH})$ is verified and that $P_0 \in \Sigma^+(H)$. Then there exists a unique optimal pair $(u^\star, x^\star)$ for the control problem* (1.3)-(1.4). *Moreover*

(i) $\lim_{\varepsilon \to 0} u_\varepsilon^\star = u^\star$ *in* $L^2(0,T;U)$,

(ii) $\lim_{\varepsilon \to 0} x_\varepsilon^\star = x^\star$ *in* $C([0,T];H)$.

Proof. We have

$$(P_\varepsilon(T)x_0, x_0) \le J_\varepsilon(u), \quad \forall\, u \in L^2(0,T;U), \tag{3.4}$$

$$\begin{aligned}(P_\varepsilon(T)x_0, x_0) \;&=\; \int_0^T \{|Cx_\varepsilon^\star(t)|^2 + |u_\varepsilon^\star(t)|^2\}dt \\ &\quad +(P_0x(T), x(T)) + \varepsilon|x_\varepsilon^\star(T)|^2.\end{aligned} \tag{3.5}$$

We notice that the sequence $\{u_\varepsilon^\star\}$ is bounded in $L^2(0,T;U)$; we have in fact

$$\int_0^T |u_\varepsilon^\star(s)|^2 ds \le J_\varepsilon(u_\varepsilon^\star) \le J_\varepsilon(0),$$

where

$$J_\varepsilon(0) = \int_0^T |Ce^{sA}x_0|^2 ds + (P_0Ce^{TA}x_0, Ce^{TA}x_0) + \varepsilon|e^{TA}x_0|^2.$$

It follows that there exists a sequence $\{\varepsilon_k\} \downarrow 0$, and elements $\widetilde{u} \in L^2(0,T;U)$, $\widetilde{x} \in C([0,T];H)$ such that

(i) $u_{\varepsilon_k} \rightharpoonup \widetilde{u}$ in $L^2(0,T;U)$,

(ii) $x_{\varepsilon_k} \rightharpoonup \widetilde{x}$ in $L^2(0,T;H)$,

(iii) $x_{\varepsilon_k}(T) \rightharpoonup \widetilde{x}(T)$ in H.

Moreover

$$\widetilde{x}(t) = e^{tA}x_0 - A\int_0^t e^{(t-s)A}E\widetilde{u}(s)\,ds, \quad t \in [0,T].$$

Now from (3.4) it follows that

$$(P_\varepsilon(T)x_0, x_0) \le J_\varepsilon(\widetilde{u}),$$

and by (3.5)

$$(P_\varepsilon(T)x_0, x_0) \ge J_\varepsilon(\widetilde{u}).$$

The conclusion follows arguing as in the proof of Theorem 3.1. ∎

Remark 3.1 We shall denote by P_{min} the mild solution of the Riccati equation corresponding to $P_0 = 0$. By Remark 2.1 it follows that P_{min} is the minimal nonnegative solution of the Riccati equation. ∎

4 Examples of controlled hyperbolic systems

Let $A : D(A) \subset H \to H$ be a linear operator, infinitesimal generator of a strongly continuous group e^{tA} and let $E : U \to H$ be a linear bounded operator. In this section we shall discuss several examples such that the key condition $(\mathcal{HH})$-(iii) is fulfilled. We start with a very simple example.

Example 4.1 *A first order problem.*

Consider a system

$$\begin{cases} \dfrac{\partial x}{\partial t}(t,\xi) = \dfrac{\partial x}{\partial \xi}(t,\xi), \quad t \in [0,T],\ \xi \in [0,2\pi], \\[2ex] x(0,\xi) = x_0(\xi), \quad \xi \in [0,2\pi], \\[2ex] x(t,2\pi) = x(t,0) + u(t), \quad t \in [0,T]. \end{cases}$$

We choose here $H = L^2_{\#}(\mathbf{R})$, the set of all 2π-periodic functions which belong to $L^2(0,2\pi;\mathbf{R})$, and $U = \mathbf{R}$. We consider the linear operator A in H

$$\begin{cases} D(A) = W^{1,2}_{\#}(\mathbf{R}) = \{x \in L^2_{\#}(\mathbf{R}) \, : \, x' \in L^2_{\#}(\mathbf{R}), \; x(0) = x(2\pi)\} \\ \\ Ax = x', \quad \forall\, x \in D(A). \end{cases}$$

As it is well known, A is the infinitesimal generator of the strongly continuous group of translations

$$e^{tA}x(\xi) = x(t+\xi), \quad t \in \mathbf{R}, \; \xi \in \mathbf{R}.$$

We proceed here as in the Example 1.1 of Chapter 2. First we assume that u is differentiable and introduce a new variable y by setting

$$y(t,\xi) = x(t,\xi) - \frac{e^{\xi}}{e^{2\pi}-1}u(t),$$

so that state equation becomes

$$\begin{cases} \dfrac{\partial y}{\partial t}(t,\xi) = \dfrac{\partial y}{\partial \xi}(t,\xi) + \dfrac{e^{\xi}}{e^{2\pi}-1}(u(t) - u'(t)), \quad t \in [0,T], \; \xi \in [0,2\pi], \\ \\ y(0,\xi) = x_0(\xi) - \dfrac{e^{\xi}}{e^{2\pi}-1}u(0) = y_0, \quad \xi \in [0,2\pi], \\ \\ y(t,2\pi) = y(t,0), \quad t \in [0,T]. \end{cases}$$

Now we can write this equation in abstract form, namely

$$y(t,\cdot) = e^{tA}y_0 + \int_0^t e^{(t-s)A}z_0(u(s) - u'(s))\,ds,$$

where z_0 is the element of $L^2_{\#}(\mathbf{R})$ defined by

$$z_0(\xi) = \frac{e^{\xi}}{e^{2\pi}-1}, \quad \xi \in [0,2\pi].$$

By integration by parts we eventually find that the state equation can be written in the required form

$$x'(t) = Ax(t) + (1-A)Eu(t), \quad x(0) = x_0,$$

where the linear operator $E \in \mathcal{L}(U; H)$ is defined as

$$E\alpha = \alpha z_0, \quad \forall \alpha \in U = \mathbf{R}.$$

Observe that the adjoint operator $E^\star \in \mathcal{L}(H; U)$ is defined as

$$E^\star x = (x, z_0), \quad \forall x \in H.$$

We now check condition $(\mathcal{HH})$-(iii). Let $t \in [0, 2\pi]$ and $x \in W^{1,2}_{\#}(\mathbf{R})$. If $s \in [0, 2\pi]$ we have

$$\begin{aligned} E^\star A^\star e^{sA^\star} x &= (A^\star x, e^{sA} z_0) \\ &= \frac{1}{e^{2\pi} - 1} \left\{ \int_0^{2\pi - s} x'(\xi) e^{s+\xi} d\xi + \int_{2\pi - s}^{2\pi} x'(\xi) e^{s+\xi-2\pi} d\xi \right\} \\ &= x(2\pi - s) - (x, e^{sA} z_0). \end{aligned}$$

It follows that

$$\begin{aligned} &\int_0^t |E^\star A^\star e^{sA^\star} x|^2 ds \\ &\leq 2 \int_0^t |x(2\pi - s)|^2 ds + 2t \int_0^t |(x, e^{sA} z_0)|^2 ds \\ &\leq 2(1 + t^2 |z_0|^2) |x|^2. \end{aligned}$$

Thus condition $(\mathcal{HH})$-(iii) is fulfilled. ∎

Example 4.2 *Abstract wave equation.*
Let Z, U be Hilbert spaces, $\Lambda : D(\Lambda) \subset Z \to Z$ a linear self–adjoint strictly positive operator, D a linear operator in $\mathcal{L}(U; Z)$ and $\lambda_0 \geq 0$. Here we are dealing with the system

$$\begin{cases} y''(t) + \Lambda y(t) + (\lambda_0^2 + \Lambda) D u(t) = 0, & t \geq 0, \\ y(0) = y_0, \quad y'(0) = y_1. \end{cases} \tag{4.1}$$

In order to write problem (4.1) in the form (1.3), we have to introduce the space $D(\Lambda^{-1/2})$ defined as the completion of Z with respect to the norm $\|x\|_{D(\Lambda^{-1/2})} = \|\Lambda^{-1/2}x\|$. Then we choose as space of states and observations $H = Z \oplus D(\Lambda^{-1/2})$, endowed with the inner product:

$$\left(\begin{bmatrix} x^0 \\ x^1 \end{bmatrix}, \begin{bmatrix} \overline{x}^0 \\ \overline{x}^1 \end{bmatrix}\right) = (x^0, \overline{x}^0)_Z + (\Lambda^{-1/2}x^1, \Lambda^{-1/2}\overline{x}^1)_Z,$$

and define the linear operator A on H

$$\begin{cases} AX = \begin{bmatrix} 0 & 1 \\ -\Lambda & 0 \end{bmatrix} \begin{bmatrix} x^0 \\ x^1 \end{bmatrix}, & \forall\, X \in D(A) \\ \\ D(A) = D(\Lambda^{1/2}) \oplus Z. \end{cases}$$

A is the infinitesimal generator of a strongly continuous group of contractions, given by the formula

$$e^{tA} = \begin{bmatrix} \cos\ (\Lambda^{1/2}t) & \Lambda^{-1/2} \sin\ (\Lambda^{1/2}t) \\ -\Lambda^{1/2} \sin\ (\Lambda^{1/2}t) & \cos\ (\Lambda^{1/2}t) \end{bmatrix}, \quad t \in \mathbf{R}.$$

Setting now

$$X(t) = \begin{bmatrix} y(t,\cdot) \\ y'(t,\cdot) \end{bmatrix}, \quad X_0 = \begin{bmatrix} y_0 \\ y_0' \end{bmatrix},$$

we can write equation (4.1) as

$$\begin{cases} X'(t) = AX(t) + (\lambda_0 - A)Eu(t), \quad t \geq 0, \\ \\ X(0) = X_0. \end{cases} \tag{4.2}$$

where $E : U \to H$ is defined by

$$E\eta = -\begin{bmatrix} D\eta \\ \lambda_0 D\eta \end{bmatrix}, \quad \eta \in U.$$

We remark that the adjoint of E, $E^\star : H \to U$ is given by

$$E^\star \begin{bmatrix} x^0 \\ x^1 \end{bmatrix} = -D^\star(x^0 - \lambda_0 \Lambda^{-1})x^1.$$

It follows that

$$E^\star A^\star e^{tA^\star} \begin{bmatrix} x^0 \\ x^1 \end{bmatrix}$$

$$= -D^\star \Lambda^{1/2} \sin\ (\Lambda^{1/2} t) x^0 + D^\star \cos\ (\Lambda^{1/2} t) x^1$$

$$+\lambda_0 D^\star \Lambda^{-1} \cos\ (\Lambda^{1/2} t) x^0 + \lambda_0 D^\star \Lambda^{1/2} \sin\ (\Lambda^{1/2} t) x^1.$$

Now it follows easily that condition $(\mathcal{HH})$-(iii) is fulfilled if and only if there exists $K_1 > 0$ such that

$$\begin{aligned} &\int_0^t |D^\star \Lambda^{1/2} \sin\ (\Lambda^{1/2} s) z|_U^2 ds \le K_1(t) |z|_Z^2, \quad \forall\, z \in D(\Lambda^{1/2}), \\ &\int_0^t |D^\star \Lambda^{1/2} \cos\ (\Lambda^{1/2} s) z|_U^2 ds \le K_1(t) |z|_Z^2, \quad \forall\, z \in D(\Lambda^{1/2}). \end{aligned} \tag{4.3}$$

■

Example 4.3 *Wave equation.*
Let Ω be a bounded set in $\mathbf{R}^n$ with a regular boundary $\partial\Omega$. Consider the problem

$$\begin{cases} y_{tt}(t,\xi) = \Delta_\xi y(t,\xi), & t \ge 0,\ \xi \in \Omega, \\ y(t,\xi) = u(t,\xi), & t \ge 0,\ \xi \in \partial\Omega, \\ y(0,\xi) = y_0(\xi), \quad y_t(0,\xi) = y_1(\xi), & \xi \in \Omega. \end{cases} \tag{4.4}$$

We set $Z = L^2(\Omega)$, $U = L^2(\partial\Omega)$ and denote by $-\Lambda$ the Laplace operator with Dirichlet boundary conditions in Ω. Then, setting $y(t) = y(t,\cdot)$, $u(t) = u(t,\cdot)$ and denoting by D the Dirichlet mapping defined in Example 1.1 of Chapter 2, we see that problem (4.4) is equivalent to problem (4.1) with $\lambda_0 = 0$.

We now prove $(\mathcal{HH})$-(iii); for this it is enough to show the first inequality (4.3), since the second one can be proven similarly. Thus we fix $z \in D(\Lambda)$ and set

$$w(t) = D^\star \Lambda^{1/2} \sin\ (\Lambda^{1/2} t) z, \quad v(t) = \Lambda^{-1/2} \sin\ (\Lambda^{1/2} t) z;$$

we remark that

$$w(t) = \frac{\partial v}{\partial \nu}(t),$$

where ν is the the outward normal to $\partial\Omega$ (since $D^\star\Lambda$ coincides with the linear operator $\frac{\partial}{\partial\nu}$). Moreover $v(t,\cdot) = v(t)$ is the classical solution to the following problem

$$\begin{cases} v_{tt}(t,\xi) = \Delta_\xi v(t,\xi), & t \geq 0,\ \xi \in \Omega, \\ v(t,\xi) = 0, & t \geq 0,\ \xi \in \partial\Omega, \\ v(0,\xi) = 0, & v_t(0,\xi) = x_0(\xi), \quad \xi \in \Omega. \end{cases} \tag{4.5}$$

Fix $T > 0$, to prove the first inequality in (4.3) it suffices to prove that there exists a constant $C_1 > 0$ such that

$$\int_0^T (T-t) \left| \frac{\partial v}{\partial \nu}(t,\cdot) \right|^2_{L^2(\partial\Omega)} dt \leq C_1 |x_0|^2_{L^2(\Omega)}. \tag{4.6}$$

For this purpose, we follow the proof in I. LASIECKA, J.-L. LIONS and R. TRIGGIANI [1]. We consider a regular vector field h in Ω that extends the outward normal ν; then we multiply the first equation in (4.5) by $(T-t)h \cdot \nabla_\xi v$ and set

$$I = \int_Q (T-t)\, v_{tt}\, h \cdot \nabla_\xi v\, dt d\xi, \quad J = \int_Q (T-t)\, (\Delta_\xi v)\, h \cdot \nabla_\xi v\, dt d\xi,$$

where $Q = [0,T] \times \Omega$ and " $\cdot$ " denotes the scalar product in $\mathbf{R}^n$.

Step 1. Estimate of I. We have

$$I = \int_\Omega [(T-t)\,(h \cdot \nabla_\xi v)\, v_t] \Big|_{t=0}^{t=T} d\xi + \int_Q (h \cdot \nabla_\xi v)\, v_t\, dt d\xi$$

$$- \int_Q (T-t)\,(h \cdot \nabla_\xi v_t)\, v_t\, dt d\xi = I_1 + I_2 + I_3.$$

Clearly, $I_1 = 0$; moreover, by the usual energy estimates, there exists $C_2 > 0$ such that

$$I_2 \leq C_2 |x_0|^2_{L^2(\Omega)}. \tag{4.7}$$

Concerning I_3 we have

$$I_3 = -\frac{1}{2}\int_Q (T-t)\,\mathrm{div}_\xi\,(hv_t^2)\,dtd\xi + \frac{1}{2}\int_Q (T-t)\,v_t^2\,\mathrm{div}_\xi h\,dtd\xi$$

$$= \frac{1}{2}\int_{[0,T]\times\partial\Omega} (T-t)\,v_t^2\,h\cdot\nu\,dtd\sigma + \frac{1}{2}\int_Q (T-t)\,v_t^2\,\mathrm{div}_\xi h\,dtd\xi.$$

The first term is equal to 0 since $v = 0$ on $\partial\Omega$; moreover the second one can be estimated as I_2. Thus, there exists a constant $C_3 > 0$ such that

$$|I| \le C_2|x_0|^2_{L^2(\Omega)}. \tag{4.8}$$

Step 2. Estimate of J.

Using Gauss-Green formulas for the term J, we obtain the boundary term in (4.6) minus the term J_1 given by the expression

$$\begin{aligned} J_1 &= \sum_{i,j=1}^n \int_Q (T-t)\,\partial_{\xi_i}v\,\partial_{\xi_i}(h_j\partial_{\xi_j}v)\,dtd\xi \\ &= \sum_{i,j=1}^n \int_Q (T-t)\,\partial_{\xi_i}v\,\partial_{\xi_i}h_j\,\partial_{\xi_j}v\,dtd\xi \\ &\quad + \sum_{i,j=1}^n \int_Q (T-t)\,h_j\,\partial_{\xi_i\xi_j}v\,\partial_{\xi_i}v\,dtd\xi, \end{aligned}$$

where ∂_{ξ_i} denotes the partial derivative with respect to ξ_i and $\partial_{ij} = \partial_i(\partial_j)$. Then

$$\begin{aligned} J_1 = &\sum_{i,j=1}^n \int_Q (T-t)\,\partial_{\xi_i}h_j\,\partial_{\xi_j}v\,\partial_{\xi_i}v\,dtd\xi \\ &+\frac{1}{2}\sum_{i,j=1}^n \int_Q (T-t)\,h_j\,\partial_{\xi_j}\,(\partial_{\xi_i}v)^2\,dtd\xi \end{aligned}$$

$$
\begin{aligned}
J_1 = & \sum_{i,j=1}^{n} \int_Q (T-t)\, \partial_{\xi_i} h_j\, \partial_{\xi_j} v\, \partial_{\xi_i} v\, dtd\xi \\
& + \frac{1}{2} \sum_{i,j=1}^{n} \int_Q (T-t)\, \partial_{\xi_j} [h_j |\partial_{\xi_i} v|^2]\, dtd\xi \\
& - \frac{1}{2} \sum_{i,j=1}^{n} \int_Q (T-t)\, \partial_{\xi_j} h_j\, |\partial_{\xi_i} v|^2\, dtd\xi \\
= & \ K_1 + \frac{1}{2} \int_{[0,T]\times\partial\Omega} (T-t)\, |\nabla_\xi v|^2\, dtd\sigma \\
= & \ K_1 + \frac{1}{2} \int_{[0,T]\times\partial\Omega} (T-t) \left|\frac{\partial v}{\partial \nu}\right|^2 dtd\sigma,
\end{aligned}
$$

since $v = 0$ on $\partial\Omega$ and

$$
h \cdot \nabla_\xi = \frac{\partial v}{\partial \nu} \text{ on } \partial\Omega.
$$

Now

$$
|K_1| \leq C_4 |x_0|^2_{L^2(\Omega)}. \tag{4.9}
$$

for some constant $C_4 > 0$. We have

$$
J = -K_1 + \frac{1}{2} \int_{[0,T]\times\partial\Omega} (T-t) \left|\frac{\partial v}{\partial \nu}\right|^2 dtd\sigma. \tag{4.10}
$$

Finally since $I = J$

$$
\int_0^T (T-t) \left|\frac{\partial v}{\partial \nu}\right|^2_{L^2(\partial\Omega)} dt = I + K_1, \tag{4.11}
$$

so that the required estimate (4.6) follows from (4.9), (4.10) and (4.11). ■

Remark 4.1 One can consider different situations such as the wave equations with Neumann Boundary conditions, or the plate equations with several boundary conditions. As for the verification of the condition $(\mathcal{HH})$-(i) the reader is referred to the review paper by I. LASIECKA and R. TRIGGIANI [7]. ■

5 Some result for general semigroups

In this section we do not assume that A generates a strongly continuous group but that a condition weaker than $(\mathcal{HH})_1$ holds.

$$(\mathcal{HH})_1' \begin{cases} (i) & A \text{ generates a strongly continuous semigroup } e^{tA} \text{ in } \\ & H \text{ of type } \omega_0 \text{ and } \lambda_0 \text{ is a real number in } \rho(A) \text{ such} \\ & \text{that } \omega_0 < \lambda_0 \\ (ii) & E \in \mathcal{L}(U;H), \\ (iii) & \exists\ K > 0 \text{ such that} \\ & \displaystyle\int_0^T |E^\star A^\star e^{sA^\star} x|^2 ds \le K^2 |x|^2, \quad \forall x \in D(A^\star). \end{cases}$$

If we replace $(\mathcal{HH})_1$ with $(\mathcal{HH})_1'$ Proposition 1.1 still holds true as can easily be seen and the state equation (1.3) is meaningful. We consider again the cost function (1.4) and assume that $(\mathcal{HH})_2$ is verified. We denote by $(\mathcal{HH})'$ the set of all hypotheses $(\mathcal{HH})_1'$ and $(\mathcal{HH})_2$. Besides the Riccati equation (1.5) and the approximating equation (1.6) we consider, as in V. Barbu and G. Da Prato [1], the dual Riccati equation

$$\begin{cases} Q' = AQ + QA^\star - QC^\star CQ + (E^\star A^\star)^\star E^\star A^\star, \\ \\ Q(0) = 0, \end{cases} \tag{5.1}$$

as well as the approximating problem

$$\begin{cases} Q_n' = AQ_n + Q_n A^\star - Q_n C^\star C Q_n + (E^\star A_n^\star)^\star E^\star A_n^\star, \\ \\ Q_n(0) = 0, \end{cases} \tag{5.2}$$

where $A_n = nAR(n,A)$ are the Yosida approximations of A.

Proposition 5.1 *Assume* $(\mathcal{HH})'$ *and let* $P_n \in C_s([0,T];\Sigma(H))$ *be the mild solution of* (1.6). *Then there is* $P \in C_s([0,T];\Sigma(H))$ *such that the following limit exists*

$$\lim_{n\to\infty} P_n = P, \qquad in\ C_s([0,T];\Sigma(H)).$$

In the sequel we shall call P the *mild solution* of (1.5).

Proof. We first remark that the hypotheses of Theorem 4.1 in Chapter 1 are verified, so that problems (5.1) and (5.2) have mild solutions, and moreover the following limit exists

$$\lim_{n\to\infty} Q_n = Q, \qquad in\ C_s([0,T];\Sigma(H)).$$

Denote by C_T a positive constant such that

$$\|Q_n(t)\| \le C_T, \quad \forall\, t \in [0,T], n \in \mathbf{N}. \tag{5.3}$$

Fix now $t \in [0,T]$; then by Theorem 7.4 in Chapter 1 we have

$$\begin{aligned} P_n(t) = \;& U_{G_{n,t}}(t,0)[I + P_0 Q_n(t)]^{-1} P_0 U^*_{G_{n,t}}(t,0) \\ & + \int_0^t U_{G_{n,t}}(t,s) C^* C U^*_{G_{n,t}}(t,s)\, ds, \end{aligned} \tag{5.4}$$

where

$$G_{n,t}(s) = A^* - C^* C Q_n(t-s), \quad s \in [0,t].$$

Therefore $U_{G_{n,t}}$ is the evolution operator with respect to $G_{n,t}$.

Step 1. There exists $C_1 > 0$ such that

$$\|U_{G_{n,t}}(\tau,s)\| \le C_1, \quad \forall\, 0 \le s \le \tau \le t \text{ and } n \in \mathbf{N}. \tag{5.5}$$

In fact let $x \in H$ and set $\varphi_n(\tau) = U_{G_{n,t}}(\tau,s)x$. Then we have

$$\varphi_n(\tau) = e^{(\tau-s)A^*}x - \int_s^\tau e^{(\tau-\sigma)A^*} C^* C Q_n(t-\sigma)\varphi_n(\sigma)\, d\sigma.$$

Let $M_T > 0$ such that

$$\|e^{sA}\| \le M_T, \quad \forall\, s \in [0,T].$$

Then, recalling (5.3), we obtain

$$|\varphi_n(\tau)| \le M_T|x| + M_T C \int_s^\tau |\varphi_n(\sigma)|\, d\sigma,$$

and the conclusion follows from Gronwall's lemma.

Step 2. We have

$$\lim_{n\to\infty} U_{G_{n,t}}(\tau,s)x = U_{G_t}(\tau,s)x, \quad \forall\, x \in H,\ 0 \le s \le \tau \le t. \tag{5.6}$$

Let $x \in H$ and set $\varphi_n(\tau) = U_{G_{n,t}}(\tau, s)x$ and $\varphi(\tau) = U_{G_n}(\tau, s)x$. Setting $\zeta_n(\cdot) = \varphi(\cdot) - \varphi_n(\cdot)$, then ζ_n is the mild solution of the following problem

$$\begin{cases} \zeta_n'(\tau) &= A^*\zeta_n(\tau) - C^*CQ_n(t-\tau)\zeta_n(\tau) \\ & \quad +C^*C(Q_n(t-\tau) - Q(t-\tau))\varphi(\tau) \\ \\ \zeta_n(0) &= 0, \end{cases}$$

that is to the integral equation

$$\zeta_n(\tau) = \int_s^\tau U_{G_{n,t}}(\tau - \sigma)C^*C(Q_n(t-\sigma) - Q(t-\sigma))\varphi(\sigma)\, d\sigma.$$

Now (5.6) follows from (5.5).

Step 3. Conclusion. It is sufficient to let n tend to infinity in (5.4). ∎

We can now proceed as in §3. Let $(u^\star, x^\star)$ (resp. $(u_n^\star, x_n^\star)$) be the optimal pair for the control problem (1.3)-(1.4) (resp. (1.7)-(1.8)).

Then the following result is proved as Theorem 3.1.

Theorem 5.1 *Assume that $(\mathcal{HH}')$ is verified and that $P_0 \in \Sigma^+(H)$. Then we have*

(i) $\lim_{n\to\infty} u_n^\star = u^\star$ *in* $L^2(0,T;U)$,

(ii) $\lim_{n\to\infty} x_n^\star = x^\star$ *in* $C([0,T];H)$.

Example 5.1 *Age dependent equations.*
Consider a dynamical system describing the evolution of a certain population (see for instance G. WEBB [1]), governed by the equations

$$\begin{cases} p_t(t,a) + p_a(t,a) + \mu(a)p(t,a) = 0, \quad a \in [0, a_+],\ t \geq 0, \\ \\ p(0,a) = p_0(a), \quad a \in [0, a_+], \\ \\ p(t,0) = \displaystyle\int_0^{a_+} \beta(b)p(t,b)db + u(t), \quad t \geq 0. \end{cases} \tag{5.7}$$

Here $p(t,a)$ is the density of the population of age a at time t, μ is the mortality rate, β the birth rate, a_+ the maximal age and u the control. We shall assume that $\mu \geq 0$, $\beta \geq 0$, $\mu, \beta \in C^1([0, a_+])$ and

$$\int_0^{a_+} \mu(b)\, db = +\infty, \quad \int_0^{a_+} \beta(b) \exp \left\{ - \int_0^{\beta} \mu(c)\, dc \right\} db \neq 1. \quad (5.8)$$

We want to minimize a quadratic cost function of the form

$$J(u) = \int_0^T dt \left[\int_0^{a_+} p^2(t, a)\, da + u^2(t) \right], \quad (5.9)$$

over all controls $u \in L^2(0, T)$ subject to state equation (5.7).

As shown in G. Da Prato and M. Iannelli [1], problem (5.7) can be written on the abstract form

$$p(t) = e^{tA} p_0 - \alpha A \int_0^t e^{(t-s)A} E u(s)\, ds, \quad t \in [0, T], \quad (5.10)$$

where A is the infinitesimal generator of a C_0 semigroup on $H = L^2(0, a_+)$ and D is a mapping from $\mathbf{R}$ into H. More precisely the operator A is given by

$$\begin{cases} D(A) = \left\{ \varphi \in H : \begin{array}{l} \varphi' + \mu\varphi \in H \\ \varphi(0) = \int_0^{a_+} \beta(b)\varphi(b)\, db \end{array} \right\} \\ \\ A\varphi = -\varphi' - \mu\varphi, \end{cases} \quad (5.11)$$

and E is defined by

$$x \mapsto E(x) = \pi x : \mathbf{R} \to H, \quad (5.12)$$

where π is the element of H given by

$$\pi(a) = e^{-\int_0^a \mu(b)\, db}, \quad a \in [0, a_+] \quad (5.13)$$

Moreover the semigroup generated by A is given by

$$e^{tA}\varphi = \begin{cases} B_\varphi(t - a)\pi(a) & \text{if } a \in [0, t[\\ \\ \varphi(a - t)\pi(t) & \text{if } a \in [t, a_+] \end{cases}, \quad \varphi \in L^2(0, a_+), \quad (5.14)$$

where B is the solution to the integral equation

$$B(a) = F_\varphi(a) + \int_0^a K(a - b) B(b)\, db, \quad a \in [0, a_+], \quad (5.15)$$

and

$$F_\varphi(a) = \int_a^{a_+} \varphi(b-a)K(b)\frac{\pi(a)}{\pi(b)}\,db, \quad a \in [0,a_+]. \tag{5.16}$$

We now check hypotheses $(\mathcal{HH})'$. It is sufficient to show that $(\mathcal{HH})_1$-(iii) holds. If $\varphi \in D(A^*)$ we have in fact

$$\begin{aligned}\int_0^{a_+} |E^*A^*e^{tA^*}\varphi|^2 dt &= \int_0^{a_+} | < A^*e^{tA^*}\varphi, \pi > |^2 dt \\ &= \int_0^{a_+} |\frac{d}{dt} < \varphi, e^{tA}\pi > |^2 dt.\end{aligned} \tag{5.17}$$

Moreover, recalling (5.14), we have

$$\begin{aligned}&\frac{d}{dt} < \varphi, e^{tA}\pi > \\ &= \frac{d}{dt}\left\{\int_0^t \varphi(a)B_\pi(t-a)\pi(a)\,da + \int_t^{a_+} \varphi(a-t)\pi(t)\,da\right\} \\ &= \varphi(t)B_\pi(0)\pi(b) + \int_0^t \varphi(a)B'_\pi(t-a)\pi(a)\,da \\ &\quad -\varphi(a_+-t)\pi(t) + \int_0^{a_+-t} \varphi(a)\pi'(t)\,da.\end{aligned} \tag{5.18}$$

Now, by plugging (5.18), into (5.17) the conclusion follows. ∎

References

P. Acquistapace [1], *On a family of generators of analytic semigroups*, in "Differential equations with applications in biology, physics, and engineeering", J.A. Goldstein, F.Kappel and W. Schappacher, eds, pp 1–6, Marcel Dekker, New York 1991.

P. Acquistapace, F. Flandoli and B. Terreni [1], *Initial boundary value problems and optimal control for non- autonomous parabolic systems*, SIAM J. Control and Optim. 29 (1991), 89–118.

P. Acquistapace and B. Terreni [1], *Some existence and regularity results for abstract non-autonomous parabolic equations*, J. Math. Anal. Appl. 99 (1984), 9–64.

[2], *A unified approach to abstract linear non-autonomous parabolic equations*, Rend. Sem. Mat. Univ. Padova 78 (1987), 47–107.

[3], *Boundary control problems for non-autonomous parabolic systems*, in "Stabilization of Flexible Structures", Lecture Notes in Control and Information Sciences, Vol. 147, J.-P. Zolésio, ed., pp. 156–166, Springer-Verlag, Berlin, Heidelberg, New York 1990.

S. Agmon [1], "Lectures on Elliptic Boundary Value Problems", Van Nostrand, New York 1965.

A.V. Balakrishnan [1], "Applied functional analysis", Springer-Verlag, New-York, Heidelberg 1976.

[2], *Boundary control of parabolic equations: L-Q-R theory*, in "Theory of non linear operators", pp. 11–13, Abh. Akad. Wiss. DDR, Abt. Math. Naturwiss. Tech., 6, Akademie-Verlag, Berlin, 1978.

H.T. Banks and K. Kunish [1], *The Linear Regulator Problem for Parabolic Systems*, SIAM J. Control and Optimization 22, no. 5 (1984), 684–699.

V. Barbu and G. Da Prato [1], *A representation formula for the solutions to the operator Riccati equation*, Differential and Integral Equations 5, no. 4 (1992), 821–829.

R.W. Brockett [1], "Finite dimensional linear systems", Wiley, New York 1970.

F. Bucci [1], *A Dirichlet boundary control problem for the strongly damped wave equation*, SIAM J. Control and Optim. 30, no. 5 (1992), 1092–1100.

P. Cannarsa, G. Da Prato and J.-P. Zolésio [1], *The damped wave equation in a moving domain*, J. Differential Equations 85 (1990), 1–16.

A. Chojnowska-Michalik [1], *A Representation theorem for general stochastic delay equations*, Bull. Pol. Acad. Sci. Ser. Math. XXVI, no 7 (1978), 634-641.

R.F. Curtain and A.J. Pritchard [1], *The infinite dimensional Riccati equation*, J. Math. Anal. and Appl. 47 (1974), 43–56.

G. Da Prato [1], *Quelques résultat d'existence, unicité et régularité pour un problème de la théorie du contrôle*, J. Maths. Pures Appl. 52 (1973), 353–375.

Cf. V. Barbu and G. Da Prato

Cf. P. Cannarsa, G. Da Prato and J.-P. Zolésio

G. Da Prato and M. Iannelli [1], *Boundary control problem for age dependent equations*, in "Third international workshop-conference on Evolution equations, control theory and biomathematics", Ph. Clément and G. Lumer, eds, Marcel Dekker, to appear.

G. Da Prato and A. Ichikawa [1], *Riccati equations with unbounded coefficients*, Ann. Mat. Pura Appl. 140 (1985), 209–221.

G. Da Prato, I. Lasiecka and R. Triggiani [1], *A direct study of the Riccati equation arising in hyperbolic boundary control problems*, J. Differential Equations 64 (1986), 26–47.

G. Da Prato and J.-P. Zolésio [1], *An optimal control problem for a parabolic equation in non-cylindrical domains*, Systems and Control Letters 11 (1988), 73–77.

[2], *Existence and optimal control for wave equation in moving domains*, in "Stabilization of Flexible Structures", Lecture Notes in Control and Information Sciences, Vol. 147, J.-P. Zolésio, ed., pp.167–190, Springer-Verlag, Berlin, Heidelberg, New York 1990.

M.C. Delfour [1], *The linear quadratic optimal control problem with delays in state and control variables: a state space approach*, SIAM J. Control and Optim. 24 (1986), 835–883.

[2], *The linear quadratic optimal control problem for hereditary differential systems: theory and numerical solution*, J. Applied Mathematics and Optimization 3 (1977), 101–162.

[3], *The linear quadratic optimal control problem over an infinite time horizon for a class of distributed parameter systems*, in "Control of distributed parameter systems", S.P. Banks and A.J. Pritchard, eds, pp. 57-66, Pergamon Press, Oxford, England 1978.

M.C. DELFOUR, E.B. LEE and A. MANITIUS [1], *F-reduction of the operator Riccati equation for hereditary differential systems*, Automatica 14 (1978), 385–395.

M.C. DELFOUR and M. SORINE [1], *The linear quadratic optimal control problem for parabolic systems with boundary control through a Dirichlet boundary condition*, in "Control of Distributed Parameter Systems", J.P. Barbary and L. Le Letty, eds., Chap. I, pp. 13-16, Pergamon Press, Oxford, England, 1982. (Proc. 3rd IFAC Symp. on Control of Distributed Parameter Systems, Toulouse, France June 29-July 2nd, 1982, see also CRMA report no. 1051, Université de Montréal, 1981).

N. DUNFORD and J.T. SCHWARTZ [1], "Linear operators, Part I: general theory", Intersciences, New York 1954.

H.O. FATTORINI [1], *Boundary control systems*, SIAM J. Control 6 (1968), 349-385.

F. FLANDOLI [1], *Riccati equation arising in a boundary control problem with distributed parameters*, SIAM J. Control Optim. 22 (1984), 76–86.

[2], *Invertibility of Riccati operators and controllability of related systems*, Systems Control Letters 9 (1987), 65–72.

[3], *A new approach to L-Q-R problem for hyperbolic dynamics with boundary control*, in "Distributed parameter systems", F. Kappel, K. Kunish and W. Schappacher, eds, pp. 89–111, Lectures Notes in Control and Information Sciences, Vol. 102 (Proceedings of the 3rd International Conference), Springer-Verlag, Berlin, Heidelberg, New York 1987.

[4], *A counterexample in the boundary control of parabolic systems*, Appl. Math. Letters 3, no. 2 (1990), 47–50.

[5], *On the semigroup approach to stochastic differential equations in Hilbert spaces*, Stochastic Analysis and Applications 10, no. 2 (1992), 181–204.

[6], *On the direct solution of Riccati equations arising in boundary control theory*, Ann. Mat. Pura Appl., to appear.

F. FLANDOLI, I. LASIECKA and R. TRIGGIANI [1], *Algebraic Riccati equations with non smooth observation arising in Hyperbolic and Euler-Bernoulli boundary control problems*, Annali di Matematica Pura e Applicata, Serie IV, Tomo CLIII (1988), 307–382.

M. Iannelli Cf. G. Da Prato and M. Iannelli

A. Ichikawa Cf. G. Da Prato and A. Ichikawa

R.E. Kalman [1], *Contributions to the theory of optimal control*, Bol. Soc. Mat. Mex. 5 (1960), 102–119.

K. Kunish Cf. H.T. Banks and K. Kunish

I. Lasiecka [1], *Unified theory for abstract parabolic boundary problems: a semigroup approach*, Appl. Math. Optim. 6 (1980), 287–333.

Cf. G. Da Prato, I. Lasiecka and R. Triggiani

Cf. F. Flandoli, I. Lasiecka and R. Triggiani

I. Lasiecka, J.-L. Lions and R. Triggiani [1], *Non homogeneous boundary value problems for second order hyperbolic operators*, J. Math. pures et appl. 65 (1986), 149–192.

I. Lasiecka and R. Triggiani [1], *A cosine operator approach to modelling $L_2(0,T;L^2(\Gamma))$-boundary input hyperbolic equations*, Appl. Math. Optim. 7 (1981), 35–93.

[2], *Regularity of hyperbolic equations under $L_2(0,T;L^2(\Gamma))$-Dirichlet boundary terms*, Appl. Math. Optim. 10 (1983), 275–286.

[3], *Dirichlet boundary control problems for parabolic equations with quadratic cost; analiticity and Riccati's feedback synthesis*, SIAM J. Control and Optim. 21 (1983), 41–67.

[4], *Riccati equations for hyperbolic partial differential equations with $L_2(0,T;L^2(\Gamma))$-Dirichlet boundary terms*, SIAM J. Control Optim. 24, no. 5 (1986), 41–68.

[5], *The regulator problem for parabolic equations with Dirichlet boundary control. Part I, Riccati's feedback synthesis*, Appl. Math. Optim. 16 (1987), 147–168.

[6], *The regulator problem for parabolic equations with Dirichlet boundary control. Part II, Galerkin approximation*, Appl. Math. Optim. 16 (1987), 187–216.

[7], "Differential and Algebraic Riccati Equations with applications to boundary/point control problems: continuous theory and approximation theory", Lecture Notes in Control and Information Sciences, Vol. 164, Springer-Verlag, Berlin, Heidelberg, New York 1991.

[8], *Riccati differential equations with unbounded coefficients and non smoothing terminal condition: the case of analytic semigroups*, SIAM J. Math. Anal. (1992), to appear.

E.B. Lee Cf. M.C. Delfour, E.B. Lee and A. Manitius

J.-L. Lions [1], "Optimal control of systems described by Partial Differential Equations", Springer-Verlag, Berlin, Heidelberg, New York 1971.

[2], *Exact controllability, stabilization and perturbations*, J. Von Neumann Lecture, SIAM Reviews 30 (1988), 1–60.

[3], "Controllabilité exacte, perturbations et stabilisation des systèmes distribués", Volumes 1, 2, Masson, Paris 1990.

Cf. I. Lasiecka, J.-L. Lions and R. Triggiani

J.-L. Lions and E. Magenes [1], "Problèmes aux limites non homogènes", Dunod, Paris 1968.

J.-L. Lions and J. Peetre [1], *Sur une classe d'espaces d'interpolation*, Institut de Hautes Etudes Scientifiques, Publ. Math. 19 (1964), 5–68.

D.L. Lukes and D.L. Russell [1], *The quadratic criterion for distributed systems*, SIAM J. Control 7 (1969), 101–121.

A. Lunardi [1], *Differentiability with respect to (t, s) of the parabolic evolution operator*, Israel J. Math. 68, no. 2 (1989), 161–184

A. Manitius Cf. M.C. Delfour, E.B. Lee and A. Manitius

L. Pandolfi [1], "The standard regulator problem for systems with input delays: an approach through singular control theory", Rapporto interno no. 17, Dipartimento di Matematica, Politecnico di Torino, 1992.

A. Pazy [1], "Semigroups of Linear Operators and Applications to Partial Differential Equations", Springer-Verlag, Berlin, Heidelberg, New York 1983.

J. Peetre Cf. J.-L. Lions and J. Peetre

A.J. Pritchard and D. Salomon [1], *The linear quadratic control problem for infinite dimensional systems with unbounded input and output operators*, SIAM J. Control Optim. 25 (1987), 121–144.

Cf. R.F. Curtain and A.J. Pritchard

D.L. Russell Cf. D.L. Lukes and D.L. Russell

J.T. Schwartz Cf. N. Dunford and J.T. Schwartz

D. Salomon Cf. A.J. Pritchard and D. Salomon

M. Sorine [1], "Sur le semigroupe non linéaire associé à l'équation de Riccati", Rapport du CRMA no. 1055, Université de Montréal 1981.

[2], "Un résultat d'existence et d'unicité pour l'équation de Riccati stationnaire", Rapport du CRMA no. 984, Université de Montréal 1980.

Cf. M.C. Delfour and M. Sorine

H. Tanabe [1], "Equations of evolution", Pitman, London 1979.

L. Tartar [1], *Sur l'étude directe d'équations non linéaires intérvenant en théorie du contrôle optimal*, J. Funct. Anal. 17 (1974), 1–46.

A.E. Taylor [1], "Introduction to Functional Analysis", J. Wiley and Sons, New York 1961.

R. Temam [1], *Sur l'équation de Riccati associée à des opérateurs non bornés en dimension infinie*, J. Funct. Anal. 7 (1971), 85–116.

B. Terreni Cf. P. Acquistapace and B. Terreni

R. Triggiani [1], *On Nambu's Boundary Stabilization Problem for Diffusion Processes*, J. Differential Equations 33 (1979), 189–200.

Cf. G. Da Prato, I. Lasiecka and R. Triggiani

Cf. F. Flandoli, I. Lasiecka and R. Triggiani

Cf. I. Lasiecka and R. Triggiani

Cf. I. Lasiecka, J.-L. Lions and R. Triggiani

D.C. Washburn [1], *A bound on the boundary input map for parabolic equations with application to time optimal control*, SIAM J. on Control 17 (1979), 652–671.

G.F. Webb [1], "Theory of nonlinear age-dependent population dynamic", Monographs and Textbooks in Pure and Applied Mathematical Series, Vol. 89, Marcel Dekker, New York and Basel 1985.

W.M. Wonham [1], *On a matrix Riccati equation of stochastic control*, SIAM J. Control 6 (1968), 681–697.

A. Yagi [1], *Coincidence entre des espaces d'interpolation et des domaines de puissances fractionnaires d'opérateurs*, C.R. Acad. Sc. Paris 299, Sér. I, no. 6 (1984), 173–176.

J. Zabczyk [1], *Infinite dimensional systems in Control Theory*, Bull. Int. Stat. Inst. XLVII, no. 2 (1977), 286–310.

J.-P. ZOLÉSIO [1], *The material derivative*, in "Optimization of Distributed Parameters Structures", E.J. Haug and J. Céa, eds, pp. 1089–1151, Sijthoff and Noordhoff, Alphen aan den Rijn 1981.

Cf. P. CANNARSA, G. DA PRATO and J.-P. ZOLÉSIO

Cf. G. DA PRATO and J.-P. ZOLÉSIO

Part III

Quadratic Optimal Control: Infinite Time Horizon

Chapter 1

Systems with Bounded Control Operators: Control Inside the Domain

1 Introduction and setting of the problem

As in Chapter 1 (Part II) we consider a dynamical system governed by the following state equation

$$\begin{cases} x'(t) = Ax(t) + Bu(t), \quad t \geq 0 \\ x(0) = x_0 \in H \end{cases} \tag{1.1}$$

and we use the notation introduced in §1 of that chapter. We assume that

$$(\mathcal{H})_\infty \quad \begin{cases} (i) \quad A \textit{ generates a } C_0 \textit{ semigroup } e^{tA} \textit{ in } H, \\ (ii) \quad B \in \mathcal{L}(U; H), \\ (iii) \quad C \in \mathcal{L}(H; Y). \end{cases}$$

Clearly, if the assumptions $(\mathcal{H})_\infty$ hold, then the assumptions $(\mathcal{H})$ of §1 in Chapter 1 (Part II) are verified with $P_0 = 0$.

We want to minimize the *cost function:*

$$J_\infty(u) = \int_0^\infty \{|Cx(s)|^2 + |u(s)|^2\} ds, \tag{1.2}$$

over all controls $u \in L^2(0,\infty;U)$ subject to the differential equation constraint (1.1). We say that the control $u \in L^2(0,\infty;U)$ is *admissible* if $J_\infty(u) < \infty$. An admissible control $u^\star \in L^2(0,\infty;U)$ is called an *optimal control* if

$$J_\infty(u^\star) \le J_\infty(u), \quad \forall u \in L^2(0,\infty;U).$$

In this case the corresponding solution of (1.1) is called an *optimal state* and the pair $(u^\star, x^\star)$ an *optimal pair.*

Simple examples in finite dimension show that admissible controls can fail to exist. When, for any $x_0 \in H$, an admissible control exists, we say that (A,B) is *stabilizable with respect to the observation operator* C, or, for brevity, that (A,B) is C*-stabilizable.* In this case it is still possible to solve problem (1.1)-(1.2) by using Dynamic Programming. In fact when there exists an admissible control, one can show that the *minimal solution* $P_{min}(t)$ to the Riccati equation (2.7) in §1 of Chapter 1 in Part II, that is the solution corresponding to the initial condition $P_0 = 0$, converges, as $t \to \infty$ to a solution P^∞_{min} of the *algebraic Riccati equation:*

$$A^\star X + XA - XBB^\star X + C^\star C = 0. \tag{1.3}$$

Moreover equation (1.3) has a nonnegative solution if and only if (A,B) is C-stabilizable. The *minimal nonnegative solution* of (1.3) P^∞_{min} is the main tool in solving problem (1.1)-(1.2). In fact one can show (see §3), that if (A,B) is C-stabilizable, then the optimal control $u^\star$ is given by

$$u^\star(t) = -B^\star P^\infty_{min} x^\star(t), \tag{1.4}$$

where $x^\star$ is the solution to the *closed loop equation:*

$$\begin{cases} x'(t) = (A - BB^\star P^\infty_{min})x(t), & t \in [0,T], \\ x(0) = x_0 \in H. \end{cases} \tag{1.5}$$

The operator

$$F = A - BB^\star P^\infty_{min} \tag{1.6}$$

is called the *feedback operator.* It is important for the applications, to know whether F is exponentially stable or not. A sufficient condition for this is that (A, C) be *detectable*, i.e. that $(A^\star, C^\star)$ be I- stabilizable. This condition is also necessary if H is finite dimensional or in special situations (see §3). In §4 we study some qualitative properties of the solutions of (1.3). In particular we find that if F is exponentially stable then P^∞_{min} is globally attractive (among all nonnegative solutions of Riccati equations), and so it is the unique solution of (1.3). We also study the existence of a maximal solution P^∞_{Max} of (1.3). We prove that a maximal solution P^∞_{Max} exists if (A, B) is I-stabilizable. In this case one can show that the infimum of $J_\infty(u)$ over all $u \in L^2(0,\infty;U)$ such that $x \in L^2(0,\infty;H)$, subject to the differential equation constraint (1.1), is precisely $(P^\infty_{Max}x_0, x_0)$.

We end §4 by studying periodic solutions of Riccati equations. Finally §5 is devoted to examples and §6 to complements.

2 The algebraic Riccati equation

We assume here that $(\mathcal{H})_\infty$ is verified and consider the system (1.1).

Definition 2.1 *(i)* (A, B) *is said to be stabilizable with respect to the observation* C*, or* C*-stabilizable, if for any* $x_0 \in H$ *there exist* $u \in L^2(0,\infty;U)$ *such that* $Cx \in L^2(0,\infty;Y)$*, where* x *is the corresponding solution to* (1.1).
(ii) (A, B) *is called feedback stabilizable with respect to the obsevation* C*, if there exists* $K \in \mathcal{L}(H;U)$*,* $N > 0$*,* $\omega > 0$ *such that*

$$\|Ce^{t(A-BK)}\| \le Ne^{-\omega t}, \quad \forall\, t \ge 0$$

We consider the Riccati equation

$$P' = A^\star P + PA - PBB^\star P + C^\star C \tag{2.1}$$

and the corresponding stationary equation

$$A^\star X + XA - XBB^\star X + C^\star C = 0. \tag{2.2}$$

In the sequel we shall consider only nonnegative solutions of (2.1) and (2.2).

Definition 2.2 *(i) We say that $X \in \Sigma^+(H)$ is a* weak solution *of* (2.2) *if*

$$(Xx, Ay) + (Ax, Xy) - (B^\star Xx, B^\star Xy) + (Cx, Cy) = 0 \tag{2.3}$$

for all $x, y \in D(A)$.
(ii) X is called a strict solution *of* (2.2) *if $X \in D(\mathcal{A})$ (the operator defined by* (3.3) *in Part II, Chapter 1) and*

$$\mathcal{A}(X) - XBB^\star X + C^\star C = 0. \tag{2.4}$$

We remark that if X is a strict solution of (2.2) then, by Part II, Chapter 1, Proposition 3.1 we have

$$A^\star Xx + XAx - XBB^\star Xx + C^\star Cx = 0 \tag{2.5}$$

for any $x \in D(A)$ (because if $x \in D(A)$ then $Xx \in D(A^\star)$).
Finally

Definition 2.3 *We say that $X \in \Sigma^+(H)$ is a* stationary solution *of equation* (2.1) *if it coincides with the mild solution of* (2.1) *with initial condition $P(0) = X$.*

Proposition 2.1 *Let $X \in \Sigma^+(H)$, then the following statements are equivalent:*
(i) X is a weak solution of (2.2).
(ii) X is a strict solution of (2.2).
(iii) X is a stationary solution of (2.1).

Proof. *(i)* $\Rightarrow$ *(ii)*. Assume that (2.3) holds for any $x, y \in D(A)$. Then we have

$$\varphi_X(x, y) = (B^\star Xx, B^\star Xy) - (Cx, Cy), \quad \forall\, x, y \in D(A),$$

where φ_X is the bilinear form defined by (3.1) in Part II, Chapter 1. Clearly φ_X is continuous in $H \times H$ so that $X \in D(\mathcal{A})$ and (2.4) holds.

(ii) $\Rightarrow$ *(i)*. Let $X \in D(\mathcal{A})$ be a strict solution of (2.1) and let $x \in D(A)$; then $Xx \in D(A^\star)$ and (2.5) holds true. Now, if $x, y \in D(A)$, (2.3) follows and X is a weak solution.

(ii) $\Rightarrow$ *(iii)*. If X is a strict solution of (2.2), then $P(t) = X$ is a strict solution. Therefore it is a stationary solution, of (2.1).

(iii) $\Rightarrow$ *(i)*. Let X be a stationary solution of (2.1), set $P(t) = X$. Then by Part II, Chapter 1, Proposition 2.1, P is a weak solution of (2.1) so that (2.3) holds and X is a weak solution of (2.2). ∎

Due to Proposition 2.1, all kinds of solutions are the same; in the sequel we shall call a *solution* both a weak and a strict solution of (2.2).

We are going to study existence of a solution of the Algebraic Riccati equation. We follow here A. PRITCHARD and J. ZABCZYK [1]. It is useful to consider the solution of the Riccati equation with initial condition 0,

$$\begin{cases} P' = A^\star P + PA - PBB^\star P + C^\star C, \\ \\ P(0) = 0. \end{cases} \tag{2.6}$$

This solution will be denoted by $P_{min}(\cdot)$.

Remark 2.1 $P_{min}(\cdot)$ is the *minimal nonnegative solution* of the Riccati equation. In fact if $P_0 \in \Sigma^+(H)$ and P is the mild solution of (2.1) such that $P(0) = P_0$, then by Part II, Chapter 1, Proposition 2.2 we have

$$P_{min}(t) \leq P(t), \quad \forall\, t \geq 0.$$

It follows that if X is a solution of (2.2) , then

$$P_{min}(t) \leq X, \quad \forall\, t \geq 0.$$

∎

We now prove the following properties.

Proposition 2.2 *The following statements hold:*
(i) for any $x \in H, (P_{min}(\cdot)x, x)$ is non decreasing,
(ii) assume that, for some $R \in \Sigma^+(H)$, we have

$$P_{min}(t) \leq R, \quad \forall\, t \geq 0.$$

Then for all $x \in H$ the limit

$$P_{min}^{\infty}x = \lim_{t\to\infty} P_{min}(t)x, \tag{2.7}$$

exists, and P_{min}^{∞} is a solution of equation (2.2).

Proof. Let $\varepsilon > 0$, $t \geq 0$ and let P be the solution of (2.1) such that $P(0) = P_{min}(\varepsilon)$. By Proposition 2.2 in Part II, Chapter 1, we have

$$P_{min}(t+\varepsilon) = P(t) \geq 3P_{min}(t)$$

and (i) is proved. Assume now $P_{min}(t) \leq R$; since $P_{min}(t)$ is non decreasing and bounded we can set

$$\gamma(x) = \lim_{t\to\infty} (P_{min}(t)x, x), \quad \forall\, x \in H.$$

For $x, y \in H$ we have

$$\begin{aligned}
&2\,\mathrm{Re}\,(P_{min}(t)x, y)\\
&= (P_{min}(t)(x+y), x+y) - (P_{min}(t)x, x) - (P_{min}(t)y, y)
\end{aligned}$$

$$\begin{aligned}
&2\,\mathrm{Im}\,(P_{min}(t)x, y)\\
&= i(P_{min}(t)(x+iy), x+iy) - (P_{min}(t)x, x) - (P_{min}(t)(iy), iy).
\end{aligned}$$

So the limit

$$\Gamma(x, y) = \lim_{t\to\infty} (P_{min}(t)x, y), \quad \forall\, x, y \in H,$$

exists and the following operator $P_{min}^{\infty} \in \Sigma^{+}(H)$ can be defined

$$\lim_{t\to\infty} (P_{min}(t)x, y) = (P_{min}^{\infty}(t)x, y), \quad \forall\, x, y \in H.$$

It follows that

$$\lim_{t\to\infty} ([P_{min}^{\infty} - P_{min}(t)]x, x) = 0, \quad x \in H,$$

which is equivalent to

$$\lim_{t\to\infty} (P_{min}^{\infty} - P_{min}(t))^{\frac{1}{2}}x = 0, \quad \forall\, x \in H.$$

This implies that

$$\lim_{t\to\infty}(P^\infty_{min} - P_{min}(t))x = 0, \quad \forall x \in H,$$

so that (2.7) holds. It remains to show that P^∞_{min} is a solution of (2.2). For this we denote by P_h the solution of (2.1) for which $P_h(0) = P_{min}(h)$, i.e. $P_h(t) = P_{min}(h+t)$. Since

$$\lim_{h\to\infty} P_{min}(h)x = P^\infty_{min}x, \quad \forall x \in H,$$

by Theorem 2.2 in Part II, Chapter 1, we have

$$\lim_{h\to\infty} P_h(\cdot)x = P^\infty_{min}x \text{ in } C([0,T];H), \quad \forall x \in H,\ \forall T > 0.$$

Moreover P^∞_{min} is a solution of (2.1) (hence stationary). ∎

Remark 2.2 Assume that there exists a solution $X \in \Sigma^+(H)$ of (2.2). Then, by Proposition 2.2 and Remark 2.1, the solution P^∞_{min} defined by (2.7) exists. By the above proposition it follows that

$$P^\infty_{min} \le X,$$

for all solution $X \in \Sigma^+(H)$ of (2.2). Thus P^∞_{min} is the *minimal solution* of the algebraic Riccati equation (2.2). ∎

We now prove that a nonnegative solution of the algebraic Riccati equation exists if and only if (A,B) is C-stabilizable.

Proposition 2.3 *Assume that* $(\mathcal{H})_\infty$ *is verified and that* (A,B) *is* C*-stabilizable. Then there exists a minimal solution* P^∞_{min} *of equation* (2.2).

Proof. We first recall that by equation (6.2) in Part II, Chapter 1, we have

$$\begin{aligned}(P_{min}(t)x_0, x_0) &+ \int_0^t |u(s) + B^\star P_{min}(t-s)x(s)|^2\, ds \\ &= \int_0^t \{|Cx(s)|^2 + |u(s)|^2\} ds,\end{aligned} \tag{2.8}$$

for any $x_0 \in H$ and any $u \in L^2(0,\infty;U)$, where x is the solution of (1.1). Let u be a control in $L^2(0,\infty;U)$ such that the corresponding solution x of (1.1) is such that Cx belongs to $L^2(0,\infty;Y)$. By (2.8) it follows that

$$\sup_{t\geq 0}\,(P_{min}(t)x_0,x_0) \leq \int_0^\infty \{|Cx(s)|^2+|u(s)|^2\}ds < \infty$$

for any $x_0 \in H$. By the Uniform Boundedness Theorem it follows that $P_{min}(t)$ is bounded, so that, by Proposition 2.2, there exists a solution of equation (2.2). ∎

In order to prove the converse result it is useful to introduce, for any $t > 0$, the following auxiliary optimal control problems over the finite time horizon $[0,t]$: to minimize

$$J_t(u) = \int_0^t \{|Cx(s)|^2+|u(s)|^2\}ds, \tag{2.9}$$

over all controls $u \in L^2(0,t;U)$ subject to the differential equation constraint (1.1). By Theorem 6.1 in Part II, Chapter 1, we know that there exists a unique optimal pair (x_t,u_t) for problem (2.9), where x_t is the mild solution to the closed loop equation:

$$\begin{cases} x_t'(s) = [A - BB^\star P_{min}(t-s)]\,x_t(s), & 0 \leq s \leq t,\\[2mm] x_t(0) = x_0, \end{cases}$$

and u_t is given by the feedback formula

$$u_t(s) = -B^\star P_{min}(t-s)x_t(s), \quad 0 \leq s \leq t.$$

Moreover the optimal cost is given by

$$(P_{min}(t)x_0,x_0) = \int_0^t \{|Cx_t(s)|^2+|u_t(s)|^2\}ds. \tag{2.10}$$

Lemma 2.1 *Assume that the minimal solution P_{min}^∞ of (2.2) exists. Denote by x_∞ the corresponding mild solution to the equation*

$$\begin{cases} x_\infty'(s) = [A - BB^\star P_{min}^\infty]x_\infty(s), & s \geq 0,\\[2mm] x_\infty(0) = x_0, \end{cases}$$

and set

$$u_\infty(s) = -B^\star P_{min}^\infty x_\infty(s), \quad s \geq 0. \tag{2.11}$$

Then we have

$$\lim_{t\to\infty} x_t(s) = x_\infty(s), \quad s \geq 0. \tag{2.12}$$

$$\lim_{t\to\infty} u_t(s) = u_\infty(s), \quad s \geq 0. \tag{2.13}$$

Proof. Fix $T > t$ and set $z_t = x_t - x_\infty$; then for all $0 \leq s \leq T$, z_t is the mild solution of the problem:

$$\begin{cases} z_t'(s) = [A - BB^\star P_{min}(t-s)]\, z_t(s) \\ \qquad\qquad + BB^\star\, [P_{min}(t-s) - P_{min}^\infty]\, x_\infty(s) \\ \\ z_t(0) = 0. \end{cases} \tag{2.14}$$

Denote by $U(r,s)$ the evolution operator corresponding to $A - BB^\star P_{min}(t-\cdot)$; then for $x \in H$

$$\begin{cases} U(r,\sigma)x = e^{(r-\sigma)A}x - \displaystyle\int_\sigma^r e^{(r-\rho)A} BB^\star P_{min}(t-\rho) U(\rho,\sigma)x\, d\rho, \\ \\ U(\sigma,\sigma) = I. \end{cases}$$

It follows that

$$\|U(r,\sigma)\| \leq Me^{(r-\sigma)\omega} + M\|B\|^2\|P_{min}^\infty\| \int_\sigma^r e^{(r-\rho)\omega}\|U(\rho,\sigma)\|\, d\rho.$$

By Gronwall's Lemma we have

$$\|U(r,\sigma)\| \leq Me^{(r-\sigma)[\omega + M\|B\|^2\|P_{min}^\infty\|]}, \quad 0 \leq \sigma \leq r \leq T. \tag{2.15}$$

We now return to problem (2.14) which we write in the form

$$z_t(s) = \int_0^s U(s,\sigma) BB^\star [P_{min}(t-\sigma) - P_{min}^\infty]\, x_\infty(\sigma)\, d\sigma.$$

By (2.15) and by the Dominated Convergence Theorem we obtain $z_t(s) \to 0$ as $t \to \infty$. So (2.12) and then (2.13) follow. ∎

We can now prove the following proposition.

Proposition 2.4 *Assume that there exists a solution of* (2.2). *Then* (A, B) *is* C*-stabilizable.*

Proof. Let x_t and u_t be defined as in the lemma. By (2.10) we have for $t \geq T$

$$(P^{\infty}_{min} x_0, x_0) \geq \int_0^T \{|Cx_t(s)|^2 + |u_t(s)|^2\} ds \tag{2.16}$$

and, as $t \to \infty$,

$$(P^{\infty}_{min} x_0, x_0) \geq \int_0^T \{|Cx_\infty(s)|^2 + |u_\infty(s)|^2\} ds. \tag{2.17}$$

But since T is arbitrary we find

$$(P^{\infty}_{min} x_0, x_0) \geq \int_0^\infty \{|Cx_\infty(s)|^2 + |u_\infty(s)|^2\} ds \tag{2.18}$$

and thus $u_\infty \in L^2(0, \infty; U)$ is an admissible control. ∎

3 Solution of the control problem

We now consider the control problem (1.1)-(1.2) and prove the central result.

Theorem 3.1 *Assume that the conditions* $(\mathcal{H})$ *are verified and that* (A, B) *is* C*-stabilizable. Then there exists a unique optimal pair* $(u^\star, x^\star)$ *for the optimal control problem* (1.1)-(1.2). *Moreover the following statements hold:*
(i) $x^\star \in C([0, \infty[; H)$ *is the mild solution to the closed loop equation* (1.5).
(ii) $u^\star \in C([0, \infty[; U)$ *is given by the feedback formula*

$$u^\star(t) = -B^\star P^{\infty}_{min} x^\star(t), \quad t \in [0, T], \tag{3.1}$$

where P^{∞}_{min} *represents the minimal solution of* (2.2).
(iii) The optimal cost $J_\infty(u^\star)$ *is given by*

$$J_\infty(u^\star) = (P^{\infty}_{min} x_0, x_0). \tag{3.2}$$

Proof. Let $u \in L^2(0, \infty; U)$ and let x be the corresponding solution of the state equation (1.1). By the identity (2.8) we have

$$(P_{min}(t) x_0, x_0) \leq \int_0^t \{|Cx(s)|^2 + |u(s)|^2\} ds \leq J_\infty(u).$$

It follows that

$$J_\infty(u) \geq (P_{min}^\infty(t)x_0, x_0), \quad \forall u \in L^2(0,\infty;U).$$

Let now u_∞ be defined by (2.11); by (2.18) we have

$$(P_{min}^\infty(t)\, x_0, x_0) \geq J_\infty(u_\infty),$$

so that u_∞ is optimal. Formula (3.1) with $u^\star = u_\infty$, $x^\star = x_\infty$ follows from (2.12)-(2.13). It remains to show uniqueness. Let $(\widehat{u}, \widehat{x})$ be another optimal pair, then $J_\infty(\widehat{u}) = (P_{min}^\infty x_0, x_0)$. Fix $T > 0$. By applying (2.8) with $t \geq T$ we obtain

$$\int_0^T |\widehat{u}(s) + B^\star P_{min}(t-s)\widehat{x}(s)|^2\, ds \leq J_\infty(\widehat{u}) - (P_{min}(t)x_0, x_0)$$

$$\leq ([P_{min}^\infty - P_{min}(T)]\, x_0, x_0),$$

As $t \to \infty$ we have

$$\int_0^T |\widehat{u}(s) + B^\star P_{min}\widehat{x}(s)|^2\, ds \leq ([P_{min}^\infty - P_{min}(T)]x_0, x_0),$$

and letting T tend to ∞, necessarily $\widehat{u}(s) = -B^\star P_{min}^\infty \widehat{x}(s)$. Consequently $\widehat{x}$ is also a solution of the closed loop equation that necessarily coincides with x_∞. Then $\widehat{u} = u_\infty$ and the proof is complete. ∎

We now give a regularity result for the optimal pair $(u^\star, x^\star)$.

Proposition 3.1 *Assume that the conditions of Theorem* 3.1 *are verified and let* $(u^\star, x^\star)$ *be the optimal pair for the control problem* (1.1)-(1.2). *Then the following statements hold:*
(i) If $x_0 \in D(A)$, *then* $x^\star$ *is a strict solution of* (1.5) *and belongs to*

$$C^1([0,\infty[\,;H) \cap C([0,\infty[\,;D(A)).$$

(ii) If e^{tA} *is an analytic semi-group then* $x^\star$ *is a classical solution of* (1.5).

Proof. (i) follows from Proposition 3.3 in Part II, Chapter 1, and (ii) from Proposition 3.9 in Part II, Chapter 1. ∎

3.1 Feedback operator and detectability

We assume here that the conditions $(\mathcal{H})$ are verified and that (A, B) is C-stabilizable. We denote by P^∞_{min} the minimal solution of the algebraic Riccati equation (2.2). Under these hypotheses, by Theorem 3.1 there exists a unique optimal pair $(u^\star, x^\star)$, and

$$\begin{cases} x^\star(t) = e^{tF}x_0, \quad t \geq 0, \\ \\ u^\star(t) = -B^\star P^\infty_{min} x^\star(t) = -B^\star P^\infty_{min} e^{tF} x_0, \quad t \geq 0, \end{cases}$$

where

$$F = A - BB^\star P^\infty_{min}$$

is called the *closed loop operator*. Moreover, by (3.2) we have

$$(P^\infty_{min}\, x_0, x_0) = \int_0^\infty \left\{ |Ce^{tF}x_0|^2 + |B^\star P^\infty_{min} e^{tF} x_0|^2| \right\} ds. \tag{3.3}$$

Remark 3.1 When $C = I$ (or if $C^{-1} \in \mathcal{L}(H)$), by (3.3) and Datko's Theorem, it follows that F is exponentially stable. In particular, if (A, B) is stabilizable with respect to the identity I then it is *feedback stabilizable.* ■

It is important for the applications to give conditions under which F is stable (even when C is not invertible). A sufficient condition is given by the *detectability* property. In order to define detectability we introduce the *dual system*

$$\begin{cases} y'(t) = A^\star y(t) + C^\star v(t), \quad t \geq 0, \\ \\ y(0) = y_0 \in H. \end{cases}$$

For this system the space of states, controls and observations are H, Y and H respectively.

Definition 3.1 *We say that (A, C) is detectable if $(A^\star, C^\star)$ is I-stabilizable.*

Remark 3.2 By Remark 3.1 it follows that (A, C) is detectable if and only if there exists $K \in \mathcal{L}(Y; H)$ such that $A - KC$ is exponentially stable. ■

The next result is due to W.M. WONHAM [1] in the finite dimensional case and to J. ZABCZYK [1] in the general case.

Proposition 3.2 *Assume that (A, C) is detectable. Then F is exponentially stable.*

Proof. Let $x_0 \in H, x(t) = e^{tF}x_0$; by Remark 3.2 there exists $K \in \mathcal{L}(Y; H)$ such that $A - KC$ is exponentially stable. Since

$$F = (A - KC) + (KC - BB^\star P^\infty_{min}),$$

we have that x is the mild solution to the problem

$$\begin{cases} x' = (A - KC)x + z, \\ \\ x(0) = x_0, \end{cases}$$

where $z = (KC - BB^\star P^\infty_{min})x$. It follows that

$$x(t) = e^{t(A-KC)}x_0 + \int_0^t e^{(t-s)(A-KC)}[KC - BB^\star P^\infty_{min}]\, x(s)\, ds.$$

By (3.3) we have that Cx and $BB^\star P^\infty_{min} x$ belong to $L^2(0, \infty; H)$; it follows that $x \in L^2(0, \infty; H)$ since $A - KC$ is exponentially stable. By Datko's Theorem this implies the conclusion. ■

Remark 3.3 The assumption that (A, C) is detectable is not a necessary condition that $F = A - BB^\star P^\infty_{min}$ be exponentially stable, as example 3.1 below shows. Notice however that this condition is necessary and sufficient in some special cases (see §3.2). ■

The following example was discussed in G. DA PRATO and M.C. DELFOUR [1], [2].

Example 3.1 Let $H = U = Y = \ell^2$, the Hilbert space of all the sequences $\{x_n\}$ of complex numbers such that $\sum_{n=1}^{\infty} |x_n|^2 < \infty$. We set

$$A\{x_n\} = \left\{\frac{n}{n+1}x_n\right\}, \quad B = I, \quad C\{x_n\} = \left\{\frac{\sqrt{2n+1}}{n+1}x_n\right\}.$$

Then equation (2.2) reduces to

$$2AX - X^2 + C^2 = 0,$$

and we have

$$P_{min}^{\infty} = A + \sqrt{A^2 + C^2} = A + I$$
$$F = A - P_{min}^{\infty} = -I$$

Thus F is exponentially stable. We now show that (A, C) is not detectable. In fact assume, by contradiction, that (A, C) is detectable. Then the algebraic Riccati equation

$$2\,AX - C^2X^2 + I = 0$$

has a positive solution X and we have

$$X\{x_n\} = \{(n+1)x_n\}.$$

However this is not a bounded operator in H. ■

3.2 Stabilizability and stability of the closed loop operator F in the point spectrum case

We assume here $(\mathcal{H})_\infty$-(i)-(ii); we are interested in the stabilizability of the system (1.1) under suitable conditions on the spectrum of A. We denote by $\sigma^-(A)$, $\sigma^+(A)$ and $\sigma^0(A)$ those elements of the spectrum of A which have negative, positive and zero real part, respectively. The elements of $\sigma^+(A) \cup \sigma^0(A)$ are called *unstable* points of $\sigma(A)$. Obviously, if $\sigma^+(A) \cup \sigma^0(A)$ is empty and if A verifies the spectral determining condition (see Volume I, Chapter 1, §2.9), then (A, B) is stabilizable. We now assume that

$$(\mathcal{P})\begin{cases} (i)\ \textit{the set } \sigma^+(A)\cup\sigma^0(A) \textit{ consists of a finite set of} \\ \quad\ \textit{eigenvalues of finite algebraic multiplicity.} \\ (ii)\ \textit{there exists } \varepsilon > 0,\ N_A > 0 \textit{ such that} \\ \quad\ \sup\limits_{\lambda\in\sigma^-(A)} \operatorname{Re}\lambda < -\varepsilon, \quad \|e^{tA}\Pi_A^-\| \le N_A e^{-\varepsilon t}, \quad \forall\, t \ge 0. \end{cases}$$

Here Π_A^- represents the projector on $\sigma^-(A)$ defined by

$$\Pi_A^- = \frac{1}{2\pi i}\int_{\gamma^-} R(\lambda, A)\, d\lambda$$

and γ^- is a simple Jordan curve around $\sigma^-(A)$. The projectors Π_A^+ and Π_A^0 are defined analogously. We set

$$H_A^- = \Pi_A^-(H), \quad H_A^+ = \Pi_A^+(H), \quad H_A^0 = \Pi_A^0(H).$$

Then H_A^-, H_A^+, H_A^0 are invariant subspace for the semigroup e^{tA}.

Remark 3.4 Assume that A fulfills $(\mathcal{P})$. Then by hypothesis (ii) H_A^+ is finite dimensional and there exists $\eta > 0$ such that

$$\operatorname{Re}\lambda > \eta, \quad \forall \lambda \in \sigma^+(A). \tag{3.4}$$

Moreover $e^{tA}H_A^+$ can be extended for $t < 0$ by the formula

$$e^{tA}\Pi_A^+ = \frac{1}{2\pi i}\int_{\gamma^+} e^{t\lambda}R(\lambda, A)\,d\lambda, \quad t \in \mathbf{R}.$$

Finally, by (3.4) there exists $N_A' > 0$ such that

$$\|e^{tA}\Pi_A^+\| \le N_A', \quad \forall t \le 0. \tag{3.5}$$

■

Remark 3.5 Assumptions $(\mathcal{P})$ are verified in each of the following cases:

(i) H is finite dimensional.
(ii) e^{tA} is compact for any $t > 0$. ■

3.3 Stabilizability

In this section we want to give a necessary and sufficient condition in order that (A, B) be stabilizable with respect to the identity I. We first prove a lemma on the existence of solutions of (1.1) in $[0, +\infty[$.

Lemma 3.1 *Assume that conditions $(\mathcal{H})_\infty$-(i)-(ii) are verified and moreover that A fulfills $(\mathcal{P})$ with $\sigma^0(A) = \emptyset$. Then the following statements are equivalent:*
(i) for all $x \in H$, there exists $u \in L^2([0, +\infty[\,; U)$ such that $x \in L^2([0, +\infty[\,; H)$, x being the solution to (1.1) corresponding to u.
(ii) Range $\gamma \supset \Pi_A^+(H)$, where γ is defined by

$$u \mapsto \gamma u = \int_0^\infty e^{-sA}\Pi_A^+ Bu(s)\,ds \;:\; L^2([0, +\infty[\,; U) \to H,$$

(iii) the mapping

$$\xi \mapsto \gamma^\star \xi = B^\star (\Pi_A^+)^\star e^{-\cdot A^\star}\xi \; : \; H \to L^2([0,+\infty[\,;U),$$

is one-to-one,

(iv) for all $\lambda \in \sigma^+(A^\star)$, $\quad Ker(B^\star) \cap Ker(\lambda - A^\star) = \{0\}$.

Proof. Let $u \in L^2([0,+\infty[\,;U)$, and let x be the solution to (1.1). Then we have

$$x(t) = e^{tA}x_0 + \int_0^t e^{(t-s)A}Bu(s)\,ds = x_-(t) + x_+(t),$$

where

$$\begin{aligned} x_-(t) \quad &= \quad e^{tA}\Pi_A^- x_0 + \int_0^t e^{(t-s)A}\Pi_A^- Bu(s)\,ds \\ &\quad - \int_t^{+\infty} e^{(t-s)A}\Pi_A^+ Bu(s)\,ds, \end{aligned}$$

$$x_+(t) = e^{tA}\left\{\Pi_A^+ x_0 + \int_0^{+\infty} e^{-sA}\Pi_A^+ Bu(s)\,ds\right\}.$$

Remark that $x_- \in L^2(0,\infty;H)$ by $(\mathcal{P})$-(ii) and (3.5) whereas $x_+ \in L^2(0,\infty;H)$ if and only if

$$\Pi_A^+ x_0 + \int_0^{+\infty} e^{-sA}\Pi_A^+ Bu(s)\,ds = 0.$$

It follows that $(i) \Leftrightarrow (ii)$. Moreover since the space $\Pi_A^+(H)$ is finite dimensional, $(ii) \Leftrightarrow (iii)$ by the Alternative Principle. Thus it remains to show that $(iii) \Leftrightarrow (iv)$. Assume (iii) and, by contradiction, that (iv) does not hold. Then there exist $\lambda \in \sigma^+(A^\star)$ and $\xi \in (\Pi_A^+(H))^\star$ different from 0 such that $B^\star\xi = 0$, $\lambda\xi = A^\star\xi = 0$. Then $e^{-sA^\star}\xi = e^{-\lambda s}\xi$ and we have

$$B^\star e^{-sA^\star}(\Pi_A^+)^\star \xi = 0, \; s \geq 0, \tag{3.6}$$

which is a contradiction.

$(iv) \Rightarrow (iii)$. Assume (iv) and, by contradiction, that (iii) does not hold. Then there exists $\xi \in (\Pi_A^+(H))^\star$ different from 0 such that

(3.6) holds. By differentiating (3.6) several times with respect to s and by setting $s = 0$ we see that

$$p(A^\star)\,\xi = 0$$

for all polynomials p. Choose p as the minimal degree polynomial such that $p(A^\star)\xi = 0$. Let λ_0 be a root of p, then clearly $\lambda_0 \in \sigma^+(A^\star)$. Set

$$q(\lambda) = \frac{p(\lambda)}{\lambda - \lambda_0} \text{ and } \psi = q(A^\star)\xi,$$

then by (3.6) it follows that $\psi \in$Ker$(B^\star)\cap$Ker$(\lambda I - A^\star)$, which is a contradiction. ∎

The following result is a straightforward generalization of a result due to M.L.J. Hautus [1].

Proposition 3.3 *Assume that conditions $(\mathcal{H})_\infty$-(i)-(ii) and $(\mathcal{P})$ are verified. Then the following statements are equivalent:*
(i) (A, B) is stabilizable with respect to I,
(ii) Ker$(\lambda - A^\star)\cap$Ker$(B^\star) = \{0\}$, $\quad \forall\, \lambda \in \sigma^0(A) \cup \sigma^+(A)$.

Proof. We can clearly choose $\varepsilon > 0$ such that, setting $A_\varepsilon = A + \varepsilon I$, then $\sigma^0(A_\varepsilon)$ is empty, so that we can apply Lemma 3.1. Setting $v(t) = e^{\varepsilon t}u(t)$, $z(t) = e^{\varepsilon t}x(t)$, problem (1.1) becomes

$$\begin{aligned} z'(t) &= A_\varepsilon z(t) + Bv(t), \\ z(0) &= x. \end{aligned}$$

Now assume that (ii) holds. By Lemma 3.1, for all $x_0 \in H$ there exists $v \in L^2([0, \infty[; U)$ such that $z \in L^2(0, \infty; H)$; thus (A, B) is I-stabilizable and (i) holds.

It remains to prove that $(i) \Rightarrow (ii)$. In fact assume (i) and that, by contradiction, (ii) does not hold. Then there exists $\lambda \in \sigma^0(A^\star) \cup \sigma^+(A^\star)$ and $\xi \in D(A^\star)$ different from 0 such that

$$B^\star\xi = 0, \quad \lambda\xi - A^\star\xi = 0. \tag{3.7}$$

Choose now $x \in H$ such that $(x_0, \xi) = 1$; by the hypothesis (i) there exists $u \in L^2(0, \infty; U)$ such that $x \in L^2(0, \infty; H)$. Since

$$(x_t, \xi) = (x_0, e^{tA^\star}\xi) + \int_0^t (u(s), B^\star e^{tA^\star}\xi)\, ds,$$

by (3.7) it follows that $(x_t, \xi) = (x_0, e^{tA^\star}\xi) = e^{\lambda t}$, which is a contradiction, since Re$\lambda \geq 0$. The proof is complete. ∎

3.4 Exponential stability of F

We assume here that (A, B) is C-stabilizable and denote by $F = A - BB^\star P_{min}^\infty$ the closed loop operator. We prove the following result.

Proposition 3.4 *Assume that (A, B) is C-stabilizable and that A and F fulfill $(\mathcal{P})$. Then the following statements are equivalent:*
(i) F is exponentially stable,
(ii) for any $\lambda \in \sigma^+(A) \cup \sigma^0(A)$, we have

$$\mathrm{Ker}\,(A - \lambda I) \cap \mathrm{Ker}\,(C) = \{0\}. \tag{3.8}$$

Proof. $(i) \Rightarrow (ii)$. Assume that F is exponentially stable and, by contradiction, that (ii) does not hold. Then there exists $x_0 \neq 0$ and $\lambda_0 \in \sigma^+(A)$ such that

$$\mathrm{Re}\,\lambda_0 \geq 0, \quad Ax_0 = \lambda_0 x_0, \quad Cx_0 = 0. \tag{3.9}$$

It follows that

$$\begin{aligned}
&\frac{d}{dt}(P_{min}(t)x_0, x_0)\\
&= 2\,\mathrm{Re}\ \lambda_0(P_{min}(t)x_0, x_0) - |B^\star P_{min}(t)x_0|^2\\
&\leq 2\,\mathrm{Re}\ \lambda_0(P_{min}(t)x_0, x_0),
\end{aligned}$$

where P_{min} is the minimal nonnegative solution of (2.1). Since $P_{min}(0) = 0$ by (3.9) it follows that $P_{min}(t)x_0 = 0$ and then $P_{min}^\infty x_0 = 0$. Thus

$$Fx_0 = Ax_0 - BB^\star P_{min}^\infty\, x_0 = Ax_0 = \lambda x_0$$

which contradicts the fact that F be exponentially stable.

$(ii) \Rightarrow (i)$ Assume now that (3.8) holds and, by contradiction, that F is not exponentially stable. Then there exists $\lambda_0 \in \mathbf{C}$ and $x_0 \neq 0$ in H such that

$$Fx_0 = \lambda_0 x_0, \quad \mathrm{Re}\ \lambda_0 \geq 0.$$

Since, as easily checked, P_{min}^∞ verifies

$$F^\star P_{min}^\infty + P_{min}^\infty F + P_{min}^\infty BB^\star P_{min}^\infty + C^\star C = 0,$$

we have

$$2\,\mathrm{Re}\ \lambda_0(P_{min}^{\infty}x_0, x_0) + |B^\star P_{min}^{\infty}x_0|^2 + |Cx_0|^2 = 0,$$

which implies $B^\star P_{min}^{\infty}x_0 = 0$, $Ax_0 = \lambda_0 x_0, Cx_0 = 0$ and this fact contradicts (ii). ■

We now give a characterization for the stability of F.

Proposition 3.5 *Assume that the conditions of Proposition* 3.4 *are verified. Then the following statements are equivalent:*
(i) F is exponentially stable,
(ii) (A, C) is detectable.

Proof. $(i) \Rightarrow (ii)$. If F is exponentially stable then, by Proposition 3.4, (3.8) holds. But this implies that $(A^\star, C^\star)$ is stabilizable with respect to I (by Proposition 3.3). Thus (ii) is proved. The implication $(i) \Rightarrow (ii)$ was proved in Proposition 3.4. ■

4 Qualitative properties of the solutions of the Riccati equation

In all this section we assume that conditions $(\mathcal{H})_\infty$ are verified and that (A, B) is stabilizable with respect to C. We again consider the equations

$$P' = A^\star P + PA - PBB^\star P + C^\star C, \tag{4.1}$$

$$A^\star X + XA - XBB^\star X + C^\star C = 0, \tag{4.2}$$

and denote by P_{min} and P_{min}^{∞} the minimal solutions of (4.1) and (4.2) respectively. We say that a solution $X \in \Sigma^+(H)$ of (4.2) is *maximal* if

$$\forall \text{ solution } Y \in \Sigma^+(H) \text{ of } (4.2) \quad \Rightarrow \quad Y \leq X.$$

If a maximal solution exists, it is clearly unique; however a maximal solution does not exist in general (see Remark 4.2 below). The existence of a maximal solution under suitable assumptions will be proved in §4.3.

4.1 Local stability results

We study here the exponential stability of any positive solution X of (4.2). This property of X is naturally related to the exponential stability of the linear operator $A - BB^\star X$.

Proposition 4.1 *Let $X \in \Sigma^+(H)$ be a solution of* (4.2). *Assume that $A - BB^\star X = K$ is exponentially stable, that is that there exist $a > 0, N > 0$ such that*

$$\|e^{tK}\| \leq Ne^{-at}, \quad t \geq 0. \tag{4.3}$$

Then there exist $r > 0$ and $a > 0$ such that, if $P_0 \in \Sigma^+(H)$ and $\|P_0 - X\| < \frac{1}{2rN^2}$, we have

$$\|P(t) - X\| \leq re^{-at}, \quad t \geq 0,$$

where P is the mild solution to (4.1) *such that $P(0) = P_0$.*

Proof. Set $Y(t) = P(t) - X$; then Y is the mild solution to the problem

$$\begin{aligned} &Y' = K^\star Y + YK - YBB^\star Y, \\ &Y(0) = P_0 - X = Y_0, \end{aligned} \tag{4.4}$$

which is equivalent to the equation $\gamma(Y) = Y$ where γ is defined by

$$\gamma(Y)(t)x = e^{tK^\star}Y_0e^{tK}x - \int_0^t e^{(t-s)K^\star}Y(s)e^{(t-s)K}x\,ds, \tag{4.5}$$

for any $x \in H$. For any $a \geq 0$ we introduce the Banach space

$$C_a([0,\infty[;\Sigma(H)) = \{Y \in C_s([0,\infty[;\Sigma(H)) \,:\, \sup_{t\geq 0}\|e^{ta}Y(t)\| < \infty\},$$

and, for any $r > 0$ we set

$$B_r = \{Y \in C_a([0,\infty[;\Sigma(H)) \,:\, \|Y\|_a \leq r\}.$$

We now want to solve (4.4) by proving that γ has a fixed point in B_r, for a suitable r. If $Y, Z \in B_r$ we have, taking into account (4.3),

$$\|\gamma(Y)\|_a \leq N^2\|Y_0\| + \frac{N^2r^2}{ea}\|B\|^2$$

$$\|\gamma(Y) - \gamma(Z)\|_a \leq N^2\|Y_0\| + 2\frac{rN^2}{ea}\|B\|^2\|Y - Z\|.$$

Now choose r such that

$$2\frac{rN^2}{ea}\|B\|^2 \le \frac{1}{2}, \quad \frac{N^2r^2}{ea}\|B\|^2 \le \frac{1}{2}$$

and then Y_0 such that $\|Y_0\| \le \frac{\rho}{2N^2}$. Then γ is a contraction on B_r and the conclusion holds. ∎

4.2 Attractivity properties of a stationary solution

We first show that the minimal solution P^∞_{min} is globally attractive *from below.*

Proposition 4.2 *Let $P_0 \in \Sigma^+(H)$ such that $P_0 \le P^\infty_{min}$ and let P be the mild solution of* (4.1) *with $P(0) = P_0$. Then we have*

$$\lim_{t\to\infty} P(t)x = P^\infty_{min}x, \quad \forall x \in H.$$

Proof. By Proposition 2.2 we have

$$P_{min}(t) \le P(t) \le P^\infty_{min},$$

so that

$$0 \le (P^\infty_{min}x - P(t)x, x) \le (P^\infty_{min}x - P_{min}(t)x, x), \quad x \in H,$$

and by (2.7)

$$\lim_{t\to\infty} (P^\infty_{min}x - P(t)x, x) = 0, \quad x \in H,$$

which implies the conclusion. ∎

We now consider a general positive solution X of (4.2) and set $K = A - BB^\star X$. We first show that if K is exponentially stable, then X is globally attractive *from above.*

Proposition 4.3 *Let $X \in \Sigma^+(H)$ be a solution of* (4.2), *assume that $K = A - BB^\star X$ is exponentially stable and that* (4.3) *holds. Then for any $P_0 \ge X$ we have*

$$\|P(t) - X\| \le N^2e^{-2at}\|P_0 - X\|, \quad t \ge 0, \tag{4.6}$$

where P is the solution to (4.1) *such that $P(0) = P_0$.*

Proof. Set $Y = P(t) - X$; since $P_0 \geq X$ we have, by Proposition 2.2 in Part II, Chapter 1, $Y(t) \geq 0$. Moreover Y is the mild solution to problem (4.4). For any $x \in D(A)$ we have (using the fact that Y is a strict solution of (4.4))

$$\frac{d}{ds}(Y(t-s)e^{sK}x, e^{sK}x) = |B^\star Y(t-s)e^{sK}x|^2, \quad \forall x \in D(A)$$

so that (since $D(A)$ is dense in H)

$$(Y(t)x, x) + \int_0^t |BY(t-s)e^{sK}x|^2\, ds = (Y(0)e^{sK}x, e^{sK}x), \quad \forall x \in H.$$

It follows that

$$(P(t)x - Xx, x) \leq (Y(0)e^{sK}x, e^{sK}x), \quad \forall x \in H,$$

which yields (4.6). ∎

The following result shows that if X is a positive solution of (4.2) such that $K = A - BB^\star X$ is exponentially stable, then X is *maximal.*

Proposition 4.4 *Let $X, Y \in \Sigma^+(H)$ be solutions of* (4.2). *Assume that $K = A - BB^\star X$ is exponentially stable. Then $X \geq Y$.*

Proof. Set $Z = X - Y$; by (4.4) it follows that

$$(Zx, Kx) + (ZKx, x) + (ZBB^\star Zx, x) = 0, \quad \forall x \in D(A),$$

which implies that

$$\frac{d}{dt}(Ze^{sK}x, e^{sK}x) = -|B^\star Ze^{sK}x|^2, \quad \forall x \in D(A).$$

Since $D(A)$ is dense in H, by integrating this between 0 and t we obtain

$$\begin{aligned}(Zx, x) &= (Ze^{tK}x, e^{tK}x) + \int_0^t |B^\star Ze^{sK}x|^2 ds \\ &\geq (Ze^{tK}x, e^{tK}x), \quad \forall x \in H.\end{aligned}$$

As $t \to \infty$ we have $(Zx, x) \geq 0, \forall\, x \in H$ so that $X \geq Y$. ∎

Corollary 4.1 *Equation* (4.2) *has at most one positive solution X such that $A - BB^\star X$ is exponentially stable.*

Proof. Let $X, Y \in \Sigma^+(H)$ be two solutions of (4.2) such that $A - BB^\star X$ and $A - BB^\star Y$ are exponentially stable. By Proposition 4.4 it follows $X \geq Y$ and $Y \geq X$ so that $X = Y$. ■

Concerning the minimal solution P_{min}^∞ we have a uniqueness result.

Corollary 4.2 *If $A - BB^\star P_{min}^\infty$ is exponentially stable, then P_{min}^∞ is the unique positive solution of* (4.2).

Proof. It is sufficient to observe that P_{min}^∞ is both the maximal and the minimal solution of (4.2). ■

We show now that if $F = A - BB^\star P_{min}^\infty$ is exponentially stable then P_{min}^∞ is *globally attractive* among all the positive solutions of (4.2). In finite dimension this result is well known, see W.M. WONHAM [1].

Proposition 4.5 *If $F = A - BB^\star P_{min}^\infty$ is exponentially stable, then for any $P_0 \in \Sigma^+(H)$ we have*

$$\lim_{t\to\infty} P(t)x = P_{min}^\infty x, \quad \forall x \in H, \tag{4.7}$$

where P is the mild solution of (4.1) *such that $P(0) = P_0$.*

Proof. Choose $n \in \mathbf{N}$ such that $P_0 \leq nI$, $P_{min}^\infty \leq nI$. Then by Proposition 2.2 in Part II, Chapter 1, we have

$$P_{min}(t) \leq P(t) \leq Q(t), \tag{4.8}$$

where Q is the mild solution of (4.1) such that $Q(0) = nI$. Now, as $t \to \infty$, $P_{min}(t)x \to P_{min}^\infty x$ and, by (4.6), $Q(t)x \to P_{min}^\infty x$; this, along with (4.8), implies (4.7). ■

Remark 4.1 By Proposition 3.5 we know that if (A, C) is detectable, then the feedback operator F is exponentially stable. Thus, by Corollary 4.2, P_{min}^∞ is the unique nonnegative solution to (4.2). However, detectability of (A, C) is not necessary for the uniqueness, as the following example shows. Let $H = U = Y = \mathbf{R}^2$,

$$A = \begin{bmatrix} 0 & 1 \\ 0 & 0 \end{bmatrix}, \quad B = I, \quad C = 0;$$

as easily seen $X = 0$ is the unique solution of (4.2), but the feedback operator $F = A$ is not stable and (A, C) is not detectable.

When H is finite dimensional M. SORINE [1] has shown that the following conditions are sufficient for the uniqueness of the solution:

(i) (A, B) is C-stabilizable,

(ii) $(A^\star, C^\star)$ is $B^\star$-stabilizable,

(iii) no eigenvalue of the closed loop operator F lies on the imaginary axis.

This result can be easily generalized to any Hilbert space H provided that A and F fulfill assumptions $(\mathcal{P})$ of §3.2. ∎

4.3 Maximal solutions

We assume here, beside $(\mathcal{H})_\infty$, that (A, B) is I-stabilizable (which, obviously is stronger than assuming that (A, B) is C-stabilizable). We follow here A. BENSOUSSAN [1].

For any $\varepsilon > 0$, we consider the problem: to minimize

$$J_\varepsilon(u) = \int_0^\infty \left\{ |Cx(s)|^2 + \varepsilon |x(s)|^2 + |u(s)|^2 \right\} ds, \tag{4.9}$$

over all controls $u \in L^2(0, \infty; U)$ subject to the differential equation constraint (1.1). Consider the observation $D_\varepsilon = \sqrt{C^\star C + \varepsilon I}$ (the observation space being H). Clearly (A, B) is D_ε–stabilizable, so that, by Theorem 3.1, there exists a unique optimal pair $(u_\varepsilon, x_\varepsilon)$ and a solution P_ε of the algebraic Riccati equation

$$A^\star X + XA - XBB^\star X + C^\star C + \varepsilon I = 0, \tag{4.10}$$

such that

$$\begin{aligned} x'_\varepsilon &= (A - BB^\star P_\varepsilon)x, \\ x_\varepsilon(0) &= x_0, \\ u_\varepsilon &= -B^\star P_\varepsilon x_\varepsilon. \end{aligned} \tag{4.11}$$

Moreover (A, D_ε) is detectable (because D_ε is invertible and $D_\varepsilon^{-1} \in \mathcal{L}(H)$). Thus by Corollary 4.2, P_ε is the unique solution in $\Sigma^+(H)$ of (4.10). Moreover, by Proposition 4.5, given $Q \in \Sigma^+(H)$ we have

$$\lim_{t \to \infty} P_Q(t) = P_\varepsilon x, \quad \forall x \in H, \tag{4.12}$$

where P_Q is the mild solution of the Riccati equation

$$P' = A^\star P + PA - PBB^\star P + C^\star C + \varepsilon I,$$

$$P(0) = Q,$$

that is P_ε is globally attractive.

Proposition 4.6 *Assume that conditions* $(\mathcal{H})_\infty$ *are verified and that* (A, B) *is I-stabilizable. Then equation* (4.2) *has a maximal solution* P^∞_{Max}. *Moreover*

$$\lim_{\varepsilon \to 0} P_\varepsilon x = P^\infty_{Max}\, x, \quad \forall x \in H. \tag{4.13}$$

Proof. We first remark that, since $\{P_\varepsilon\}$ is nonincreasing (by Proposition 2.2 in Part II, Chapter 1) and bounded below by 0 the limit

$$\lim_{\varepsilon \to 0} P_\varepsilon x = P^\infty_{Max} x, \quad \forall x \in H, \tag{4.14}$$

exists. We have to show that P^∞_{Max} is the required maximal solution.

Step 1. P^∞_{Max} is a solution to (4.2).
In fact for all $x, y \in D(A)$, we have

$$(P_\varepsilon Ax, y) + (P_\varepsilon x, Ay) - (B^\star P_\varepsilon x, B^\star P_\varepsilon x) + (Cx, Cy) + \varepsilon(x, y) = 0;$$

letting ε tend to 0 and using (4.14) we see that P^∞_{Max} is a solution of (4.2).

Step 2. P^∞_{Max} is maximal.
Let Q be any solution of (4.2), then, by Proposition 2.2 in Part II, Chapter 1, we have $P_Q(t) \geq Q$ (where P_Q is the solution to (4.9)). Thus, by (4.12) and (4.14) it follows that $P^\infty_{Max} \geq Q$, so that P^∞_{Max} is maximal. ∎

We will now give the variational interpretation of the maximal solution.

Theorem 4.1 *Assume* $(\mathcal{H})_\infty$ *and that* (A, B) *is I–stabilizable. Let* P^∞_{Max} *be the maximal solution of* (4.2) *and let* $x_0 \in H$. *Then*

$$\begin{aligned}&(P^\infty_{Max} x_0, x_0)\\ &= \inf \{J_\infty(u) \,:\, u \in L^2(0, \infty; U), x \in L^2(0, \infty; H)\},\end{aligned} \tag{4.15}$$

where (x, u) are subject to the differential equation constraint (1.1). *Moreover*

$$(P_{Max}^{\infty}x_0, x_0) \geq J_{\infty}(\widehat{u}) \tag{4.16}$$

where $\widehat{u} = \lim_{\varepsilon \to 0} u_{\varepsilon}$ and u_{ε} is defined by (4.11).

Proof. Set

$$U_{ad} = \inf\{J_{\infty}(u) \,:\, u \in L^2(0,\infty;U),\, x \in L^2(0,\infty;H)\},$$

where x is the solution of (1.1).

We first remark that, recalling (4.13) and arguing as we did in the proof of Lemma 2.1, we can show that the following limits

$$\widehat{x}(t) = \lim_{\varepsilon \to 0} x_{\varepsilon}(t), \quad \widehat{u}(t) = \lim_{\varepsilon \to 0} u_{\varepsilon}(t), \quad t \geq 0 \tag{4.17}$$

exist and are uniform on the bounded subsets of $[0, \infty[$. Let now $u \in U_{ad}$. By (6.2) in Chapter 1 of Part II, we have

$$(P_{Max}^{\infty}x_0, x_0) + \int_0^t \left\{|u(s) + B^{\star}P_{Max}^{\infty}x(s)|^2\right\} ds$$

$$= \int_0^t \left\{|Cx(s)|^2 + |u(s)|^2\right\} ds + (P_{Max}^{\infty}x(t), x(t)).$$

Since $x \in L^2(0,\infty;H)$ there exists a sequence $t_n \nearrow \infty$ such that $x(t_n) \to 0$. Therefore

$$(P_{Max}^{\infty}x_0, x_0) + \int_0^t |u(s) + B^{\star}P_{Max}^{\infty}x(s)|^2\, ds = J_{\infty}(u), \tag{4.18}$$

which implies that

$$(P_{Max}^{\infty}x_0, x_0) \leq J_{\infty}(u), \quad \forall\, u \in U_{ad}.$$

Letting ε tend to zero in the equality

$$(P_{\varepsilon}x_0, x_0) = J_{\infty}(u_{\varepsilon}) + \varepsilon\|x_{\varepsilon}\|^2_{L^2(0,\infty;H)},$$

one obtains (4.16); also, since $u_{\varepsilon} \in U_{ad}$, the conclusion follows. ∎

Proposition 4.7 *Under the conditions of Theorem 4.1, the following statements are equivalent.*

(i) The operator $F_M = A - BB^\star P^\infty_{Max}$ is exponentially stable.

(ii) $(P^\infty_{Max}x_0, x_0) = \min\{J_\infty(u) \,:\, u \in L^2(0,\infty;U), x \in L^2(0,\infty;H)\}$, where x is given by (1.1).

Proof. *(i)* $\Rightarrow$ *(ii)*. Let $\widehat{x}(t)$ and $\widehat{u}(t)$ be defined by (4.17). If F_M is exponentially stable, we have $\widehat{x} \in L^2(0,\infty;H)$ so that $\widehat{u} \in U_{ad}$ and *(ii)* holds.

(ii) $\Rightarrow$ *(i)*. Let $u^\star \in L^2(0,\infty;U)$ such that the corresponding solution $x^\star$ of (1.1) belongs to $L^2(0,\infty;H)$. Setting $u = u^\star$ in (4.18), we obtain

$$\int_0^\infty |u^\star(s) + B^\star P^\infty_{Max} x^\star(s)|^2\, ds = 0$$

which implies that

$$x^\star(t) = e^{tF_M}x_0, \quad u^\star(t) = -B^\star P^\infty_{Max} e^{tF_M}x_0,$$

and the conclusion follows from Datko's Theorem. ∎

Corollary 4.3 *Assume that the conditions $(\mathcal{H})_\infty$ are verified and that (A, B) is I-stabilizable. Let P^∞_{min} and P^∞_{Max} be the minimal and maximal solutions of the algebraic Riccati equation. Then*

(i) $P^\infty_{min} \le P^\infty_{Max}$,

(ii) P^∞_{min} is globally attractive from below,

(iii) P^∞_{Max} is globally attractive from above.

Proof. *(i)* is clear, *(ii)* follows from Proposition 4.2 and *(iii)* from Theorem 4.1 and Proposition 4.3. ∎

Remark 4.2 (i) The maximal solution does not exist in general. Let in fact

$$H = U = Y = \mathbf{R}^2, \quad A = \begin{bmatrix} 0 & 1 \\ -1 & 0 \end{bmatrix}, \quad B = 0, \quad C = 0.$$

Then (4.2) reduces to $A^\star X + XA = 0$ and $X = \lambda I$ is a solution for all $\lambda \ge 0$.

(ii) In (4.15) the infimum is not a minimum in general. In fact, let $A = 0$, $U = H$, $B = I$, $C = 0$. Then the algebraic Riccati equation reduces to $P^2 = 0$ and one has $P^{\infty}_{min} = P^{\infty}_{Max} = 0$. Thus the infimum in (4.15) is 0; however the control $u = 0$ does not belong to U_{ad} if $x_0 \neq 0$. Remark also that F_M is not exponentially stable. ∎

Remark 4.3 For the existence of the maximal solution it is not necessary that (A, B) be I-stabilizable. It is sufficient to assume that there exists $D \geq C^\star C$ such that

(i) (A, B) is $D^{\frac{1}{2}}$*-stabilizable.*

(ii) $(A, D^{\frac{1}{2}})$ is *detectable.*

In fact, under these assumptions, one can easily repeat the previous arguments. ∎

4.4 Continuous dependence of stationary solutions with respect to the data

We consider here a sequence of Riccati equations

$$
\begin{aligned}
&P_k' = A_k^\star P_k + P_k A_k - P_k B_k B_k^\star P_k + C_k^\star C_k, \\
&A_k^\star X_k + X_k A_k - X_k B_k B_k^\star X_k + C_k^\star C_k = 0.
\end{aligned}
\tag{4.19}
$$

We assume (2.18) of Part II, Chapter 1, and we set

$$
J_\infty^k(u) = \int_0^\infty \{|C_k x_k(s)|^2 + |u(s)|^2\}\, ds, \quad u \in L^2(0,\infty;U),
$$

where x_k is the mild solution of the system

$$
\begin{aligned}
x_k' &= A_k x_k + B_k u, \\
x_k(0) &= x_{k0}.
\end{aligned}
\tag{4.20}
$$

We say that (A_k, B_k) is *stabilizable with respect to* C_k *uniformly in* k, if for any $x_0 \in H$ there exists $u \in L^2(0,\infty;U)$ such that

$$
J_\infty^k(u) < \infty, \quad \forall\, k \in \mathbf{N}. \tag{4.21}
$$

If this assumption is verified, then, according to Theorem 3.1, the minimal solution of (2.2) exists. We denote it by $P^{\infty}_{k,min}$. Finally we set

$$
F_k = A_k - B_k B_k^\star P^{\infty}_{k,min}.
$$

We now prove the following theorem (see also J.S. Gibson [1]).

Theorem 4.2 *Assume that assumptions* (2.18) *in* Chapter 1 *of* Part II *are verified and that* (A_k, B_k) *is stabilizable with respect to* C_k *uniformly in* k. *If, in addition, there exist* $N > 0$ *and* $a > 0$ *such that*

$$\|e^{tF_k}\| \le Ne^{-at}, \quad t \ge 0, \tag{4.22}$$

then we have

$$\lim_{k\to\infty} P^\infty_{k,min}\, x = P^\infty_{min}\, x, \quad \forall x \in H.$$

Proof. We have

$$(P^\infty_{k,min} x_0, x_0) \le \int_0^\infty \left\{|x_k(s)|^2 + |u(s)|^2\right\} ds,$$

where x_k is the solution of (4.20). Choose u such that (4.21) holds; then, by using the Uniform Boundedness Theorem, it is easy to show that there exists $c > 0$ such that

$$P^\infty_{k,min} \le cI, \quad \forall k \in \mathbf{N}. \tag{4.23}$$

Now set $V_k = Q_k - P^\infty_{k,min}$ (resp. $V = Q - P^\infty_k$) where Q_k (resp. Q) is the solution of (4.19) (resp. (4.1)) such that $Q_k(0) = cI$ (resp. $Q(0) = cI$). Then V_k is the mild solution of the problem

$$V_k' = F_k^\star V_k + V_k F_k - V_k B_k B_k^\star V_k,$$

$$V_k(0) = cI - P^\infty_{k,min}.$$

Remark that $V_k(0) \ge 0$ in virtue of (4.23). By (4.22) it follows that

$$\|V_k(t)\| \le N^2 e^{-2at}, \quad t \ge 0.$$

Moreover

$$|P^\infty_{min}x - P^\infty_{k,min}x|$$

$$\le |P^\infty_{min}x - Q(t)x| + |Q(t)x - Q_k(t)x| + |P^\infty_{k,min}x - Q_k(t)x|$$

$$\le 2cN^2e^{-2at}|x| + |Q(t)x - Q_k(t)x|,$$

and the result follows from Theorem 2.2 of Chapter 1 (Part II). ∎

4.5 Periodic solutions of the Riccati equation

In this section we study periodic solutions of Riccati equations. For the finite dimensional case see M.A. SHAYMAN [1]. Here we follow G. DA PRATO [1], [2].

Let $T > 0$ be fixed. We say that $P \in C_s(\mathbf{R}; \Sigma(H))$ is a *T-periodic solution* of the Riccati equation

$$P' = A^\star P + PA - PBB^\star P + C^\star C, \tag{4.24}$$

if

(i) $P(t+T) = P(t), \quad \forall\, t \in \mathbf{R}$.

(ii) P *is a mild solution of* (4.24), that is, for any $t, s \in \mathbf{R}$ with $t < s$ we have

$$\begin{aligned} P(t)x &= e^{(t-s)A^\star} P(s) e^{(t-s)A} x + \int_t^s e^{(t-\tau)A^\star} C^\star C e^{(t-\tau)A} x \, d\tau \\ &\quad - \int_t^s e^{(t-\tau)A^\star} P(\tau) BB^\star P(\tau) e^{(t-\tau)A} x \, d\tau. \end{aligned}$$

If, in addition, P is not constant we say that P is a *non trivial periodic solution* of (4.24).

Proposition 4.8 *Assume that $(\mathcal{H})_\infty$ is verified, then the following statements are equivalent:*

(i) (A, B) is C-stabilizable,

(ii) there exists a positive T-periodic solution (possibly trivial) of equation (4.24).

Proof. $(i) \Rightarrow (ii)$ follows from Proposition 2.3 since any stationary solution is also periodic.

$(ii) \Rightarrow (i)$. Assume, by contradiction, that there exists a T-periodic solution of (4.24) whereas (i) does not hold. Then, by Proposition 2.2, there exists $x_0 \in H$ such that $(P_{min}(t)x_0, x_0) \to \infty$ as $t \to \infty$, where $P_{min}(\cdot)$ is the minimal positive solution of (4.24). Since $P(t) \geq P_{min}(t)$ we have $(P(t)x_0, x_0) \to \infty$ as $t \to \infty$ which is a contradiction since P is periodic. ■

Lemma 4.1 *Assume that (A, B) is C-stabilizable and let P be a positive T-periodic solution of* (4.24). *Then we have $P(t) \geq P_{min}^\infty$,* *where* P^∞_{min} *is the minimal solution of* (2.2). *If, in addition,* (A, B) *is* I*-stabilizable we have* $P(t) \leq P^\infty_{Max}$.

Proof. For the first statement, it is sufficient to let n tend to infinity in the inequality

$$P(t) = P(t + nT) \geq P_{min}(t + nT),$$

whereas the second one follows from Corollary 4.3. ■

Let now P be a positive T-periodic solution of (4.24) and set $Q = P - P^\infty_{min}$ and $F = A - BB^\star P^\infty_{min}$. By Lemma 4.1 we have $Q \geq 0$; moreover, as easily checked, Q is the mild solution of (4.24) such that $Q(0) = P(0) - P^\infty_{min} = Q_0$, that is

$$\begin{cases} Q' = F^\star Q + QF - QBB^\star Q, \\ \\ Q(0) = P(0) - P^\infty_{min} = Q_0. \end{cases} \tag{4.25}$$

Problem (4.25) can be explicitely solved; we could use formula (2.22) in Part II, Chapter 1, but we prefer to give a simpler proof in the following lemma.

Lemma 4.2 *The solution of problem* (4.25) *is given by the formula*

$$Q(t) = e^{tF^\star} Q_0 (I + \Omega(t) Q_0)^{-1} e^{tF}, \tag{4.26}$$

where

$$\Omega(t)x = \int_0^t e^{sF} BB^\star e^{sF^\star} x \, ds, \quad x \in H.$$

Proof. First of all we remark that formula (4.26) is meaningful since $\Omega(t) \in \Sigma(H)$ so that the inverse of $I + \Omega(t)Q_0$ belongs to $\mathcal{L}(H)$ (see Proposition 1.1 in Appendix A to this volume). Now let Q_n be the solution to the approximating problem

$$\begin{aligned} &Q_n' = F_n^\star Q_n + Q_n F_n - Q_n BB^\star Q_n, \\ &Q_n(0) = Q_0 + \frac{1}{n} I, \end{aligned}$$

where F_n are the Yosida approximation of F. Setting $V_n = Q_n^{-1}$, this problem reduces to the following linear one:

$$\begin{aligned} &V_n' = -F_n V_n - V_n F_n^\star + BB^\star \\ &V_n(0) = (Q_0 + \frac{1}{n} I)^{-1}. \end{aligned}$$

Thus we have

$$V_n(t) = e^{-tF_n}(Q_0 + \frac{1}{n}I)^{-1}e^{-tF_n^\star} + \int_0^t e^{(t-s)F_n}BB^\star e^{(t-s)F_n^\star}ds$$

$$= e^{-tF_n}\left\{I + \int_0^t e^{sF_n}BB^\star e^{sF_n^\star}dsQ_0 + \frac{1}{n}I\right\}(Q_0 + \frac{1}{n}I)^{-1}e^{-tF_n^\star}.$$

which implies that

$$Q_n(t) = e^{tF_n^\star}(Q_0 + \frac{1}{n}I)\left\{I + \int_0^t e^{sF_n}BB^\star e^{sF_n^\star}dsQ_0 + \frac{1}{n}I\right\}^{-1} e^{tF_n}.$$

Now the conclusion follows from Theorem 2.1 of Chapter 1 in Part II. ∎

From Lemma 4.1 we have the following result.

Proposition 4.9 *Assume that the conditions $(\mathcal{H})_\infty$ are verified and that (A, B) is stabilizable with respect to C. Let $P_0 \geq P_{min}^\infty$ and let P be the solution of (4.24) such that $P(0) = P_0$. Then P is T-periodic if and only if $X = P_0 - P_{min}^\infty$ verifies the equation*

$$e^{TF^\star}X(I + \Omega(T)X)^{-1}e^{TF} = X, \tag{4.27}$$

where

$$\Omega(T)x = \int_0^T e^{sF}BB^\star e^{sF^\star}x\,ds, \quad x \in H,$$

and

$$F = A - BB^\star P_{min}^\infty.$$

We will now study a special case of equation (4.27). We remark that it is also important to decide whether a given solution of (4.27) is a non trivial periodic solution or not. Assume that

$$\begin{cases} (i) \quad A = D + \frac{\alpha}{2}I, \quad C = 0, \quad \alpha > 0, \\ (ii) \quad D + D^\star = 0, \quad e^{2\pi D} = I, \quad \|e^{tD}\| \leq 1, \\ (iii) \quad \textit{there exists } \delta > 0 \textit{ such that} \\ \qquad \int_0^{2\pi} e^{\alpha s}e^{sD}BB^\star e^{sD^\star}ds \geq \delta. \end{cases} \tag{4.28}$$

Then equation (4.27) is equivalent to

$$e^{2\pi\alpha}X(I + \Omega X)^{-1} = X, \tag{4.29}$$

where

$$\Omega = \Omega(2\pi) = \int_0^{2\pi} e^{\alpha s} e^{sD} BB^\star e^{sD^\star} ds$$

and the Riccati equation (4.24) becomes

$$P' = D^\star P + PD + \alpha P - PBB^\star P. \tag{4.30}$$

Proposition 4.10 *Assume that the hypotheses of Proposition 4.9 hold and that, in addition, the conditions (4.28) are verified. Then all solutions of equation (4.29) are given by the formula*

$$X = (e^{2\pi\alpha} - 1)\Omega^{-\frac{1}{2}}\,\Xi\,\Omega^{-\frac{1}{2}}$$

where Ξ is any hermitian projector operator in $\Sigma^+(H)$.

Proof. Equation (4.29) is equivalent to $(e^{2\pi\alpha} - 1)\,X = X\,\Omega\,X$ and setting $Y = \Omega^{\frac{1}{2}}\,X\,\Omega^{\frac{1}{2}}$ to $(e^{2\pi\alpha} - 1)\,Y = Y^2$; thus the conclusion follows. ∎

Let X be a solution of (4.29) and let P be the solution of (4.30) such that $P(0) = X$; P is 2π–periodic. We now want to see if P is non trivial. Let us consider the special case when $B = I$. In this case we have $X = \alpha\Xi$ and moreover

$$P(t) = \alpha e^{\alpha t} e^{tD^\star}\Xi[I + (e^{\alpha t} - 1)\Xi]^{-1} e^{tD}.$$

Since Ξ is a projector we have $\Xi(I + b\Xi)^{-1} = (1 + b)^{-1}\Xi$ for any $b > 0$, so that

$$P(t) = \alpha e^{tD^\star}\Xi e^{tD}.$$

Thus, if $\Xi = 0$ or I, P(t) is constant whereas if Ξ is different from 0 or I,$P(t)$ is not trivial. ∎

Remark 4.4 *Unbounded observation operator*
Assume here that $(\mathcal{H}_\infty)$-(i)-(ii) and $(\mathcal{H}_\infty)$-(v) are verified and consider the optimal control (1.1)-(1.2). Obviously, in the definition of the cost functional J_∞, Cx must be defined as in Part II, Chapter 1, §6. The definitions of optimal pair, stabilizability and weak and stationary solutions (but not strict solutions) of the Riccati equation are now the same as before. However several of the previous results can be generalized, with the exception of strict solutions of the algebraic Riccati equation and of the detectability, since it is related to the dual system which has unbounded control operator.

∎

5 Some generalizations and complements

5.1 Non homogeneous state equation

We consider, as in Part II, Chapter 1, §7, a system governed by a non homogeneous state equation and the following infinite horizon problem: to minimize

$$J_\infty(u) = \int_0^\infty \left\{ |Cx(s)|^2 + |u(s)|^2 \right\} ds, \tag{5.1}$$

over all controls $u \in L^2(0,\infty;U)$ subject to the differential equation constraint

$$\begin{array}{l} x'(t) = Ax(t) + f(t) + Bu(t), \quad t \geq 0, \\ x(0) = x_0. \end{array} \tag{5.2}$$

We assume that

$$\left\{ \begin{array}{ll} (i) & f \in L^2(0,\infty;H), \\ (ii) & \textit{hypotheses } (\mathcal{H}) \textit{ hold}, \\ (iii) & (A,B) \textit{ is } C\textit{-stabilizable}, \\ (iv) & F = A - BB^\star P_{min}^\infty \textit{ is exponentially stable.} \end{array} \right. \tag{5.3}$$

Under conditions (5.3) it is easy to check that the following problem

$$\begin{array}{l} r'(t) + F^\star r(t) + P_{min}^\infty f(t) = 0, \quad t \geq 0, \\ r(\infty) = 0 \end{array}$$

has a unique mild solution given by

$$r(t) = \int_t^\infty e^{(s-t)F^\star} P_{min}^\infty f(s)\, ds.$$

We can now prove the following result.

Lemma 5.1 *Assume* (5.3) *and let* $x_0 \in H$ *and* $u \in L^2(0,\infty;U)$, *then we have*

$$\begin{array}{rcl} J_\infty(u) & = & (P_{min}^\infty x_0, x_0) + 2(r(0), x_0) \\ & & + \displaystyle\int_0^\infty \left\{ 2(r(s), f(s)) - |B^\star r(s)|^2 \right\} ds \\ & & + \displaystyle\int_0^\infty |u(s) + B^\star r(s) + B^\star P_{min}^\infty x(s)|^2 ds. \end{array}$$

Proof. It is sufficient to set $P = P_{min}$ (the minimal solution of the Riccati equation) in equation (7.4) of Part II, Chapter 1, and to let T tend to infinity. ∎

Now the following result is easily proved.

Theorem 5.1 *Assume* (5.3) *and let* $x_0 \in H$. *Then there exists a unique optimal pair* $(u^\star, x^\star)$ *for problem* (5.1)-(5.2). *Moreover the following statements hold.*

(i) $x^\star$ *is the mild solution to the closed loop equation*

$$\begin{cases} x'(t) = [A - BB^\star P_{min}^\infty]\, x(t) - BB^\star r(t) + f(t), & t \geq 0 \\ x(0) = x_0. \end{cases}$$

(ii) $u^\star$ *is given by the feedback formula*

$$u^\star(t) = -B^\star[P_{min}^\infty\, x^\star(t) + r(t)].$$

(iii) The optimal cost is given by

$$\begin{aligned} J_\infty(u) &= (P_{min}^\infty x_0, x_0) + 2(r(0), x_0) \\ &\quad + \int_0^\infty \left\{2(r(s), f(s)) - |B^\star r(s)|^2\right\} ds. \end{aligned}$$

Remark 5.1 If $f \notin L^2(0, \infty; H)$ it can happen that no admissible control exists. Consider in fact the following example. Let $H = U = Y = \mathbf{C}$, $A = 0$, $B = C = 1$, $f = 1$. It is readily seen that

$$J_\infty(u) = \int_0^\infty \left\{ \left| t + \int_0^t u(s)ds \right|^2 + |u(t)|^2 \right\} dt = \infty.$$

In this case the following cost function seems more appropriate (see G. Da Prato and A. Ichikawa [4]):

$$J_\infty(u) = \limsup_{T \to \infty} \frac{1}{T} \int_0^T \left\{ |Cx(s)|^2 + |u(s)|^2 \right\} ds.$$

5.2 Time dependent state equation and cost function

We consider here the system

$$\begin{cases} x'(t) = A(t)x(t) + B(t)u(t), & t \geq 0 \\ x(0) = x_0 \end{cases} \tag{5.4}$$

where $A(t) : D(A(t)) \subset H \to H$, $t \geq 0$ and $B(t) \in \mathcal{L}(U; H)$, $t \geq 0$, are linear operators. We assume that condition (7.8) in Chapter 1 of Part II is verified. We want to mimimize the cost function

$$J_\infty(u) = \int_0^\infty \left\{ |C(s)x(s)|^2 + |u(s)|^2 \right\} ds$$

over all controls $u \in L^2(0, \infty; U)$ subject to the differential equation constraint (5.4).

The definitions of admissible controls and optimal pair are the same as before. Also when, for any $x_0 \in H$, an admissible control exists, we say that (A, B) is C-*stabilizable.* In the autonomous case this problem was studied with the tool of the minimal positive solution of the algebraic Riccati equation. Now the same role is played by the *minimal positive bounded solution* of the Riccati equation

$$Q' + A^\star Q + QA - QBB^\star Q + C^\star C = 0 \tag{5.5}$$

in $[0, +\infty[$. In the remaining of this subsection we follow G. Da Prato and A. Ichikawa [4].

Definition 5.1 *We say that* $Q \in C_s([0, +\infty[; \Sigma^+(H))$ *is a (mild) solution to equation* (5.5)*, if for any* $a > 0$, $t \in [0, a]$ *and* $x \in H$*, we have*

$$Q(t)x = U^*(a,t)Q(a)U(a,t)x + \int_t^a U^*(s,t)C^*(s)C(s)U(s,t)x\, ds$$

$$- \int_t^a U^*(s,t)Q(s)B(s)B^*(s)Q(s)U(s,t)x\, ds.$$

Theorem 5.2 *Assume that conditions (7.8) and (7.9) of Chapter 1 in Part II are verified, and that* (A, B) *is* C-stabilizable. *Then equation* (5.5) *has a nonnegative bounded solution* Q. *This solution is minimal among all nonnegative bounded solutions of* (5.5).

Proof. For any $\lambda > 0$ and $n > 0$ sufficiently large, we introduce the Riccati equations

$$\begin{cases} \dfrac{dQ^\lambda}{dt} + A^\star Q^\lambda + Q^\lambda A - Q^\lambda BB^\star Q^\lambda + C^\star C = 0, \\ \\ Q^\lambda(\lambda) = 0, \quad t \in [0, \lambda], \end{cases} \tag{5.6}$$

$$\begin{cases} \dfrac{dQ_n^\lambda}{dt} + A_n^\star Q_n^\lambda + Q_n^\lambda A_n - Q_n^\lambda BB^\star Q_n^\lambda + C^\star C = 0, \\ \\ Q_n^\lambda(\lambda) = 0, \quad t \in [0, \lambda], \end{cases} \tag{5.7}$$

where A_n are the Yosida approximations of A. By Theorem 7.2 of Chapter 1 in Part II , problems (5.6) and (5.7) have unique solutions Q^λ and Q_n^λ respectively, and moreover for any $x \in H$

$$\lim_{n\to\infty} Q_n^\lambda(t)x = Q^\lambda(t)x, \tag{5.8}$$

uniformly in $t \in [0, \lambda]$. Also, by Proposition 7.1 of Chapter 1 in Part II,

$$Q_n^\lambda(t) \le Q_n^\mu(t), \quad t \in [0, \lambda] \text{ if } \mu > \lambda. \tag{5.9}$$

We now prove that there exists $C_1 > 0$ such that

$$\|Q^\lambda(t)\| \le C_1, \quad \forall\, \lambda > 0, \forall\, t \in [0, \lambda]. \tag{5.10}$$

For this it is sufficient to prove that

$$\|Q_n^\lambda(t)\| \le C_1, \quad \forall\, \lambda > 0, \forall\, t \in [0, \lambda], \forall\, n > 0. \tag{5.11}$$

Let $x \in H$ and let $u \in L^2(0, \infty; U)$ such that $J_\infty(u) < +\infty$. Let moreover x_n be the solution of the initial value problem:

$$\begin{aligned} x_n' &= A_n x_n + Bu, \\ x_n(0) &= x. \end{aligned}$$

Then we have

$$\begin{aligned} &\frac{d}{dt}(Q_n^\lambda(t)x_n(t), x_n(t)) \\ &= |u(t) + B^\star(t)Q_n^\lambda(t)x_n(t)|^2 - |C(t)x_n(t)|^2 - |u(t)|^2. \end{aligned}$$

Integrating this from t and λ, we find

$$< Q_n^\lambda(t)x, x > \leq \int_t^\lambda \{|C(s)x_n(s)|^2 + |u(s)|^2\} ds \leq \text{ Const. } |x|^2$$

and the inequalities (5.10), (5.11) follow.

From the estimates (5.9) and (5.10) the following limit exists

$$Q_\infty(t)x = \lim_{\lambda \to \infty} Q_n^\lambda(t)x, \quad \forall\, x \in H.$$

By letting λ tend to infinity in the equality

$$\begin{aligned} Q^\lambda(t)x &= U^*(a,t)Q^\lambda(a)U(a,t)x \\ &+ \int_t^a U^*(s,t)C^*(s)C(s)U(s,t)x\, ds \\ &- \int_t^a U^*(s,t)Q^\lambda(s)B(s)B^*(s)Q^\lambda(s)U(s,t)x\, ds \end{aligned}$$

we see that Q_∞ is a bounded solution of (5.5). It remains to prove minimality. Let R be a bounded nonnegative solution of (5.5). Then by Proposition 7.1 of Chapter 1 in Part II and the fact that $Q^\lambda(\lambda) = 0$,

$$R(t) \geq Q^\lambda(t), \quad \forall\, t \in [0, \lambda],$$

and so $R \geq Q_\infty$. ∎

Corollary 5.1 *Assume that the conditions of Theorem* 5.2 *are verified and that, in addition, A, B and C are τ–periodic functions for some $\tau > 0$. Then equation* (5.5) *has a nonnegative τ–periodic solution Q. This solution is minimal among all nonnegative τ-periodic solutions of* (5.5).

Proof. Let Q_∞ be the minimal non negative bounded solution of the Riccati equation. Then we have

$$Q^\lambda(t+\tau) = Q^{\lambda-\tau}(t), \quad \forall\, t \in [0, \lambda - \tau],$$

as $\lambda \to \infty$, we find that Q_∞ is τ–periodic. ∎

Remark 5.2 This result was proved in G. DA PRATO and A. ICHIKAWA [3]. If A, B, C are almost periodic and H is a finite dimensional space, then Q_∞ is almost periodic (see T. MOROZAN [1]). ∎

5.3 Periodic control problems

Consider a dynamical system governed by a linear equation

$$x'(t) = A(t)x(t) + f(t) + B(t)u(t) \tag{5.12}$$

with periodic coefficients $f(\cdot), A(\cdot), B(\cdot)$ of period 2π.

Then it is natural to consider 2π-periodic controls u. For any Banach space E we denote by $L^2_\#(E)$ the space of all 2π- periodic functions $u : \mathbf{R} \to E$ which belong to $L^2(0, 2\pi; E)$.

We assume that

$$\left\{\begin{array}{ll} (i) & \textit{Conditions (7.8) and (7.9) of Chapter 1 in Part II} \\ & \textit{hold.} \\ (ii) & A, B, C \ \textit{are} \ 2\pi \ \textit{periodic.} \\ (iii) & f \in L^2_\#(H). \end{array}\right. \tag{5.13}$$

Obviously equation (5.12) does not necessarily have a $2\pi-$ periodic solution for any control $u \in L^2_\#(U)$. This happens when no Floquet exponent of A is equal to 1. In such a case we say that A is *non resonant.* As it is well known, this is equivalent to require that 1 belongs to the resolvent set of $U_A(2\pi, 0)$, where $U_A(t, s)$ is the evolution operator relative to A. In this case equation (5.12) has a unique mild solution given by the formula

$$x(t) = U_A(t,0)[I - U_A(2\pi,0)]^{-1} \int_0^{2\pi} U_A(2\pi, s)[Bu(s) + f(s)]\, ds$$
$$+ \int_0^t U(t,s)[Bu(s) + f(s)]\, ds.$$

We are interested in the general case, that is when A is possibly resonant. For any $u \in L^2_\#(U)$ we set

$$\Lambda(u) = \{y \in L^2_\#(H) \ : \ y \text{ fulfills (5.12)}\} \tag{5.14}$$

The control u is said to be *admissible* if $\Lambda(u) \neq \emptyset$. The set of all admissible controls will be denoted by U_{ad}.

We want to minimize the *cost function*

$$J(u,y) = \int_0^{2\pi} \left\{|C(t)y(t)|^2 + |u(t)|^2\right\} dt \tag{5.15}$$

over all $u \in U_{ad}$ and $y \in \Lambda(u)$.

If there exist $u^\star \in U_{ad}$ and $x^\star \in \Lambda(u^\star)$ such that

$$J(u^\star, x^\star) \le J(u,x), \quad \forall\, u \in U_{ad}, \forall\, x \in \Lambda(u),$$

$u^\star$ is called an *optimal control*, $x^\star$ an *optimal state* and $(u^\star, x^\star)$ an *optimal pair*.

We will study this minimization problem by using again the Dynamic Programming approach and by proceeding in the following steps.

(i) We consider a periodic solution to the Riccati equation

$$Q' + A^\star Q + QA - QBB^\star Q + C^\star C = 0. \tag{5.16}$$

(ii) We look for a periodic solutions to the dual equation

$$r'(t) + (A - BB^\star Q)^\star r(t) + Qf(t) = 0 \tag{5.17}$$

and to the closed loop equation

$$x'(t) = (A - BB^\star Q)x(t) - BB^\star r(t) + f(t) \tag{5.18}$$

Then, we show that x is an optimal state and u, given by

$$u(t) = -B^\star(Qx(t) + r(t)), \tag{5.19}$$

is an optimal control.

Concerning the point (i), the existence of a nonnegative periodic solution of (5.16) was proved in Corollary 5.1 when (A, B) is C-stabilizable. The approach described below was introduced in G. DA PRATO and A. ICHIKAWA [3] where the existence of an optimal pair was proved with the additional condition that (A, C) be detectable. In fact in this case one can show that the Floquet exponents of the closed loop operator $F = A - BB^\star Q$ have all modulus less than 1 and then F is non resonant and equations (5.17) and (5.18) have periodic solutions.

Here we do not assume detectability, but we give, following G. DA PRATO [2], a characterization of the Floquet exponents of F with modulus greater or equal to one. In the following theorem U_F represents the evolution operator associated with F (U_F is well-defined since F is a bounded perturbation of A) and Q_∞ the minimal non negative periodic solution of the Riccati equation (5.16)

Theorem 5.3 *Assume that condition* (5.13) *is verified and that* (A, B) *is* C*–stabilizable. Let* Q_∞ *be the minimal nonnegative* 2π*–periodic solution of Riccati equation* (5.16) *and* $F = A - BB^*Q_\infty$*. Let* $\mu \in \mathbf{C}$ *such that* $|\mu| \geq 1$ *and let* $x_0 \in H$ *be different from* 0. *Then the following statements are equivalent.*

(i) $U_F(2\pi, 0)x_0 = \mu x_0$.

(ii) $U_A(2\pi, 0)x_0 = \mu x_0$ *and* $C(t)U_A(t, 0)x_0 = 0$, *for all* $t \geq 0$.

Moreover if either (i) or (ii) holds true, we have

$$U_F(t, 0)x_0 = U_A(t, 0)x_0, \quad \forall t \geq 0 \tag{5.20}$$

Proof. $(i) \Rightarrow (ii)$. Let $x_0 \in H$ and let x be the mild solution of the problem

$$\begin{aligned} &x' = Fx, \quad t \geq 0, \\ &x(0) = x_0, \end{aligned}$$

then the following identity holds

$$\begin{aligned} &(Q_\infty(t)x(t), x(t)) - (Q_\infty(0)x_0, x_0) \\ &+ \int_0^t \{|B^*(s)Q_\infty(s)x(s)|^2 + |C(s)x(s)^2|ds\} = 0. \end{aligned} \tag{5.21}$$

If $x_0 \in D(A)$ this follows easily by integrating the identity

$$\frac{d}{dt}(Q_\infty(t)x(t), x(t)) = -|B^\star(t)Q_\infty(t)x(t)|^2 - |C(t)x(t)|^2$$

between 0 and t. For general x_0 equation (5.21) follows by density.

Let now $x_0 \in H$ be such that $U_F(2\pi, 0)x_0 = x(2\pi) = \mu x_0$ with $|\mu| \geq 1$. Then, setting $t = 2\pi$ in (5.21) we find

$$(|\mu|^2 - 1)(Q_\infty(0)x_0, x_0) + \int_0^{2\pi} \{|B^*Q_\infty x|^2 + |Cx^2|ds\} = 0,$$

which yields

$$B^\star(t)Q_\infty(t)x(t) = 0, \ C(t)x(t) = 0, \quad \forall t \geq 0.$$

Consequently $x' = Ax$, and

$$x(t) = U_F(t, 0)x_0 = U_A(t, 0)x_0, \quad t \geq 0,$$

which implies (ii).

$(ii) \Rightarrow (i)$. Given $x_0 \in H$, let y be the mild solution of the system $y' = Ay,\ y(0) = x_0$. Moreover let Q^λ be the mild solution of the Riccati equation

$$\begin{aligned} &Q' + A^*Q + QA - QBB^*Q + C^*C = 0, \\ &Q(2\pi\lambda) = 0. \end{aligned}$$

Then the following identity holds

$$\begin{aligned} &(Q^\lambda(0)x_0, x_0) + \int_0^{2\pi\lambda} |B^*(s)Q^\lambda(s)x(s)|^2 ds \\ &= \int_0^{2\pi\lambda} |C(s)x(s)^2| ds. \end{aligned} \tag{5.22}$$

In fact if $x_0 \in D(A)$ this result readily follows by integrating the identity

$$\frac{d}{dt}(Q^\lambda(t)y(t), y(t)) = |B^\star(t)Q^\lambda(t)y(t)|^2 - |C(t)y(t)|^2$$

between 0 and $2\pi\lambda$. For general x_0 equation (5.22) follows by density.

Let now x_0 be such that

$$y(2\pi) = U_A(2\pi, 0)x_0 = \mu x_0, \quad C(t)y(t) = 0, \quad t \geq 0.$$

Then, by letting λ tend to infinity in (5.22) we find

$$(Q_\infty(0)x_0, x_0) + \int_0^\infty |B^*(t)Q_\infty(t)\, x(t)|^2 dt = 0,$$

which implies $B^*(t)Q_\infty(t) = 0,\ t \geq 0$, so that $y' = Fy$ and $U_A(t,0)x_0 = U_F(t,0)x_0$ and (i) holds. ■

Remark 5.3 Assume that the spectra of $U_A(2\pi, 0)$ and $U_F(2\pi, 0)$ consist only of eigenvalues, (this is for instance the case for parabolic state equations in bounded domains, see Example 6.1 below). By the above theorem it follows that F is non resonant if one of the following conditions holds

(i) A is non resonant.
(ii) A is resonant but the following implication holds

$$x_0 \in H,\ x_0 \neq 0,\ U_A(2\pi,0)x_0 = x_0 \quad \Rightarrow \quad C(t_0)U_A(t_0,0)\,x_0 \neq 0,$$

for at least one $t_0 \geq 0$.

For a case where the eigenvalues of $U_A(2\pi,0)$ and $U_F(2\pi,0)$ have limit points, see G. DA PRATO, [4]. ∎

We can prove now the result.

Theorem 5.4 *Assume that condition* (5.13) *is verified, that A, B is C-stabilizable and that $F = A - BB^\star Q_\infty$ is non resonant, where Q_∞ is the minimal non negative 2π-periodic solution of the Riccati equation* (5.16). *Then there exists a unique optimal pair $(u^\star, x^\star)$ for problem* (5.12)-(5.15) *and the following conditions are satisfied:*

(i) $x^\star$ is the unique periodic solution to the closed loop equation (5.18),

(ii) $u^\star$ is given by the feedback formula

$$u^\star(t) = -B^\star[Q_\infty(t)x^\star(t) + r^\star(t)], \tag{5.23}$$

where r is the unique periodic solution of equation (5.17),

(iii) the optimal cost $J(u^\star, x^\star)$ is given by

$$J(u^\star, x^\star) = \int_0^{2\pi} \{(r(t), f(t)) - |B^\star(t)r(t)|^2\}dt. \tag{5.24}$$

Proof. Let $u \in U_{ad}$ and $y \in \Lambda(u)$. By computing

$$\frac{d}{dt}\{(Q_\infty(t)y(t), y(t))) + 2(r(t), f(t))\}$$

and by integrating between 0 and 2π, we find the identity

$$J(u,y) = J^\star + \int_0^{2\pi} \|R(t)\|^2 dt, \tag{5.25}$$

where

$$J^\star = \int_0^{2\pi} \{(r(t), f(t)) - |B^\star(t)r(t)|^2\}dt$$

and

$$R(t) = B^*(t)[Q_\infty(t)y(t) + r(t)] + u(t).$$

We remark that the computation can be made rigorous by approximating $A(t)$ by their Yosida approximations. It follows that

$$J(u^\star, y^\star) \geq J^\star, \quad \forall\, u \in U_{ad}, \forall\, y \in \Lambda(u)$$

Now, let x^* be the solution of (5.18) and let u^* be given by (5.23). Setting in (5.25), $u = u^\star$ and $y = y^\star$ we obtain $J(u^\star, y^\star) = J^\star$ so that the pair $u^\star, y^\star$ is optimal. Finally uniqueness of the optimal pair is proved as in Theorem 6.1 of Chapter 1 in Part II. ∎

Remark 5.4 It is also possible to study almost periodic control problems (see G. DA PRATO and A. ICHIKAWA [2].) ∎

6 Examples of controlled systems

6.1 Parabolic equations

We shall continue here the Example 8.1 in Chapter 1 of Part II. Moreover we denote by $\{\varphi_k\}$ a complete set of eigenvector of A and by $\{\lambda_k\}$ the corresponding sequence of eigenvalues. We assume that $\{\lambda_k\}$ is non increasing. We consider the infinite horizon control problem: to minimize

$$J_\infty(u) = \int_0^\infty \int_\Omega \left\{|(Cx(t,\cdot))(\xi)|^2 + |u(t,\xi)|^2\right\} dt d\xi, \tag{6.1}$$

over all $u \in L^2([0,\infty] \times \Omega)$ subject to condition (1.48) of Chapter 1 in Part II.

We have to discuss the existence of admissible controls. We consider two cases.

First Case. $\lambda_1 < 0$.
In this case A is exponentially stable so that (A, B) is C-stabilizable. Thus by Theorem 3.1 and Corollary 4.2, the algebraic Riccati equation

$$AP + PA - PBB^\star P + C^\star C = 0, \tag{6.2}$$

has a unique solution $P_{min}^\infty = P_{Max}^\infty$ and the feedback operator $F = A - BB^\star P_{min}^\infty$ is exponentially stable.

Second Case. $\lambda_1 \geq 0, \lambda_2 < 0$.
In this case we have $\sigma^+(A^\star) \cup \sigma^0(A^\star) = \{\lambda_1\}$; thus, by Proposition 3.3, (A, B) is I-stabilizable if and only if $B^\star\varphi_1 \neq 0$. Under this assumption equation (6.2) has a minimal and a maximal nonnegative solution. Moreover by Proposition 3.4, F is exponentially stable if and only if $C\varphi_1$ is not identically zero. It is easy to generalize the previous discussion when $\lambda_m \geq 0$ and $\lambda_{m+1} < 0$ for some $m \in \mathbf{N}$.

6.2 Wave equation

We continue here Example 8.2 in Part II, Chapter 1. Consider the infinite horizon problem: to minimize

$$J_\infty(u) = \int_0^\infty \int_\Omega \left\{ |\nabla_\xi x(t,\xi))|^2 + \left|\frac{\partial}{\partial t}(t,\xi)\right|^2 + |u(t,\xi)|^2 \right\} dt d\xi, \tag{6.3}$$

over all $u \in L^2([0,\infty] \times \Omega)$ subject to equation (8.2) in Chapter 1 of Part II. In this case the algebraic Riccati equation reads as follows

$$\begin{bmatrix} 0 & -1 \\ \Lambda & 0 \end{bmatrix} X + X \begin{bmatrix} 0 & 1 \\ -\Lambda & 0 \end{bmatrix} - X \begin{bmatrix} 0 & 0 \\ 0 & 1 \end{bmatrix} X + \begin{bmatrix} 1 & 0 \\ 0 & 1 \end{bmatrix} = 0.$$

We prove now that (A, B) is I-stabilizable; for this it is sufficient to show that $A - 2\alpha BB^\star$ is exponentially stable if $0 < \alpha^2 < \lambda_0$ where λ_0 is the principal eigenvalue of the Laplace operator in Ω, with Dirichlet boundary conditions. By a direct computation we find

$$e^{t(A-2\alpha BB^\star)} =$$

$$e^{-\alpha t} \begin{bmatrix} \cos\ (Et) + \frac{\alpha}{E} \sin\ (Et) & \frac{1}{E} \sin\ (Et) \\ \frac{\alpha^2+E^2}{E} \sin\ (Et) & -\frac{\alpha}{E} \sin\ (Et) + \cos\ (Et) \end{bmatrix}$$

where $E = \sqrt{\Lambda - \alpha^2 I}$. Since $(A, C) = (A, I)$ is detectable, the Riccati equation has a unique nonnegative solution

$$P_{min}^\infty = P_{Max}^\infty = \begin{bmatrix} P_{11} & P_{12} \\ P_{21} & P_{22} \end{bmatrix}.$$

Then, by Theorem 3.1, there exists a unique optimal pair $(u^\star, x^\star)$ with:

$$u^\star(t,\xi) = -(P_{21}x^\star(t,\cdot))(\xi) - (P_{22}\frac{\partial x^\star}{\partial t}(t,\cdot)(\xi).$$

Remark 6.1 Similar consideration apply to the wave equation with Neumann boundary conditions. See Remark 8.1 in Part II, Chapter 1

■

Remark 6.2 For a situation in which there exist periodic non trivial solutions see G. Da Prato [2]. ■

6.3 Strongly damped wave equation

Let Ω be an open bounded set of $\mathbf{R}^n$ with regular boundary $\partial\Omega$. Consider the equation

$$\begin{cases} \dfrac{\partial^2 x}{\partial t^2}x(t,\xi) = \Delta_\xi(t,\xi) + \rho\Delta_\xi\dfrac{\partial x}{\partial t}(t,\xi) \\ \qquad\qquad + (Bu(t,\cdot))(\xi) \text{ in }]0,T] \times \Omega \\ \dfrac{\partial x}{\partial \nu}(t,\xi) = 0 \text{ on }]0,T] \times \partial\Omega \\ x(0,\xi) = x_0(\xi), \ \dfrac{\partial x}{\partial t}(0,\xi) = x_1(\xi) \text{ in } \Omega, \end{cases} \tag{6.4}$$

where ν is the outward normal to $\partial\Omega$, and ρ is a given positive number.

As in Chapter 1 of Part II, we choose the spaces H, Y, U as in (8.10), the scalar product in H as in (8.11) and consider the positive self-adjoint operator Λ_1 defined by (8.12). Since Ω is bounded there exists a complete orthonormal system in $L^2(\Omega)$, $\{e_k\}$, and a sequence of real numbers

$$\mu_1 = 0 < \mu_2 \leq \mu_2 \leq \ldots\ldots \to +\infty$$

such that

$$\Lambda_1 e_k = \mu_k e_k, \ k = 1, 2, \ldots.$$

Define the linear operator A_1 on H

$$\begin{cases} A_1 X = \begin{bmatrix} 0 & 1 \\ -\Lambda_1 & -\rho\Lambda_1 \end{bmatrix} \begin{bmatrix} x^0 \\ x^1 \end{bmatrix}, \quad \forall X \in D(A_1) \\ \\ D(A_1) = \{x \in H^2(\Omega) : \dfrac{\partial}{\partial \nu} x = 0 \text{ on } \partial\Omega\} \oplus H_0^1(\Omega)\}. \end{cases} \tag{6.5}$$

It is an easy exercise to prove that the spectrum of A_1 is given by

$$\sigma(A_1) = \{-1/\rho\} \cup \left\{ -\frac{1}{2}(\sigma\lambda_k \pm \sqrt{(\rho^2\lambda_k^2 - 4\lambda_k)}; k = 1, 2, \ldots \right\}$$

and the resolvent by

$$R(\lambda, A_1) = \begin{bmatrix} \lambda + \rho\Lambda_1 & 1 \\ -\Lambda_1 & \lambda \end{bmatrix} [\lambda^2 + \rho\Lambda_1\lambda + \Lambda_1]^{-1}, \quad \forall \lambda \in \sigma(A_1).$$

From the above formula, it is not difficult to prove that A_1 is the infinitesimal generator of an analytic semigroup in H.

Now set

$$Bu = \begin{bmatrix} 0 \\ u \end{bmatrix}, \quad u \in U.$$

We want to check that condition (ii) of Proposition 3.5 (Hautus condition) is fulfilled, so that (A_1, B) is stabilizable with respect to the observation I. Remark that A_1 does not have a pure point spectrum, since $-\frac{1}{\rho}$ is not an eigenvalue. However $\sigma(A_1)$ consists of 0 and elements of real part less than a fixed negative number. Thus, in order to check the Hautus condition, we only have to prove the implication

$$A_1^* \begin{bmatrix} x^0 \\ x^1 \end{bmatrix} = 0 \quad \Rightarrow \quad B_1^* \begin{bmatrix} x^0 \\ x^1 \end{bmatrix} \neq 0. \tag{6.6}$$

In fact, since

$$A_1^* X = \begin{bmatrix} 0 & -1 \\ \Lambda_1 & -\rho\Lambda_1 \end{bmatrix} \begin{bmatrix} x^0 \\ x^1 \end{bmatrix}$$

the condition $A_1^* \begin{bmatrix} x^0 \\ x^1 \end{bmatrix} = 0$ is equivalent to $x^0 = 0$, $\Lambda_1 x^1 = 0$. So it holds if and only if $\begin{bmatrix} x^0 \\ x^1 \end{bmatrix} = \begin{bmatrix} 0 \\ pe^1 \end{bmatrix}$ for any real number p. Since $B^* \begin{bmatrix} 0 \\ pe^1 \end{bmatrix} = pe_1$ the implication (6.6) is verified.

Now we can consider the cost functional (6.3) and solve the corresponding minimization problem.

Remark 6.3 If one replaces in (6.4) Neumann with Dirichlet boundary conditions, the situation is easier to describe, since in that case one obtains a linear operator A of negative type. ∎

Remark 6.4 Strongly damped wave equation with time–dependent periodic coefficients has been studied in G. DA PRATO and A. LUNARDI [1]. ∎

Chapter 2

Systems with Unbounded Control Operators: Parabolic Equations with Control on the Boundary

1 Introduction and setting of the problem

As in Part II Chapter 2 we consider a dynamical system governed by the following equation

$$\begin{cases} x'(t) = Ax(t) + (\lambda_0 - A)Du(t), & t \geq 0, \\ x(0) = x_0, \end{cases}$$

or equivalently

$$x(t) = e^{tA}x_0 + (\lambda_0 - A)\int_0^t Du(s)\,ds, \tag{1.1}$$

where $x_0 \in H$ and $u \in L^2(0,\infty;U)$. We assume that

$$(\mathcal{HP})_\infty \begin{cases} (i) & A \text{ generates an analytic semigroup } e^{tA} \text{ of type } \omega_0 \\ & \text{and } \lambda_0 \text{ is a real number in } \rho(A) \text{ such that } \omega_0 < \lambda_0 \\ (ii) & \exists\, \alpha \in]0,1[\text{ such that } D \in \mathcal{L}(U; D([\lambda_0 - A]^\alpha)), \\ (iii) & C \in \mathcal{L}(H;Y). \end{cases}$$

Clearly, if hypotheses $(\mathcal{HP})_\infty$ hold, then the hypotheses $(\mathcal{HP})$ of Part II, Chapter 2 are fulfilled with $P_0 = 0$. If $\alpha \le \frac{1}{2}$, we will choose once and for all a number β belonging to $]1 - \frac{\alpha}{2}, \frac{1-\alpha}{2}[$. We want to minimize the cost function:

$$J_\infty(u) = \int_0^\infty \{|Cx(s)|^2 + |u(s)|^2\} ds \tag{1.2}$$

over all controls $u \in L^2(0, \infty; U)$ subject to the differential equation constraint (1.1). We say that the control $u \in L^2(0, \infty; U)$ is *admissible* if $J_\infty(u) < \infty$. The definitions of optimal control, optimal state and optimal pair are the same as in Chapter 1. When, for any $x_0 \in H$, an admissible control exists, we say that (A, AD) is C-stabilizable.

In this Chapter, we want to generalize the results of Chapter 1. Several of these generalizations are straightforward and the corresponding proofs will only be sketched. We will start by proving the existence of a minimal solution P_{min}^∞ of the algebraic Riccati equation

$$\begin{cases} A^\star X + XA - Y^\star EE^\star Y + C^\star C = 0, \\ Y = [\lambda_0 - A^\star]^{1-\alpha} X, \end{cases} \tag{1.3}$$

where $E = (\lambda_0 - A)^\alpha D$, under the hypothesis that (A, AD) is C-stabilizable. We remark that in order that (1.3) be meaningful, we need a regularity property of X, namely

$$X \in \Sigma^+(H) \cap \mathcal{L}(H; D([\lambda_0 - A^\star]^{1-\alpha})).$$

We shall set $V_{min}^\infty = [\lambda_0 - A^\star]^{1-\alpha} P_{min}^\infty$. The above requirement will make the proof of existence much more involved than in Chapter 1, (see §2 below). Once the existence is proved, one can show, quite easily (see §3), that the optimal control $u^\star$ exists and is given by the feedback formula

$$u^\star(t) = -E^\star V_{min}^\infty x^\star(t),$$

where $x^\star$ is the solution to the closed loop equation

$$x(t) = e^{tA} x_0 + \int_0^t [\lambda_0 - A]^{1-\alpha} e^{(t-s)A} EE^\star V_{min}^\infty\, x(s)\, ds. \tag{1.4}$$

Another new difficulty arises in the study of the closed loop operator F (see §3), formally defined by

$$F = A - EE^{\star}V^{\infty}_{min}.$$

In §3 we prove a characterization of F which enables us to generalize all results concerning detectability and Hautus conditions of Chapter 1. Also we leave the reader to extend all results on the qualitative behaviour of the solutions of Riccati equations proved in Chapter 1.

2 The algebraic Riccati equation

We assume here that the assumptions $(\mathcal{HP})_{\infty}$ are verified and consider system (1.1) and the Riccati equation

$$P' = A^{\star}P + PA - V^{\star}EE^{\star}V + C^{\star}C, \tag{2.1}$$

along with the Algebraic Equation (1.3). We say that P (belonging to $C_{s,\alpha}([0,\infty[\,;\Sigma^{+}(H))$ if $\alpha > \frac{1}{2}$ and to $C_{s,\alpha,\beta}([0,\infty[\,;\Sigma^{+}(H))$ if $\alpha \leq \frac{1}{2}$) is a mild solution of (2.1), if

$$\begin{aligned} P(t)x &= e^{tA^{\star}}P(0)e^{tA}x + \int_0^t e^{(t-s)A^{\star}}C^{\star}Ce^{(t-s)A}x\,ds \\ &\quad + \int_0^t e^{(t-s)A^{\star}}V^{\star}(s)EE^{\star}V(s)e^{(t-s)A}x\,ds. \end{aligned} \tag{2.2}$$

Moreover, X is said to be a *solution* of (1.3) if

(i) $X \in \Sigma^{+}(H) \cap \mathcal{L}(H; D([\lambda_0 - A^{\star}]^{1-\alpha}))$.

(ii) $(Xx, Ay) + (Ax, Xy) - (E^{\star}Yx, E^{\star}Yy) + (Cx, Cy) = 0$
for all $x, y \in D(A)$, where $Y = [\lambda_0 - A^{\star}]^{1-\alpha}X$.

As in the previous chapter, we shall denote by P_{min} the minimal nonnegative solution of Riccati equation (2.1); that is, the solution of (2.1) such that $P(0) = 0$. (see Propositions 2.3 and 2.4 of Part II, Chapter 2).

We start by proving existence in the easier case $\alpha > \frac{1}{2}$.

Proposition 2.1 *Assume $(\mathcal{HP})_{\infty}$ with $\alpha > \frac{1}{2}$ and that (A, AD) is C-stabilizable. Then there exists*

$$P^{\infty}_{min} \in \Sigma^{+}(H) \cap \mathcal{L}(H; D([\lambda_0 - A^{\star}]^{1-\alpha}))$$

such that

(i) for each $x \in H$ *the following limit exists*

$$\lim_{t \to \infty} P_{min}(t)x = P_{min}^{\infty}x, \tag{2.3}$$

(ii) $P_{min}^{\infty} \in \Sigma^{+}(H) \cap \mathcal{L}(H; D([\lambda_0 - A^{\star}]^{1-\alpha}))$.

(iii) P_{min}^{∞} *is a solution of* (1.3).

Proof. Since (A, AD) is C–stabilizable, we can prove, by repeating the proof of Proposition 1.2.3, that there exists $K > 0$ such that

$$\|P_{min}(t)\| \leq K, \quad t \geq 0. \tag{2.4}$$

Now, arguing as we did in the proof of Proposition 2.2 in Chapter 1, we see that there exists P_{min}^{∞} such that (2.3) holds true. Let Q and $Q_n, n = 0, 1, \ldots$, be the mild solutions (granted by Theorem 2.1 in Part II, Chapter 2) of the Riccati equations

$$\begin{cases} Q' = A^{\star}Q + QA - V_Q^{\star}EE^{\star}V_Q + C^{\star}C, \\ \\ Q(0) = P_{min}^{\infty}, \end{cases} \tag{2.5}$$

and

$$\begin{cases} Q_n' = A^{\star}Q_n + Q_nA - V_{Q,n}^{\star}EE^{\star}V_{Q,n} + C^{\star}C, \\ \\ Q_n(0) = P_{min}(n), \end{cases} \tag{2.6}$$

where $V_Q = [\lambda_0 - A^{\star}]^{1-\alpha}Q$ and $V_{Q,n} = [\lambda_0 - A^{\star}]^{1-\alpha}Q_n$. We clearly have

$$Q_n(t) = P_{min}(t + n), \quad t > 0.$$

Moreover, by Proposition 2.2 in Part II, Chapter 2, since

$$Q_n(0)x \to Q(0)x, \quad \forall\, x \in H,$$

we have

$$\begin{aligned} Q(t)x &= \lim_{n \to \infty} Q_n(t)x = \lim_{n \to \infty} P_{min}(t + n)x \\ &= P_{min}^{\infty}x, \quad \forall\, t > 0, x \in H. \end{aligned}$$

Thus $Q(t)$ is constant and coincides with P^{∞}_{min}. Since, by Theorem 2.1 in Part II, Chapter 2, $Q \in C_{s,\alpha}([0,\infty[\,;\Sigma^+(H))$, we see that (ii) holds true. Finally, (iii) follows from Proposition 2.1 in Part II, Chapter 2, since if $x, y \in D(A)$, we have

$$\begin{aligned} 0 &= \frac{d}{dt}(Q(t)x, y) \\ &= (P^{\infty}_{min}x, Ay) + (P^{\infty}_{min}Ax, y) + (Cx, Cy) - (E^\star V^{\infty}_{min}x, E^\star V^{\infty}_{min}y). \end{aligned}$$

The proof is complete. ∎

Next, we consider the case $\alpha \leq \frac{1}{2}$. We first recall that, by Theorem 2.2 in Part II, Chapter 2, for any $x \in D([\lambda_0 - A]^\beta)$ we haven$P_{min}(t)\,[\lambda_0 - A]^\beta x \in D([\lambda_0 - A^\star]^\beta)$, and the linear operator $[\lambda_0 - A^\star]^\beta P_{min}(t)\,[\lambda_0 - A]^\beta$ is closable. We denote by $(P_{min}(t))_\beta$ its closure.

Proposition 2.2 *Assume $(\mathcal{HP})_\infty$ with $\alpha \leq \frac{1}{2}$ and that (A, AD) is C-stabilizable. Then there exists*

$$P^{\infty}_{min} \in \Sigma^+(H) \cap \mathcal{L}(H; D([\lambda_0 - A^\star]^{1-\alpha}))$$

such that for all $x \in H$

(i) the limits

$$\lim_{t\to\infty} P_{min}(t)x = P^{\infty}_{min}x, \quad \forall\, x \in H. \tag{2.7}$$

$$\lim_{t\to\infty} (P_{min}(t))_\beta x = R^{\infty}_{min}x, \quad \forall\, x \in H. \tag{2.8}$$

exist, where $R^{\infty}_{min} = (P^{\infty}_{min})_\beta$ and

$$(P^{\infty}_{min})_\beta = \textit{closure of } [\lambda_0 - A^\star]^\beta\, P^{\infty}_{min}\,[\lambda_0 - A^\star]^\beta.$$

(ii) $P^{\infty}_{min} \in \Sigma^+(H) \cap \mathcal{L}(H; D([\lambda_0 - A^\star]^{1-\alpha}))$.

(iii) P^{∞}_{min} is a solution of (1.3).

Proof. By proceeding as in the proof of Proposition 2.1, we can prove (2.7) and that there exists a constant L_1 such that

$$\|P_{min}(t))_\beta\| \leq L_1, \quad t \geq 0. \tag{2.9}$$

We follows here an argument of F. FLANDOLI [2]; since the proof of (2.9) is easier for $\lambda_0 < 0$, we introduce a shifting $L = A - \lambda_0 I$ of A. Clearly L is of negative type and, as easily checked, $P(t) = P_{min}(t)$ is the solution to Riccati equation

$$\begin{cases} P' = L^\star P + PL + \Delta^2 - V^\star EE^\star V, \\ \\ P(0) = 0, \end{cases} \tag{2.10}$$

where

$$\Delta^2 = C^\star C + (2\lambda_0 + 2)P.$$

Now, if $x \in D([\lambda_0 - A]^\beta) = D([I - L]^\beta)$, we have

$$(P(t)(\lambda_0 - A)^\beta x, (\lambda_0 - A)^\beta x) = (P(t)(I - L)^\beta x, (I - L)^\beta x)$$

$$\leq \int_0^t |\Delta(I - L)^\beta e^{(t-s)L} x|^2 ds$$

$$\leq L^2[\,\|C\|^2 + 2K(\lambda_0 + 1]\int_0^\infty e^{-\lambda_0 s} s^{-2\beta} ds\,|x|^2,$$

which implies (2.9). Since the family $\{(P_{min}(t))_\beta\}$ is bounded and non decreasing in β, (2.8) follows.

We now prove (ii). Let Q and Q_n be the mild solutions (established in Theorem 2.2 in Part II, Chapter 2) of the Riccati equations (2.5) and (2.6). We clearly have

$$Q_n(t) = P_{min}(t + n), \quad t > 0.$$

Moreover, by Proposition 2.2 in Part II, Chapter 2, since $Q_n(0)x \to Q(0)x$, $\forall\, x \in H$, we have

$$\begin{aligned} Q(t) &= \lim_{n\to\infty} Q_n(t)x \\ &= \lim_{n\to\infty} P_{min}(t+n)x = P^\infty_{min}x, \quad \forall\, t > 0,\, x \in H. \end{aligned}$$

Thus $Q(t)$ is constant and coincides with P^∞_{min}. Since, by Theorem 2.2, in Part II, Chapter 2, $Q \in C_{s,\alpha,\beta}([0,\infty[\,;\Sigma^+(H))$, we see that (ii) holds true. Finally, the proof of (iii) is similar to that of the previous proposition. So it will be omitted. The proof is complete. ■

3 Dynamic programming

3.1 Existence and uniqueness of the optimal control

In this section, we consider the control problem (1.1)-(1.2). We assume $(\mathcal{HP})_\infty$ and that (A, AD) is C-stabilizable, and denote by

$$P_{min}^\infty \in \Sigma^+(H) \cap \mathcal{L}(H; D([\lambda_0 - A^\star]^{1-\alpha}))$$

the minimal solution of the Algebraic Equation (1.3) and we set $V_{min}^\infty = [\lambda_0 - A^\star]^{1-\alpha} P_{min}^\infty$.

We first recall that, in virtue of (3.1) in Chapter 1, we have

$$\begin{aligned}(P_{min}(t)x_0, x_0) &+ \int_0^t |u(s) + E^\star V_{min}(t-s)x(s)|^2\, ds \\ &= \int_0^t \left\{|Cx(s)|^2 + |u(s)|^2\right\} ds,\end{aligned} \tag{3.1}$$

for any $x_0 \in H$ and any $u \in L^2(0, \infty; U)$, where x is the solution of (1.1). Now we study the closed loop equation (1.4) which is meaningful since $P_{min}^\infty \in \Sigma^+(H) \cap \mathcal{L}(H; D([\lambda_0 - A^\star]^{1-\alpha}))$ and $\|[\lambda_0 - A]^{1-\alpha}\, e^{t(A-\lambda_0)}\| \leq L(t-s)^{\alpha-1}$.

Proposition 3.1 *There exists a unique solution $x \in C([0, \infty[; H)$ of* (1.4).

Proof. For any $x \in C([0, \infty[; H)$ we set

$$(\lambda(y))(t) = \int_0^t [\lambda_0 - A]^{1-\alpha}\, e^{(t-s)A} EE^\star\, V_{min}^\infty x(s)\, ds,$$

for all $t \in [0, T]$. Then we have

$$\|(\lambda(y))(t)\| \leq L^2 \|V_{min}^\infty\| \int_0^t (t-s)^{\alpha-1} ds\, \|y\|_{C([0,t];H)}$$

$$\leq \frac{L^2}{2\alpha - 1} \|V_{min}^\infty\|\, t^\alpha\, \|y\|_{C([0,t];H)},$$

so that $\lambda \in \mathcal{L}(C([0, t]; H))$ for any $t \in]0, T]$ and

$$\|\lambda\|_{\mathcal{L}(C([0,t];H))} \leq \frac{L^2}{2\alpha - 1} \|V_{min}^\infty\|\, t^\alpha.$$

Thus, if t is small, λ is a contraction and equation (1.4) has a unique solution in $C([0,t];H)$. Now this argument can be repeated in the interval $[t,2t]$ and so on, giving the conclusion. ∎

We prove now the main result of this section.

Theorem 3.1 *Assume that assumption $(\mathcal{HP})_\infty$ is verified and that (A,B) is C-stabilizable. Then there exists a unique optimal pair $(u^\star,x^\star)$ for the optimal control problem* (1.1)-(1.2). *Moreover the following statements hold:*

(i) $x^\star \in C([0,\infty[\,;H)$ *is the mild solution to the closed loop equation* (1.4).

(ii) $u^\star \in C([0,\infty[\,;U)$ *is given by the feedback formula*

$$u^\star(t) = -E^\star V^\infty_{min} x^\star(t), \tag{3.2}$$

where P^∞_{min} represents the minimal solution of the Algebraic Riccati equation (1.3),

(iii) the optimal cost $J_\infty(u^\star)$ is given by

$$J_\infty(u^\star) = (P^\infty_{min} x_0, x_0). \tag{3.3}$$

Proof. In view of Proposition 3.1, the proof is similar to that of Theorem 1.3.1. So it will be omitted. ∎

3.2 Feedback operator and detectability

Assume that the hypotheses of Theorem 3.1 hold and let $(u^\star,x^\star)$ be the optimal pair for problem (1.1)-(1.2). We want here to construct, following G. Da Prato and A. Ichikawa [1], a closed loop operator F for the system (3.1); that is a linear operator, infinitesimal generator of a strongly continuous semigroup e^{tF}, such that

$$x^\star(t) = e^{tF} x_0, \quad t \geq 0. \tag{3.4}$$

Proposition 3.2 *Assume $(\mathcal{HP}_\infty)$ and that (A,B) is C-stabilizable. For any $x_0 \in H$ set*

$$S(t)x = x^\star(t)$$

where $x^\star(t)$ is the optimal state corresponding to x_0. Then $S(\cdot)$ is an analytic semigroup in H and its infinitesimal generator F is given by

$$\begin{aligned} Fx &= (\lambda_0 - A)\,(AR(\lambda_0, A)x + DE^\star V_{min}^\infty x) \\ D(F) &= \{x \in H \,:\, AR(\lambda_0, A)x + DE^\star V_{min}^\infty x \in D(A)\}. \end{aligned} \tag{3.5}$$

Finally, if the resolvent of A is compact, so is the resolvent of F.

Proof. We first remark that for any $x_0 \in H$, $x(t) = S(t)x_0$ is precisely the solution of the closed loop equation (1.4). Thus, x is continuous, by Proposition 3.1, and $S(t)$ is a strongly continuous semigroup in H. Let $F : D(F) \subset H \to H$ be its infinitesimal generator; we want to prove that F is given by (3.5). To this aim, let $x_0 \in D(F)$, $x(t) = S(t)x_0$. We have

$$R(\lambda_0, A)x(t) = e^{tA}R(\lambda_0, A)x_0 + \int_0^t e^{(t-s)A}DE^\star V_{min}^\infty x(s)\,ds.$$

Since $x(t)$ is continuously differentiable, we have

$$R(\lambda_0, A)x'(t) = AR(\lambda_0, A)x(t) + DE^\star V_{min}^\infty x(t).$$

Setting $t = 0$, it follows that

$$R(\lambda_0, A)Fx_0 = AR(\lambda_0, A)x_0 + DE^\star V_{min}^\infty x_0, \tag{3.6}$$

so $AR(\lambda_0, A)x_0 + DE^\star V_{min}^\infty x_0 \in D(A)$ and

$$Fx_0 = (\lambda_0 - A)(AR(\lambda_0, A)x_0 + DE^\star V_{min}^\infty x_0). \tag{3.7}$$

Conversely assume that,

$$AR(\lambda_0, A)x_0 + DE^\star V_{min}^\infty x_0 \in D(A). \tag{3.8}$$

Then, by (3.6) and the density of $D(F)$, the linear operator $R(\lambda_0, A)F$ is closable and its closure $N = \overline{R(\lambda_0, A)F}$ is bounded. Now, if (3.8) holds we have $Nx_0 \in D(A)$ and it is not difficult to show that this implies $x_0 \in D(F)$ and $(\lambda_0 - A)Nx_0 = Fx_0$.

It remains to show that e^{tF} is an analytic semigroup. Let

$$S_{\omega,\theta_0} = \{\lambda \in \mathbf{C} \,:\, |\arg\ (\lambda - \omega)| \le \theta_0\}$$

be a sector contained in the resolvent set $\rho(A)$ and assume that

$$\|R(\lambda, A)\| \leq \frac{M(\theta)}{|\lambda - \omega|}, \quad \lambda \in S_{\omega,\theta_0}, \tag{3.9}$$

for some constants ω and $M(\theta) = M(-\theta)$ (see Volume I, Chapter 1, §2.7). Let $\lambda \in S_{\omega,\theta_0}, \eta \in H$; consider the equation $\lambda\xi - F\xi = \eta$, which is equivalent to

$$\lambda\xi - [\lambda_0 - A]\{AR(\lambda_0, A)\xi + DE^\star V_{min}^\infty \xi\} = \eta. \tag{3.10}$$

Multiplying (3.10) by $R(\lambda, A)$, gives

$$\xi - [\lambda_0 - A]^{1-\alpha} R(\lambda, A) EE^\star V_{min}^\infty \xi) = R(\lambda, A)\eta. \tag{3.11}$$

Since $[\lambda_0 - A]^\alpha D$ is bounded and $\|[\lambda_0 - A]^{1-\alpha} R(\lambda, A)\| \leq$const. $|\lambda|^{-\alpha}$, there exists $\rho > 0$ such that, for any $\lambda \in \{\mu \in S_{\omega,\theta_0} \ : \ |\mu| \geq \rho\}$, equation (3.11) has a unique solution $\xi = R(\lambda, F)\eta$, where

$$\begin{aligned} &R(\lambda, F) \\ &= \left\{I - [\lambda_0 - A]^{1-\alpha} R(\lambda, A) \left([\lambda_0 - A]^\alpha D\right) DE^\star V_{min}^\infty\right\}^{-1} R(\lambda, A). \end{aligned} \tag{3.12}$$

From (3.9) and (3.12) it follows that the semigroup generated by F is analytic and that if $R(\lambda, A)$ is compact so is $R(\lambda, F)$. ∎

The operator F is said to be the *closed loop* operator for the optimal control problem (1.1)-(1.2). Remark that, by (3.3) we know that the function $Cx^\star(t) = Ce^{tF}x_0$ belongs to $L^2(0, \infty; H)$. Thus, when $C = I$ (or if $C^{-1} \in \mathcal{L}(H)$), by Datko's Theorem, it follows that F is exponentially stable. Thus, in this case, (A, B) is feedback stabilizable.

We give now a generalization of Proposition 1.3.4.

Proposition 3.3 *Assume that (A, C) is detectable. Then F is exponentially stable.*

Proof. Let $x_0 \in H$, and let $(u^\star, x^\star)$ be the optimal pair corresponding to x_0. By Remark 3.2 in Chapter 1, there exists $K \in \mathcal{L}(Y; H)$ such that $A - KC$ is exponentially stable. Now it is easy to check that

$$x^*(t) = e^{t(A-KC)}x_0 + \int_0^t (\lambda_0 - A)e^{(t-s)(A-KC)}Du(s)\,ds$$
$$+\int_0^t e^{(t-s)(A-KC)}KCx(s)\,ds.$$

Since $Cx^\star, u^\star \in L^2(0,\infty;U)$ and $A - KC$ is exponentially stable it follows that $x^\star \in L^2(0,\infty;H)$. Datko's Theorem yelds the conclusion. ■

3.3 Stabilizability and stability of F in the point spectrum case

We consider here the system (1.1) under hypothesis $(\mathcal{HP}_\infty)$ and we assume that A verifies the *point spectrum* hypotheses $(\mathcal{P})$ (see §3.2 in Chapter 1). We want to give a necessary and sufficient condition in order that (A, AD) be stabilizable with respect to the identity I. We need a lemma whose proof is similar to that of Lemma 3.2 (Part I, Chapter 1).

Lemma 3.1 *Assume $(\mathcal{HP}_\infty)$ and $(\mathcal{P})$ with $\sigma^0(A) = \emptyset$. Then the following statements are equivalent:*

(i) for all $x \in H$, there exists $u \in L^2(0,\infty;U)$ such that the solution x of (1.1) belongs to $L^2(0,\infty;H)$,

(ii) the mapping

$$u \mapsto \gamma u = (\lambda_0 - A)\int_0^\infty e^{-sA}\Pi_A^+ Du(s)\,ds \;:\; L^2(0,\infty;U) \to H$$

is onto,

(iii) the mapping

$$\gamma^\star : H \to L^2(0,\infty;U),$$
$$(\gamma^\star\xi)(s) = D^\star[\lambda_0 - A^\star](\Pi_A^+)^\star e^{-sA^\star}\xi, \quad \xi \in H$$

is one-to-one,

(iv) $Ker(D^\star(\lambda - A^\star)\cap Ker(\lambda - A^\star) = \{0\}$, for all $\lambda \in \sigma^+(A^\star)$.

Proof. It is completely similar to the proof of Lemma 3.2 in Chapter 1. ∎

We now generalize Propositions 3.5, 3.6 and 3.7. of Chapter 1.

Proposition 3.4 *Assume $(\mathcal{HP})_\infty$ and $(\mathcal{P})$. Then the following statements are equivalent:*

(i) (A, AD) is stabilizable with respect to I,

(ii) for any $\lambda \in \sigma^0(A^\star) \cup \sigma^+(A^\star)$,

$$\text{Ker }(D^\star(\lambda - A^\star) \cap \text{Ker }(\lambda - A^\star)) = \{0\}.$$

Proposition 3.5 *Assume that (A, AD) is C-stabilizable and that A and F fulfill $(\mathcal{P})$. Then the following statements are equivalent:*

(i) F is exponentially stable,

(ii) for any $\lambda \in \sigma^+(A) \cup \sigma^0(A)$, we have

$$\text{Ker }(A - \lambda) \cap \text{Ker }(C) = \{0\}. \tag{3.13}$$

Proposition 3.6 *Assume that the hypotheses of Proposition* 3.5 *hold. Then the following statements are equivalent:*

(i) F is exponentially stable.

(ii) (A, C) is detectable.

Example 3.1 Let Ω be an open bounded set of $\mathbf{R}^n$ with regular boundary $\partial\Omega$. Consider the state equation

$$\begin{cases} \dfrac{\partial x}{\partial t}(t,\xi) = (\Delta_\xi + c)x(t,\xi) \ \text{ in } \ ]0,T[\times\Omega, \\ \\ x(t,\xi) = u(t,\xi) \ \text{ on } \ ]0,T[\times\partial\Omega, \\ \\ x(0,\xi) = x_0(\xi) \ \text{ in } \ \Omega, \end{cases} \tag{3.14}$$

where $c > 0$, $x_0 \in L^2(\Omega)$ and $u \in L^2(\partial\Omega)$. We choose $H = Y = L^2(\Omega)$ as space of states and observations and $U = L^2(\partial\Omega)$ as space of controls. We denote by A the linear self-adjoint operator in H:

$$\begin{cases} Ax = \Delta_\xi x + cx, \quad \forall\, x \in D(A) \\ \\ D(A) = H^2(\Omega) \cap H_0^1(\Omega). \end{cases}$$

We consider the following problem: to minimize

$$J(u) = \int_0^\infty \int_\Omega |x(t,\xi)|^2 \, dt d\xi + \int_0^\infty \int_{\partial\Omega} |u(t,\xi)|^2 \, dt d\xi.$$

Let $D : L^2(\partial\Omega) \to L^2(\Omega)$ the Dirichlet mapping introduced in Example 1.1 of Part II, Chapter 2; then problem (3.14) can be written as

$$x(t) = e^{tA}x_0 + (cI - A) \int_0^t e^{(t-s)A} Du(s) \, ds,$$

and hypotheses $(\mathcal{HP}_\infty)$ hold with $\lambda_0 = c$. We denote by $\{\varphi_k\}$ a complete set of eigenvector of A and by $\{-\lambda_k\}$ the corresponding sequence of eigenvalues. We assume that $\{\lambda_k\}$ is a non decreasing sequence and that

$$0 < \lambda_1 < c, \quad \lambda_k > c, \quad k = 2, 3, \dots$$

and we prove that (A, AD) is stabilizable. Recalling Proposition 3.4 it is enough to observe that

$$D^\star(A^\star - c)\varphi_1 = \frac{\partial \varphi_1}{\partial \nu}$$

is not identically zero; this in fact a consequence of a result in E.J.P.G. SCHMIDT and N. WECK [1].

Remark 3.1 Stabilizability for parabolic boundary control problems was studied by several authors, see T. NAMBU [1], I. LASIECKA and R. TRIGGIANI [1], [2] and H. AMANN [1].

These results have been generalized by A. LUNARDI [1], [2] for non autonomous state equations with periodic coefficients (see also P. KOCH MEDINA [1]). ∎

Chapter 3

Systems with Unbounded Control Operators: Hyperbolic Equations with Control on the Boundary

1 Introduction and setting of the problem

We use here the notation of Chapter 3 in Part II. We assume that

$$(\mathcal{HH})_\infty \begin{cases} (i) & A \textit{ generates a strongly continuous group } e^{tA} \textit{ in } H \\ & \textit{of type } \omega_0 \textit{ and } \lambda_0 \textit{ is a real number in } \rho(A) \textit{ such} \\ & \textit{that } \omega_0 < \lambda_0, \\ (ii) & E \in \mathcal{L}(U;H), \\ (iii) & \forall T > 0,\ \exists\ K_T > 0 \textit{ such that} \\ & \displaystyle\int_0^t |E^\star A^\star e^{sA^\star} x|^2\, ds \le K_T^2 |x|^2, \quad \forall x \in D(A^\star),\ t \ge 0, \\ (iv) & C \in \mathcal{L}(H;Y). \end{cases}$$

Clearly, if $(\mathcal{HH})_\infty$ hold, then the hypotheses $(\mathcal{HH})$ of Chapter 3 in Part II are fulfilled with $P_0 = 0$.

We want to minimize the cost function:

$$J_\infty(u) = \int_0^\infty \{|Cx(s)|^2 + |u(s)|^2\}ds, \tag{1.1}$$

over all controls $u \in L^2(0,\infty;U)$ subject to the equation constraint

$$\begin{aligned} x(t) &= e^{tA}x_0 + G(u)(s) \\ G(u)(s) &= (\lambda_0 - A)\int_0^t e^{(t-s)A}Eu(s)\,ds. \end{aligned} \tag{1.2}$$

Moreover, $x_0 \in H$ and $u \in L^2(0,\infty;U)$. We recall that by Proposition 3.1 in Part II , Chapter 3, $x \in C([0,T];H)$ for all $\in L^2(0,T;U)$, more precisely

$$G \in \mathcal{L}(L^2(0,T;U);C([0,T];H)), \quad \forall T > 0.$$

We say that the control $u \in L^2(0,\infty;U)$ is *admissible* if $J_\infty(u) < \infty$. The definitions of optimal control, optimal state and optimal pair are the same as in Chapters 1 and 2. When, for any $x_0 \in H$, an admissible control exists, we say that (A, AE) is C-stabilizable.

In order to solve the control problem (1.1)-(1.2), we consider the optimization problems: to minimize

$$J_t(u) = \int_0^t \{|Cx(s)|^2 + |u(s)|^2\}\,ds, \tag{1.3}$$

over all controls $u \in L^2(0,t;U)$ subject to the equation constraint (1.1), and denote by $(u_t^\star, x_t^\star)$ the corresponding optimal pair, then we prove the convergence of $(u_t^\star, x_t^\star)$ to an optimal pair for the problem (1.1)-(1.2).

For a different approach to the study of the Algebraic Riccati Equation and more results see the paper by F. Flandoli, I. Lasiecka and R. Triggiani [1], and the Lecture Notes by I. Lasiecka and R. Triggiani [3]. The situation where assumption $(\mathcal{HH})_\infty$-(iii) is not fulfilled is considered in I. Lasiecka and R. Triggiani [4], where specific examples are provided.

2 Main results

The main result of this Chapter is the following.

Theorem 2.1 *Assume* $(\mathcal{HH})_\infty$ *and that* (A, AE) *is* C*–stabilizable. Let* $(u_t^\star, x_t^\star)$ *be the optimal pair corresponding to problem* (1.2)–(1.3); *then there exist* $u^\star \in L^2(0,\infty;U)$ *and* $x^\star \in L^2(0,\infty;H)$ *such that*

(i) $\lim_{t\to\infty} u_t^\star = u^\star$ *in* $L^2(0,T;U)$, *for all* $T > 0$.

(ii) $\lim_{t\to\infty} x_t^\star = x^\star$ *in* $L^2(0,T;H)$, *for all* $T > 0$.

(iii) $(u^\star, x^\star)$ *is an optimal pair for the problem* (1.1)-(1.2).

Proof. Denote by $P_{min}(\cdot)$ the mild solution to equation (1.5) of Part II, Chapter 3, with $P_0 = 0$. Let $x_0 \in H$, $u \in L^2(0,\infty;U)$ and let x be the corresponding solution to (1.1); we have

$$\begin{aligned}(P_{min}(t)x_0, x_0) &\le \int_0^t \{|Cx(s)|^2 + |u(s)|^2\}\,ds \\ &\le J_\infty(u), \quad \forall\, u \in L^2(0,\infty;U).\end{aligned} \tag{2.1}$$

Since (A, AE) is C–stabilizable, there exists $M(x_0) > 0$ such that

$$(P_{min}(t)x_0, x_0) \le M(x_0), \quad \forall\, t > 0.$$

By the Uniform Boundedness Theorem it follows that there exists $M > 0$ such that

$$\|P_{min}(t)\| \le M, \quad \forall\, t \ge 0. \tag{2.2}$$

Then there exists the limit

$$P_{min}^\infty x_0 = \lim_{t\to\infty} P_{min}(t)x_0, \quad \forall\, x_0 \in H.$$

From (2.1) it follows that

$$(P_{min}^\infty x_0, x_0) \le J_\infty(u), \quad \forall\, u \in L^2(0,\infty;U). \tag{2.3}$$

On the other hand we have

$$(P_{min}(t)x_0, x_0) = \int_0^t \{|Cx_t^\star(s)|^2 + |u_t^\star(s)|^2\}\,ds, \quad \forall\, t > 0. \tag{2.4}$$

and, taking into account (2.2), we see easily that the set $\{\widehat{u}_t\}_{t\geq 0}$, where

$$\widehat{u}_t(s) = \begin{cases} u_t^\star(s), & \text{if } s \in [0,t], \\ 0, & \text{if } s \geq t, \end{cases}$$

is bounded in $L^2(0,\infty;U)$. Thus there exists a sequence $t_n \uparrow \infty$ and a function $\widehat{u} \in L^2(0,\infty;U)$ such that

$$\widehat{u}_{t_n} \rightharpoonup \widehat{u}, \text{ in } L^2(0,\infty;U) \text{ as } n \to \infty. \tag{2.5}$$

Set

$$\widehat{x}(t) = e^{tA}x_0 + G(u)(t).$$

Fix now $T > 0$. Obviously

$$u^*_{t_n} \rightharpoonup \widehat{u}, \text{ in } L^2(0,T;U) \text{ as } n \to \infty,$$

and consequently,

$$x^*_{t_n} \rightharpoonup \widehat{x}, \text{ in } L^2(0,T;U) \text{ as } n \to \infty.$$

From (2.4) it follows

$$(P^\infty_{min}x_0, x_0) \geq \int_0^T \{|\widehat{x}(s)|^2 + |\widehat{u}(s)|^2\}ds, \quad \forall t > 0.$$

Since T is arbitrary we can conclude that $C\widehat{x} \in L^2(0,\infty;H)$ and

$$(P^\infty_{min}x_0, x_0) \geq J_\infty(\widehat{u}), \tag{2.6}$$

which, along with (2.3), implies that the pair $(\widehat{u}, \widehat{x})$ is optimal. It remains to prove (i) and (ii). We first remark that, by the Lebesgue dominated convergence theorem

$$Cx^*_{t_n} \rightharpoonup \widehat{x}, \text{ in } L^2(0,T;\infty) \text{ as } n \to \infty.$$

Letting n tend to infinity in the equality

$$\int_0^{T_n} \{|Cx^*_{T_n}(s)|^2 + |u^*_{T_n}(s)|^2\}\, ds = (P_{min}(T_n)x_0, x_0),$$

we find

$$\lim_{n\to\infty} \int_0^\infty \{|Cx^*_{T_n}(s)|^2 + |u^*_T(s)|^2\}\, ds = (P^\infty_{min}x_0, x_0).$$

This implies that the convergences of $\{u^*_T(s)\}$ to $\widehat{u}$ and of $\{Cx^*_T(s)\}$ to $C\widehat{x}$ are strong. Since the optimal pair is unique, by the strict convexity of the cost J_∞, (i) and (ii) follows.

Example 2.1 We consider here the system described in Example 4.3 of Part II, Chapter 3, but with $T = +\infty$. In order to apply Theorem 2.1 it is enough to check that (A, AE) is exactly controllable. As remarked at the end of §2 in Chapter 2 of Part I, the relevant estimate to prove is that, for T sufficiently large, there exists a constant $C(T) > 0$ such that

$$\int_0^T \left|\frac{\partial v}{\partial \nu}\right|^2_{L^2(\partial\Omega)} dt \geq C(T)E(0), \tag{2.7}$$

where v is the solution to the problem

$$\begin{cases} v_{tt}(t,\xi) = \Delta_\xi v(t,\xi), \quad t \geq 0,\ \xi \in \Omega, \\ v(t,\xi) = 0, \quad t \geq 0,\ \xi \in \partial\Omega, \\ v(0,\xi) = v_0 \in H_0^1(\Omega), \\ v_t(0,\xi) = v_1 \in L^2(\Omega), \end{cases} \tag{2.8}$$

and

$$E(t) = \frac{1}{2}\int_\Omega [|\nabla v(t,\xi)|^2 + |v_t(t,\xi)|^2]\, d\xi\, dt.$$

To prove (2.7) we follows J.-L. Lions [1]. We introduce the following notation:

$$X(t) = (v_t, \xi \cdot \nabla v), \quad Y(t) = (v_t, v),$$

where $(\cdot,\cdot)$ denotes the scalar product and $|\cdot|$ the norm in $L^2(\Omega)$. We first prove the identity

$$Y(T) - Y(0) = \int_{\Omega\times[0,T]} [v_t^2 - |\nabla v|^2]\, d\xi\, dt. \tag{2.9}$$

In fact, multiplying both sides of the first equation in (2.8) by v and taking into account that

$$(v^2)_{tt} = 2v_t^2 + 2\, v\, v_{tt},$$

we find

$$v\, v_{tt} = \frac{1}{2}(v^2)_{tt} - v_t^2 = v\,\Delta v.$$

Integrating on Ω

$$\frac{1}{2}\int_\Omega (v^2)_{tt}\,d\xi - \int_\Omega v_t^2\,d\xi = \int_\Omega v\,\Delta v\,d\xi = -\int_\Omega |\nabla v|^2\,d\xi,$$

and so

$$\frac{1}{2}\frac{d}{dt}\int_\Omega (v^2)_t\,d\xi = \int_\Omega [v_t^2 - |\nabla v|^2]\,d\xi,$$

which implies (2.9).

Now we multiply both sides of the first equation in (2.8) by $\xi \cdot \nabla v$ and integrate in $\Omega \times [0,T]$. We find

$$I = J$$

where

$$I = \int_{\Omega\times[0,T]} (\xi\cdot\nabla v)\,v_{tt}\,d\xi\,dt$$

and

$$J = \int_{\Omega\times[0,T]} (\xi\cdot\nabla v)\,\Delta v\,d\xi\,dt$$

We proceed now in three steps.

Step 1. *Estimate of I.*

By integrating by parts successively in t and in ξ, we find

$$\begin{aligned} I &= X(T) - X(0) - \int_{\Omega\times[0,T]} (\xi\cdot\nabla v_t)\,v_t\,d\xi\,dt \\ &= X(T) - X(0) - \frac{1}{2}\int_{\Omega\times[0,T]} \xi\cdot\nabla(v_t^2)\,d\xi\,dt \\ &= X(T) - X(0) + \frac{n}{2}\int_{\Omega\times[0,T]} v_t^2\,d\xi\,dt \end{aligned}$$

Step 2. *Estimate of J.*

By integrating by parts in ξ, we find

$$\begin{aligned}
J &= \sum_{h,k=1}^{n} \int_{\Omega\times[0,T]} \xi_h \frac{\partial v}{\partial \xi_h} \frac{\partial^2 v}{\partial \xi_k^2}\, d\xi\, dt \\
&= \sum_{h,k=1}^{n} \int_{\partial\Omega\times[0,T]} \xi_h\, \nu_k \frac{\partial v}{\partial x_h} \frac{\partial v}{\partial \xi_k}\, d\sigma\, dt \\
&\quad - \sum_{h,k=1}^{n} \int_{\Omega\times[0,T]} \frac{\partial}{\partial \xi_k}\left(\xi_h \frac{\partial v}{\partial \xi_h}\right) \frac{\partial v}{\partial \xi_k}\, d\xi\, dt \\
&= \int_{\partial\Omega\times[0,T]} (\xi \cdot \nabla v) \left|\frac{\partial v}{\partial \nu}\right|^2 d\sigma\, dt \\
&\quad -\frac{1}{2} \sum_{h,k=1}^{n} \int_{\Omega\times[0,T]} \xi_h \frac{\partial}{\partial \xi_k} \left|\frac{\partial v}{\partial \xi_h}\right|^2 d\xi\, dt - \int_{\Omega\times[0,T]} |\nabla v|^2\, d\xi\, dt.
\end{aligned}$$

We observe now that, since $v = 0$ on $\partial\Omega$ we have $\nabla v = \frac{\partial v}{\partial \nu}\nu$. It follows

$$\begin{aligned}
J &= \int_{\partial\Omega\times[0,T]} (\xi \cdot \nabla v) \left|\frac{\partial v}{\partial \nu}\right|^2 d\sigma\, dt \\
&\quad -\frac{1}{2} \int_{\Omega\times[0,T]} \xi \cdot \nabla\left(|\nabla v|^2\right) dx\, dt - \int_{\Omega\times[0,T]} |\nabla v|^2\, d\xi\, dt \\
&= \int_{\partial\Omega\times[0,T]} (\xi \cdot v) \left|\frac{\partial v}{\partial \nu}\right|^2 d\sigma\, dt + \left(\frac{n}{2} - 1\right) \int_{\Omega\times[0,T]} |\nabla v|^2\, d\xi\, dt.
\end{aligned}$$

Step 3. *Conclusion.*
Since $I = J$ we find

$$\begin{aligned}
&\int_{\partial\Omega\times[0,T]} (\xi \cdot v) \left|\frac{\partial v}{\partial \nu}\right|^2 d\sigma \\
&= X(T) - X(0) + \frac{n}{2} \int_{\Omega\times[0,T]} v_t^2\, d\xi\, dt - \left(\frac{n}{2} - 1\right) \int_{\Omega\times[0,T]} |\nabla v|^2\, dx\, dt \\
&= X(T) - X(0) + \frac{n}{2} \int_{\Omega\times[0,T]} [v_t^2 - |\nabla v|^2]\, d\xi\, dt + \int_{\Omega\times[0,T]} |\nabla v|^2\, d\xi\, dt.
\end{aligned}$$

In conclusion

$$\int_{\partial\Omega\times[0,T]} (\xi \cdot v) \left|\frac{\partial v}{\partial \nu}\right|^2 d\sigma\, dt = \left[X + \frac{n-1}{2} Y\right]_0^T + TE(0).$$

Now, let $R > 0$ such that Ω is included in the ball $B(0, R)$ and let $\alpha_0 > 0$ such that $|v| \leq \alpha_0 |\nabla v|$; then

$$|X| \leq R|v_t||\nabla v| \leq RE(0)$$

and

$$|Y| \leq |v||v_t| \leq \alpha_0 |v_t||\nabla v| \leq \alpha_0 E(0),$$

and so

$$\int_{\partial\Omega\times[0,T]} (\xi \cdot v) \left|\frac{\partial v}{\partial \nu}\right|^2 d\sigma\, dt \geq E(0)[T - 2R - 3\alpha_0],$$

and the required estimate (2.7) holds for T large. ∎

3 Some result for general semigroups

In this section we do not assume that A generates a strongly continuous group but

$$(\mathcal{HH})'_\infty \begin{cases} (i) & A \textit{ generates a strongly continuous semigroup } e^{tA} \\ & \textit{in } H \textit{ of type } < \lambda_0, \textit{ for some real number } \lambda_0, \\ (ii) & E \in \mathcal{L}(U; H), \\ (iii) & \forall T > 0,\ \exists K_T > 0 \textit{ such that} \\ & \int_0^t |E^\star A^\star e^{sA^\star} x|^2 ds \leq K_T^2 |x|^2, \quad \forall x \in D(A^\star), t \geq 0. \\ (iv) & C \in \mathcal{L}(H; Y). \end{cases}$$

By the results of Part II §3.5 we are able to show the existence of an optimal pair (u_t^*, x_t^*) of the control problem (1.2)-(1.3) under hypotheses $(\mathcal{HH})'_\infty$. Then the following theorem is proved as Theorem 2.1.

Theorem 3.1 *Assume* $(\mathcal{HH})'_\infty$ *and that* (A, AE) *is* C*–stabilizable. Let* $(u_t^\star, x_t^\star)$ *be the optimal pair corresponding to problem* (1.2)–(1.3). *Then there exist* $u^\star \in L^2(0, \infty; U)$ *and* $x^\star \in L^2(0, \infty; H)$ *such that* *(i)* $\lim_{t\to\infty} u_t^\star = u^\star$ *in* $L^2(0,T;U)$*, for all* $T > 0$.

(ii) $\lim_{t\to\infty} x_t^\star = x^\star$ *in* $L^2(0,T;H)$*, for all* $T > 0$.

(iii) $(u^\star, x^\star)$ *is an optimal pair for the problem* (1.1)-(1.2).

Example 3.1 *Age dependent equations*
We continue here Example 5.1 of Part II, Chapter 3, but taking $T = +\infty$. In order to apply Theorem 2.2 we have only to check stabilizability. Set

$$\sigma_1(A) = \{\lambda \in \sigma(A) \ : \ \operatorname{Re}\lambda < 0\},$$
$$\sigma_2(A) = \{\lambda \in \sigma(A) \ : \ \operatorname{Re}\lambda \geq 0\},$$

and denote by Q_1 and Q_2 the spectral projectors on $\sigma_1(A)$ and $\sigma_2(A)$ respectively. As well known Q_2 is a finite dimensional projector, moreover, since the semigroup $S(\cdot)$ is differentiable for $t > a_+$ the spectral determining condition holds true, and there exists $M > 0$ and $r > 0$ such that

$$\|S(t)Q_1\| \leq Me^{-rt}, \quad t \geq 0. \tag{3.1}$$

We set $p_1(t) = Q_1 p(t)$, $t \geq 0$ and $p_2(t) = Q_2 p(t)$, $t \geq 0$ and denote by A_1 and A_2 the restrictions of A to $Q_1(H)$ and $Q_2(H)$ respectively. By (3.1) it follows easily that $p_1 \in L^2(0,\infty;H)$ for any $u \in L^2(0,\infty)$. So, we have only to check that, for any $p_0 \in H$ there exists $u \in L^2(0,\infty)$ such that $p_2 \in L^2(0,\infty;Q_2(H))$. Now p_2 is the solution of the finite dimensional Cauchy problem

$$\begin{cases} p_2'(t) = A_2 p_2(t) - \alpha A_2 D u(t) \\ \\ p_2(0) = Q_2 p_0. \end{cases} \tag{3.2}$$

By a well known result due to Hautus (see Part II, §3.2) the system (3.2) is stabilizable if and only if

$$\operatorname{Ker}(\lambda - A_2^*) \cap \operatorname{Ker}(D^* A_2^*) = \{0\}, \quad \forall\, \lambda \in \sigma_2(A). \tag{3.3}$$

Now we can easily check that the spectrum $\sigma(A)$ of consists of simple eigenvalues λ that are exactly all the solutions to the equation

$$1 = \int_0^{a_+} e^{-\lambda b} K(b)\, db. \tag{3.4}$$

If λ is an eigenvalue of A then a corresponding eigenvector is given by

$$\varphi_\lambda(a) = e^{-\lambda a}\pi(a), \quad a \in [0, a_+]. \tag{3.5}$$

Moreover the adjoint A^* of A is given by

$$\begin{cases} D(A^*) = \{\psi \in H^1(0, a_+) : \ \psi(T) = 0\}, \\ \\ A^*\psi = \psi' - \mu\psi + \beta\psi(0). \end{cases} \tag{3.6}$$

Finally $\sigma(A^*) = \sigma(A)$ and if $\lambda \in \sigma(A^*)$ then a corresponding eigenvector is given by

$$\psi_\lambda(a) = \frac{1}{\pi(a)} \int_a^{a_+} e^{\lambda(a-b)} K(b)\, db, \quad a \in [0, a_+]. \tag{3.7}$$

Let now $\lambda \in \sigma_2(A)$ and $\psi \in \text{Ker } (D^* A_2^*)$, then we have

$$1 = \int_0^{a_+} e^{-\lambda b} K(b)\, db. \tag{3.8}$$

and a corresponding eigenvector to λ is given by

$$\psi_\lambda(a) = \frac{1}{\pi(a)} \int_a^{a_+} e^{\lambda(a-b)} K(b)\, db, \quad a \in [0, a_+]. \tag{3.9}$$

It follows, recalling hypotheses (5.8) of Part II, Chapter 3,

$$D^* A_2^* \psi = <\psi, \pi> = \frac{1}{\lambda}\left[\int_0^{a_+} e^{-\lambda b} K(b)\, db - 1\right] \neq 0.$$

Thus (A, AE) is I-stabilizable. ∎

References

P. ACQUISTAPACE and B.TERRENI [1], *A unified approach to abstract linear non-autonomous parabolic equations*, Rend. Sem. Mat. Univ. Padova 78 (1987), 47–107.

H. AMANN [1], *Feedback Stabilization of Linear and Semilinear Parabolic Systems*, in "Proceedings of Trends in Semigroup Theory and Applications", pp. 21–57, (Trieste, 1987), Lecture Notes in Pure and Applied Mathematics, Marcel Dekker, New York 1989.

V. BARBU and G. DA PRATO [1], *A representation formula for the solutions to the operator Riccati equation*, Differential and Integral Equations 5, no. 4 (1992), 821–829.

A. BENSOUSSAN [1], *Observateurs et stabilité*, Colloque CNES, Paris 1987.

G. DA PRATO [1], *Synthesis of optimal control for an infinite dimensional periodic problem*, SIAM J. Control and Optim. 25 (1987), 706–714.

[2], *Periodic solutions of an infinite dimensional Riccati equation*, in "System modelling and optimization", Lecture Notes in Control and Information Sciences, Vol. 84, A. Prékopa, J. Szeleszán and B. Strazicky, eds, pp. 714–722, Springer-Verlag, Berlin, Heidelberg, New York 1986.

[3], *Some remarks on the periodic linear quadratic regulator problem*, in "Modelling and Inverse Problems of Control for Distributed Parameter Systems", A. Kurzhanski and I. Lasiecka, eds, pp. 28-42, Lecture Notes in Control and Information Sciences, Vol. 154, Springer-Verlag, Berlin, Heidelberg, New York 1991.

[4], *Spectral properties of the closed loop operator and periodic control problems*, Journal of Mathematical Systems, Estimation and Control 3, no. 1 (1993), 41–50.

Cf. V. BARBU and G. DA PRATO

G. DA PRATO and M.C. DELFOUR [1], *Stabilization and unbounded solutions of the Riccati equation*, in "Proc. 27th IEEE Conference on Decision and Control", pp. 352–357, IEEE Publications, New York 1988.

[2], *Unbounded solutions to the linear quadratic control problem*, SIAM J. Control and Optim. 30 (1992), 31–48.

G. DA PRATO and M. IANNELLI [1], *Boundary control problem for age dependent equations*, in "Third international workshop-conference on

Evolution equations, control theory and biomathematics", Ph. Clément and G. Lumer, eds, Marcel Dekker, to appear.

G. DA PRATO and A. ICHIKAWA [1], *Riccati equations with unbounded coefficients*, Ann. Mat. Pura Appl. 140 (1985), 209-221.

[2], *Optimal control of linear systems with almost periodic inputs*, SIAM J. Control and Optim. 25 (1987), 1007–1019.

[3], *Quadratic Control for Linear Periodic Systems*, Appl. Math. Optim. 18 (1988), 39–66.

[4], *Quadratic control for linear time varying systems*, SIAM J. Control and Optim. 28, no. 2 (1990), 359–381.

[5], *Optimal control for integrodifferential equations of parabolic type*, SIAM J. Control and Optim., to appear.

G. DA PRATO and A. LUNARDI [1], *Floquet exponents and stabilizability in time periodic parabolic systems*, Appl. Math. Optim 22 (1990), 91–113.

[2], *Stabilizability of integrodifferential parabolic equations*, J. Integral Equations Appl. 2 (1990), 281–304.

M.C. DELFOUR Cf. G. DA PRATO and M.C. DELFOUR

F. FLANDOLI [1], *Algebraic Riccati equation arising in a boundary control problem with distributed parameters*, SIAM J. Control and Optim. 25 (1987), 612–636.

[2], *On the direct solution of Riccati equations arising in boundary control theory*, Annali di Matematica Pura e Applicata, to appear.

F. FLANDOLI, I. LASIECKA and R. TRIGGIANI [1], *Algebraic Riccati equations with non smooth observation arising in Hyperbolic and Euler-Bernoulli boundary control problems*, Annali di Matematica Pura e Applicata, Serie IV, Tomo CLIII (1988), 307–382.

J.S. GIBSON [1], *The Riccati integral equations for optimal control problems on Hilbert spaces*, SIAM J. on Control and Optim. 17 (1979), 537–565.

M.L.J. HAUTUS [1], *Stabilization, controllability and observability of linear autonomous systems*, Indagationes Math. 32 (1970), 448–455.

M. IANNELLI Cf. G. DA PRATO and M. IANNELLI

A. ICHIKAWA Cf. G. DA PRATO and A. ICHIKAWA

P. KOCH MEDINA [1], "Feedback stabilizability of time–periodic parabolic equations", University of Zürich, Doctoral Thesis, 1991.

I. LASIECKA [1], *Unified theory for abstract parabolic boundary problems: a semigroup approach*, Appl. Math. Optim. 6 (1980), 287–333.

Cf. F. FLANDOLI, I. LASIECKA and R. TRIGGIANI

I. LASIECKA and R. TRIGGIANI [1], *Stabilization and structural assignement of Dirichlet boundary feedback parabolic equations*, SIAM J. Control and Optim. 21 (1983), 766–803.

[2], *Stabilization of Neumann boundary feedback parabolic equations: the case of trace in the feedback loop*, Appl. Math. Optim. 10 (1983), 307–350.

[3], "Differential and Algebraic Riccati Equations with applications to boundary/point control problems: continuous theory and approximation theory", Lecture Notes in Control and Information Sciences, Vol. 164, Springer-Verlag, Berlin, Heidelberg, New York 1991.

[4], *Differential Riccati equations with unbounded coefficients: applications to boundary control/boundary observation of hyperbolic problems*, Nonlinear Analysis 17, no. 7 (1991), 655–682.

J.-L. LIONS [1], *Exact controllability, stabilization and perturbations*, J. Von Neumann Lecture, SIAM Review 30 (1988), 1–60.

[2], "Controllabilité exacte, perturbations et stabilisation des systèmes distribués", Volumes 1, 2, Masson, Paris 1990.

J.-L. LIONS and E. MAGENES [1], "Problèmes aux limites non homogènes", Dunod, Paris 1968.

A. LUNARDI [1], *Bounded solutions of linear periodic abstract parabolic equations*, Proc. Royal Soc. Edinburgh 110 A (1988), 135–159.

[2], *Stabilizability of time parabolic equations*, SIAM J. Control and Optim. 29, no. 4 (1991), 810–828.

[3], *Neumann boundary stabilization of structurally damped time periodic wave and plate equations*, in "Differential equations with applications in biology, physics, and engineeering", J. A. Goldstein, F.Kappel and W. Schappacher, eds, pp. 241–257, Marcel Dekker, New York 1991.

Cf. G. DA PRATO and A. LUNARDI

E. MAGENES Cf. J.-L. LIONS and E. MAGENES

T. Morozan [1], *Almost periodic solutions for Riccati equations of stochastic control*, Appl. Math. Optim., to appear.

T. Nanbu [1], *On the stabilization of diffusion equations*, J. Differential Equations 52 (1884), 204–233.

A. Pritchard and J. Zabczyk [1], *Stability and stabilizability of infinite dimensional systems*, SIAM Rev. 23 (1981), 25–52.

E.J.P.G. Schmidt and N. Weck [1], *On the boundary behaviour of solutions to elliptic and parabolic equations, with applications to boundary control for parabolic equations*, SIAM J. Control and Optim. 16 (1978), 593–598.

M.A. Shayman [1], *Phase portrait of the matrix Riccati equation*, SIAM J. Control and Optim. 24 (1986), 1–64.

M. Sorine [1], "Un résultat d'existence et d'unicité pour l'équation de Riccati stationnaire", INRIA Report 55, Rocquencourt, France 1981 (also CRMA report no. 984, Université de Montréal, 1980).

[2], "Sur le semigroupe non linéaire associé à l'équation de Riccati", CRMA report no. 1055, Université de Montréal, 1981.

B. Terreni Cf. P. Acquistapace and B. Terreni

R. Triggiani [1], *Boundary Feedback Stabilizability of Parabolic Equations*, Appl. Math. Optim. 6 (1980), 201–220.

Cf. F. Flandoli, I. Lasiecka and R. Triggiani

Cf. I. Lasiecka and R. Triggiani

N. Weck Cf. E.J.P.G. Schmidt and N. Weck

W.M. Wonham [1], *On a matrix Riccati equation of stochastic control*, SIAM J. Control 6 (1968), 681–697.

J. Zabczyk [1], *Remarks on the Algebraic Riccati Equation in Hilbert spaces*, Appl. Math. Optim. 2 (1976), 251–258.

Cf. A. Pritchard and J. Zabczyk

Appendix A

An Isomorphism Result

We present two proofs of an isomorphism result which plays a key role in the solution of the linear quadratic optimal control problem. Proposition 1.1 gives a direct proof, while Proposition 1.2 gives a proof based on the variational characterization of the minimum of a linear quadratic cost function under a linear state equation constraint. This second proof also yields a general result on the invertibility of a 2×2 matrix of operators which corresponds to the usual optimality system obtained from the coupled system in the state and adjoint state variables.

In this appendix X represents a Hilbert space, with inner product $(\cdot, \cdot)$ and norm $|\cdot|$ and T and S linear bounded operators in X.

Proposition 1.1 *Assume that T and S are symmetric and nonnegative, then $I + TS$ is one–to–one and onto. Moreover*

$$\|S(I+TS)^{-1}\| \leq \|S\|$$

and

$$\|(I+TS)^{-1}\| \leq 1 + \|T\| \, \|S\|.$$

Proof. **Step 1.** $I + TS$ is one–to–one.
Let $x_0 \in X$ such that $x_0 + TSx_0 = 0$. Then we have $Sx_0 + STSx_0 = 0$, and so

$$(Sx_0, x_0) + (TSx_0, Sx_0) = 0,$$

which implies $(Sx_0, x_0) = 0$, and consequently $Sx_0 = 0$ and finally $0 = x_0 + TSx_0 = x_0$.

Step 2. $R := (I + TS)(X)$ is dense in X.
Let $y_0 \in X$ be such that

$$((I + TS)x, y_0) = 0, \ \forall\, x \in X.$$

Then

$$(x, (I + ST)y_0) = 0, \ \forall\, x \in X,$$

which implies $y_0 = 0$ by Step 1 (exchanging T and S).

Step 3. $0 \le (S(I + TS)^{-1}x, x) \le (Sx, x), \quad \forall\, x \in R.$
Let $x \in R$ and let $z = (I + TS)^{-1}x$. We have, recalling that $T \ge 0, S \ge 0$,

$$\begin{aligned}
(Sx, x) &= (S(I + TS)z, (I + TS)z) \\
&= (Sz, z) + (Sz, TSz) \\
&\quad +(STSz, z) + (STSz, TSz) \\
&\ge (Sz, z) + (Sz, TSz) = (Sz, (I + TS)z) \\
&= (S(I + TS)^{-1}x, x) \ge 0
\end{aligned}$$

Step 4. $(S(I + TS)^{-1}x, y) = (x, S(I + TS)^{-1}y), \quad \forall\, x, y \in R.$
Let $x, y \in R$ and let $z = (I + TS)^{-1}x, \ w = (I + TS)^{-1}y$. We have

$$\begin{aligned}
(S(I + TS)^{-1}x, y) &= (Sz, w + TSw) = ((I + TS)z, Sw) \\
&= (x, S(I + TS)^{-1}y)
\end{aligned}$$

Step 5. Conclusion.
By steps 2,3,4 the symmetric operator $S(I + TS)^{-1}$ has a unique extension to a symmetric bounded operator that we denote again by $S(I + TS)^{-1}$. Now for any $x \in R$ we have

$$(I + TS)^{-1}x = x - TS(I + TS)^{-1}x,$$

thus, by Step 3

$$|(I + TS)^{-1}x| \leq |x| + \|T\|\,|Sx|.$$

This implies that R is closed; since it is dense by Step 2, the conclusion follows. ∎

The above result was first proved in an early version of this book via an optimal control argument. We give it below since this type of result is useful in many contexts.

Consider the real Hilbert spaces U and H, the continuous linear operators

$$B \in \mathcal{L}(U; H), \quad Q \in \mathcal{L}(H; H)$$

and assume that

$$\begin{aligned} &\forall\, x \in H, \quad (Qx, x) \geq 0 \\ &\forall\, x,\, y \in H, \quad (Qx, y) = (Qy, x). \end{aligned}$$

Define the following optimal control problem: to find $u \in U$ such that

$$\inf_{u \in U} J(u)$$

where

$$J(u) = \frac{1}{2}(Qx + 2q, x) + \frac{1}{2}|u|^2$$

and

$$x = Bu + f$$

for some given pair $(f, g) \in H \times H$.

This problem has a unique solution $u \in U$ which is completely characterized by the following optimality condition

$$dJ(u; v) = 0, \quad \forall v \in U,$$

where

$$dJ(u; v) = (Qx + q, y) + (u, v)$$

and

$$x = Bu + f, \quad y = Bv.$$

Introduce the adjoint variable $p \in H$

$$p = Qx + q$$

and rewrite the optimality condition

$$\begin{aligned} &(p, Bv) + (u, v) = 0, \quad \forall v \in U \\ &\Rightarrow \quad u + B^* p = 0. \end{aligned}$$

Using this characterization we can now write the coupled system in the form

$$\begin{bmatrix} I & BB^* \\ -Q & I \end{bmatrix} \begin{bmatrix} x \\ p \end{bmatrix} = \begin{bmatrix} f \\ q \end{bmatrix}$$

We conclude that for any given pair (f, q), there exists a pair (x, p) for which the above identity is verified. Moreover when $f = g = 0$,

$$\begin{aligned} x + BB^* p = 0 \quad &\Rightarrow \quad (x, p) + |B^* p|^2 = 0 \\ -Qx + p = 0 \quad &\Rightarrow \quad -(Qx, x) + (p, x) = 0 \end{aligned}$$

and $x = p = 0$. Therefore the matrix of operators is invertible.

Finally set $q = 0$ and eliminate p in the first equation of the coupled system. Then

$$\begin{aligned} &x + BB^* Qx = f \\ &\Rightarrow \quad [I + BB^* Q]x = f. \end{aligned}$$

Again for any f in H, there exists an x in H for which the above identity is verified. For $f = 0$

$$\begin{aligned} &x + BB^* Qx = 0 \\ &\Rightarrow \quad (Qx, x) + |B^* Qx|^2 = 0 \\ &\Rightarrow \quad Q^{\frac{1}{2}} x = 0 \quad \Rightarrow \quad x = 0. \end{aligned}$$

Hence $I + BB^* Q$ is invertible.

Proposition 1.2 *Given two real Hilbert spaces U and H and two linear bounded operators*

$$B \in \mathcal{L}(U, H) \quad \textit{and} \quad Q \in \mathcal{L}(H, H)$$

such that

$$\forall\, x \in H, \quad (Qx, x) \geq 0$$
$$\forall\, x,\, y \in H, \quad (Qx, y) = (Qy, x),$$

then the operators

$$\begin{bmatrix} I & BB^* \\ -Q & I \end{bmatrix} \quad \textit{and} \quad [I + BB^*Q]$$

are (algebraic and topological) isomorphisms.

This last proposition was also proved by M. SORINE [1].

Proposition 1.1 is now obtained with $X = H = U$, $S = Q$ and $B = T^{\frac{1}{2}}$.

Index

Corrections to Volume I

p. 18, $\uparrow \ell\, 13$. J. ZABCZYK instead of J. ZABCZYCK.

p. 20, $\uparrow \ell\, 8$. ...for all ω ...instead of ...for all w ...

p. 29, $\uparrow \ell\, 6$.
such that $\sigma(A) \supset \{\lambda \in \mathbf{C} : \mathrm{Re}\lambda > \omega\}$ *and the following inequalities hold* ...instead of ...
such that $\sigma(A) \subset \{\lambda \in \mathbf{C} : \mathrm{Re}\lambda > \omega\}$ *and the following inequalities hold*

p. 53, $\uparrow \ell\, 10$. (2.41) ...instead of ...(2.42).

p. 54, $\uparrow \ell\, 10$. R. TRIGGIANI [1] ...instead of ...R. TRIGGIANI [4].

p. 108, $\downarrow \ell\, 9$.

$$D_A\left(\frac{1}{2},2\right) \cong (D(A),H)_{\frac{1}{2},2} = [D(A),H]_{\frac{1}{2}}$$

instead of

$$D_A\left(\frac{1}{2},2\right) \cong [D(A),H]_{\frac{1}{2}} = [D(A),H]_{\frac{1}{2}}.$$

p. 108, $\downarrow \ell\, 10$. Example 4.3 instead of Example 4.6.

p. 112, $\uparrow \ell\, 8$. Equation (6.6).

$$D(A^\alpha) \subset [D(A),H]_{1-\alpha} \text{ and } D(A^{*\beta}) \subset [D(A^*),H]_{1-\beta}$$

instead of

$$D(A^\alpha) \subset [D(A),H]_{1-\alpha} \text{ and } D(A^{*\beta}) \subset [D(A^*),H]_{1-\alpha}.$$

p. 115, $\uparrow \ell\, 9$. The title of that reference is incorrect. It should read: *Hölder classes with boundary conditions as interpolation spaces.*

p. 120, $\downarrow \ell\, 17$. ... *Coïncidence entre des espaces d'interpolation* ...instead of ... *Concidence entre des espaces d/interpolation* ....

p. 190, $\downarrow \ell\, 3$. ...SIAM J. Control 6 ...instead of ...SIAM J. Control C.

p. 210, $\downarrow \ell\, 14$. delete ...vspace1x ...

Systems & Control: Foundations & Applications

Series Editor
Christopher I. Byrnes
School of Engineering and Applied Science
Washington University
Campus P.O. 1040
One Brookings Drive
St. Louis, MO 63130-4899
U.S.A.

Systems & Control: Foundations & Applications publishes research monographs and advanced graduate texts dealing with areas of current research in all areas of systems and control theory and its applications to a wide variety of scientific disciplines.

We encourage the preparation of manuscripts in TeX, preferably in Plain or AMS TeX—LaTeX is also acceptable—for delivery as camera-ready hard copy which leads to rapid publication, or on a diskette that can interface with laser printers or typesetters.

Proposals should be sent directly to the editor or to: Birkhäuser Boston, 675 Massachusetts Avenue, Cambridge, MA 02139, U.S.A.

Estimation Techniques for Distributed Parameter Systems
H.T. Banks and K. Kunisch

Set-Valued Analysis
Jean-Pierre Aubin and Hélène Frankowska

Weak Convergence Methods and Singularly Perturbed Stochastic Control and Filtering Problems
Harold J. Kushner

Methods of Algebraic Geometry in Control Theory: Part I
Scalar Linear Systems and Affine Algebraic Geometry
Peter Falb

H^∞-Optimal Control and Related Minimax Design Problems
Tamer Başar and Pierre Bernhard

Identification and Stochastic Adaptive Control
Han-Fu Chen and Lei Guo

Viability Theory
Jean-Pierre Aubin

Representation and Control of Infinite Dimensional Systems, Vol. I
A. Bensoussan, G. Da Prato, M.C. Delfour and S.K. Mitter

Mathematical Control Theory: An Introduction
Jerzy Zabczyk

H_∞-Control for Distributed Paramater Systems: A State Space Approach
Bert van Keulen

Representation and Control of Infinite Dimensional Systems, Vol. II
A. Bensoussan, G. Da Prato, M.C. Delfour and S.K. Mitter

GPSR Compliance

The European Union's (EU) General Product Safety Regulation (GPSR) is a set of rules that requires consumer products to be safe and our obligations to ensure this.

If you have any concerns about our products, you can contact us on ProductSafety@springernature.com

In case Publisher is established outside the EU, the EU authorized representative is:

Springer Nature Customer Service Center GmbH
Europaplatz 3
69115 Heidelberg, Germany

Batch number: 10577878

Printed by Printforce, the Netherlands